T0319258

DEVELOPMENTS IN
ENVIRONMENTAL MODELLING, 28

ECOLOGICAL MODEL TYPES

DEVELOPMENTS IN ENVIRONMENTAL MODELLING, 28

ECOLOGICAL MODEL TYPES

Edited by

SVEN ERIK JØRGENSEN*

Emeritus Professor, Copenhagen University, Denmark

ELSEVIER

AMSTERDAM • BOSTON • HEIDELBERG • LONDON
NEW YORK • OXFORD • PARIS • SAN DIEGO
SAN FRANCISCO • SINGAPORE • SYDNEY • TOKYO

*Deceased.

Elsevier
Radarweg 29, PO Box 211, 1000 AE Amsterdam, Netherlands
The Boulevard, Langford Lane, Kidlington, Oxford OX5 1GB, United Kingdom
50 Hampshire Street, 5th Floor, Cambridge, MA 02139, United States

Library of Congress Cataloging-in-Publication Data
A catalog record for this book is available from the Library of Congress

British Library Cataloguing in Publication Data
A catalogue record for this book is available from the British Library

ISBN: 978-0-444-63623-2
ISSN: 0167-8892

For information on all Elsevier publications
visit our website at https://www.elsevier.com/

Working together
to grow libraries in
developing countries

www.elsevier.com • www.bookaid.org

Publisher: Candice Janco
Acquisition Editor: Laura Kelleher
Editorial Project Manager: Emily Thomson
Production Project Manager: Anitha Sivaraj
Designer: Maria Ines Cruz

Typeset by TNQ Books and Journals

Contents

List of Contributors

Z.P. Du Institute of Geographical Science and Natural Resources Research, Beijing, China

S.M. Duke-Sylvester University of Louisiana, Lafayette, LA, United States

W. He Peking University, Beijing, China

S.E. Jørgensen Emeritus Professor, Copenhagen University, Denmark

D. Justic Louisiana State University, Baton Rouge, LA, United States

X.-Z. Kong Peking University, Beijing, China

T. Legović Ruđer Bošković Institute, Zagreb, Croatia

S. Lek UMR CNRS-Université Paul Sabatier, Université de Toulouse, Toulouse, France

J. Liang Louisiana State University, Baton Rouge, LA, United States

W.-X. Liu Peking University, Beijing, China

Y.-S. Park Kyung Hee University, Seoul, Republic of Korea

T. Prato Professor Emeritus, Department of Agricultural and Applied Economics, University of Missouri, Columbia, MO, United States

J.M. Visser University of Louisiana, Lafayette, LA, United States

F.-L. Xu Peking University, Beijing, China

Z. Xue Louisiana State University, Baton Rouge, LA, United States

B. Yang Peking University, Beijing, China

T.X. Yue Institute of Geographical Science and Natural Resources Research, Beijing, China; University of Chinese Academy of Sciences, Beijing, China

N. Zhao Institute of Geographical Science and Natural Resources Research, Beijing, China; University of Chinese Academy of Sciences, Beijing, China

Acknowledgements

During the development of this book, Dr. Sven Erik Jørgensen passed away on 5 March 2016. Beyond being an award winning scientist, he was a champion for ecology and environmental management, and overall a remarkable man. He will truly be missed.

After his passing, a number of friends and colleagues stepped up to help us in completing this book. We extend a special thank you to Dr. Dubravko Justic and to Dr. Soeren Nors Nielsen. Without their help, this book would not have been completed.

Laura Spence Kelleher
Acquisitions Editor − Ecology and
Environmental Sciences
Elsevier

1

Introduction: An Overview

S.E. Jørgensen

Emeritus Professor, Copenhagen University, Denmark
E-mail: soerennorsnielsen@gmail.com

1.1 APPLICATION OF ECOLOGICAL MODELING

The field of ecological modeling has developed rapidly during the last four decades due to essentially three factors:

1. The development of computer technology, which has enabled us to handle very complex mathematical systems.
2. A general increased knowledge about environmental problems, including that a complete elimination of pollution is not feasible (denoted zero discharge), but that a proper pollution control with limited economical resources available requires serious considerations of the influence of pollution impacts on ecosystems. Models are indispensable tools in this context.
3. Our knowledge about environmental and ecological problems has increased significantly. We have particularly gained more knowledge of the quantitative relations in the ecosystems and between the ecological properties and the environmental factors.

Good models are syntheses of our knowledge about ecosystems and their environmental problems, in contrast to a statistical analysis, which only will reveal the relationships between the data. A model is able to include our entire knowledge about the system, if required for a proper solution of the environmental problem:

- which components interact with which other components, for instance that zooplankton grazes on phytoplankton,
- our knowledge about the processes often formulated as mathematical equations which have been proved valid generally,
- the importance of the processes with reference to the problem;

These are a few examples of knowledge which may often be incorporated in an ecological model. This implies that a model can offer a deeper understanding of the system than a statistical analysis. It is therefore a stronger tool in research and can result in better management plans, on how to solve environmental problems. This does of course not mean that statistical analytical results are not applied in the development of models. On the contrary, models are build on all available knowledge, including knowledge gained by statistical analyses of data, physical-chemical-ecological knowledge, the laws of nature, common sense, and so on. That is in short the advantage of modeling.

The idea behind the use of ecological management models is demonstrated in Fig. 1.1. Urbanization and technological development has had an increasing impact on the environment. Energy and pollutants are released into ecosystems, where they may cause more rapid growth of algae or bacteria, damage species, or alter the entire ecological structure. An ecosystem is extremely complex, and it is therefore an overwhelming task to predict the environmental effects that such emissions may have. It is here that the model comes into the picture. With sound ecological knowledge, it is possible to extract the components and processes of the ecosystem that are particularly involved in a specific pollution problem to form the basis of the ecological model (see also the discussion in Chapter 2). As indicated in Fig. 1.1, the

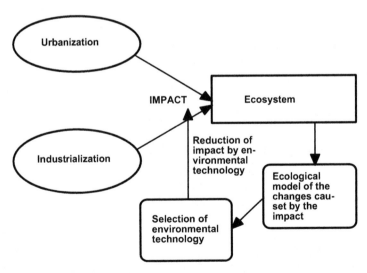

FIGURE 1.1   The environmental problems are rooted in the emissions resulting from industrialization and urbanization. Sound ecological knowledge is used to extract the components and processes of the ecosystem that are particularly involved in a specific pollution problem to form the ecological model applied in environmental management to select good solutions to the focal problem.

resulting model can be used to select the environmental technology eliminating the emission most effectively.

Fig. 1.1 represents the idea behind the introduction of ecological modeling as a management tool around years 1970–1975. The environmental management of today is more complex and applies therefore a wider spectrum of tools.

Today, we have as alternative and supplement to environmental technology, cleaner technology, ecotechnology, environmental legislation, international agreements, and sustainable management plans. Ecotechnology is mainly applied to solve the problems of nonpoint or diffuse pollution, often originating from agriculture. The significance of nonpoint pollution was hardly acknowledged before around 1980. The global environmental problems play furthermore a more important role today than 20 or 30 years ago, for instance the reduction of the ozone layer and the climatic changes due to the greenhouse effect. The global problems can hardly be solved without international agreements and plans. Fig. 1.2 attempts to illustrate the more complex picture of environmental management today.

Mathematical models are not only applied in environmental management but are widely applied in science, too. Newton's laws are for instance relatively simple mathematical models of the influence of gravity on bodies, but they do not account for frictional forces, influence of wind, etc. Ecological models do not differ essentially from other scientific models not even by their complexity, as many models used in nuclear physics today may be even more complex than ecological models. The application of models in ecology is almost compulsory, if we want to understand the function of such a complex system as an ecosystem. It is simply not possible to survey the many components and their reactions in an ecosystem without the use of a model as a holistic tool. The reactions of the system might not necessarily

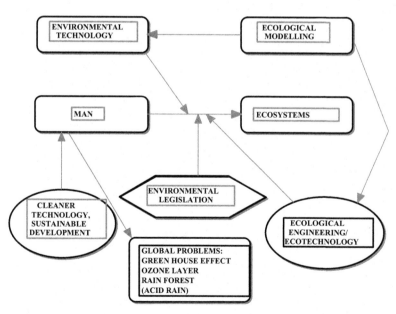

FIGURE 1.2 The idea behind the use of environmental models in environmental management. The environmental management of today is very complex and must apply environmental technology, alternative technology, and ecological engineering or ecotechnology. In addition the global environmental problems play an increasing role. Environmental models are used to select environmental technology, environmental legislation, and ecological engineering.

be the sum of all the individual reactions, which implies that the properties of the ecosystem as a system cannot be revealed without the use of a model of the entire system.

It is therefore not surprising that ecological models have been used increasingly in ecology as an instrument to understand the properties of ecosystems as systems. This application has clearly revealed the advantages of models as a useful tool in ecology, which may be summarized in the following points (Jørgensen and Fath (2011)):

1. Models are useful instruments in **survey** of complex systems.
2. Models can be used to reveal **system properties**.
3. Models reveal the weakness in our knowledge and can therefore be used to set up **research priorities**.
4. Models are useful in tests of **scientific hypotheses**, as the model can simulate ecosystem reactions, which can be compared with observations.

And we would add here:

5. Models are useful **experimental tools**. If a model is developed and it is working reasonably meaning that the calibration and validation are acceptable, it is possible to ask the model several questions. A few examples are given here, while a couple of more detailed illustrations are presented in Chapter 11, as models have been used as experimental tools frequently in the development of toxic substance models. The model can be used as an experimental tool by answering for instance the following illustrative questions that are easily answered by the model:
   a. an alternative equation has been proposed for a certain specific process and therefore it is natural to ask: would it be beneficial to use this alternative equation or is the previously used equation giving more acceptable results?
   b. the model is working fully acceptable for a number of cases (for instance for several compounds in ecotoxicological modeling) but not for other cases (or other compounds in ecotoxicological modeling). Would it be possible to explain this discrepancy by the lack of some processes in the nonworking cases? Let us try and see how the model reacts to addition of one or more processes in the nonworking cases.
   c. if the model is not working in a minor number of cases, could it be explained by the influence of a factor which has another value in the nonworking cases or maybe another equation is valid in these cases.

Hence, be open to ask the model questions and to test the model and its components. A model is a very useful experimental tool in this context. It is easy to test additions or changes, and performances of these tests imply that we use the model as an experimental tool.

1.2 MODEL CLASSIFICATION AND MODEL TYPES

Table 1.1 shows a general perception of ecological model classification 40 years ago. The differences among the three model types are the choice of the ecological components that are used as state variables. If the model aims for a description of a number of individuals, species, or classes of species, then the model will be called **biodemographic**. A model that describes the energy flows is named **bioenergetic** and the state variables will typically be

TABLE 1.1 Model Identification

Model Types	Organization	Pattern	Measurements
Biodemographic	Conservation of genetic information	Life cycles of species	Number of species or individuals
Bioenergetic	Conservation energy	Energy flow	Energy
Biogeochemical	Conservation of mass	Element cycles	Mass or concentrations

expressed in kW or kW per unit of volume or area. **Biogeochemical models** consider the flow of material, and the state variables are indicated as kg or kg per unit of volume or area. These model classes were extensively used in ecological modeling 40 years ago and are still widely applied today, because ecological models have, indeed, focus on the mass flows, energy flows, and the changes of populations, as they describe the dynamics of ecosystems.

The title of this book is ecological model types, because the development in ecological modeling has to a high extent been: How can we consider the many other additional factors in our model? For instance:

- How can we include the spatial distribution which is often significant for the ecological reactions and dynamics? or
- How can we consider the differences among individuals in a population? or
- How to account for adaptations and shifts in species compositions?
- Can we build models if we have heterogeneous databases? Or only qualitative data?
- How do we tackle the problem of ecological changes caused by the expected climatic changes?

These kinds of problems that were emerging in ecological modeling as the field developed called for solutions, and the problems have been solved during the last 30–40 years. We have today 14 model types meaning that it is possible to distinguish between 14 different modeling approaches that are able to solve the model problems that emerged during the development of ecological modeling during the last 40 years. Models of today are, however, often using two or more of these approaches simultaneously because two or more of the factors mentioned above have impact on the ecosystem dynamics at the same time. Therefore, we should maybe rather distinguish between 14 different tool boxes that are available to develop ecological models today, considering *all* the possible factors, influencing the ecological processes in ecosystems. Throughout the book we will use the expression model type, but it will be understood that the 14 types represent 14 solutions to processes and factors that are important to consider if we want to cover almost all possible ecosystem dynamic cases.

The 14 different model approaches are denoted model types, but it is acknowledged that they are 14 different tool boxes that can be applied to solve the modeling problem that confronts us, when we want to develop a model that is able to solve a specific environmental or scientific problem. We do acknowledge that in a practical and specific model situation, we need to apply two or more—sometimes several—of the modeling tool boxes.

The 14 tool boxes or model types can be described as follows:

1. Biogeochemical models are models that describe the flows of mass and/or energy in ecosystems by using the mass and energy conservation principles. The models are often based on differential equations, expressing accumulation per unit of time = inputs − outputs. As flows of energy and matter are important dynamic processes in ecosystems, it is not surprising that this model type is the most applied of the 15 types.
2. Population dynamic models describe the changes of the number of individuals in populations. This model type applies differential equations. Increasing numbers are explained by growth and decreasing numbers by mortality. Interactions between populations are often considered in this type and the age structure of the population is often considered in the model, as the properties of a population are dependent on the age class.
3. Steady state models are as the name indicates not able to consider the dynamics of ecosystems, but often are these simpler models very useful to apply, because the models are simpler than type 1 and 2, and often is a steady state sufficient information for environmental management or for understanding of ecological reactions?
4. Spatial models give, in addition to the flows of matter and energy and the changes of populations, information about the spatial distribution of matter, energy, or populations. The spatial distribution in ecosystems is often crucial for the ecological reactions to ecosystem impacts. We distinguish between surface models that describe the spatial distribution and
5. Spatial models, geo process models, that not only describe the spatial distribution but also the spatial movements. It is a further development of surface models. Generally, advanced software has been developed to cope with the spatial distribution of both forcing functions and state variables, as for instance a geographical information system, GIS.
6. Structural dynamic models that include a description of the changes of the species' properties due to adaptation and the shifts in the species composition due to changes of the prevailing conditions. These models attempt therefore to describe the changes in the ecological structure due to changes in the impacts (described by the forcing functions) on the ecosystems. The abbreviation SDM is often applied. SDM can be developed either by use of a goal function and use of knowledge about which species are best fitted under which circumstances. The latter approach is sometimes denoted artificial intelligence.
7. Individual-based models (IBM) consider that individuals in a population may be different and have different properties. Obviously, differences are often crucial for the ecological processes and reactions to changed impacts and may be used to cover the selection of the best fitted properties—it means that the adaptation can be described by IBM.
8. Artificial Neural Network (ANN), is mathematical modeling tool that is able to extract model information from a large database even if it is heterogeneous. A direct use of ANN leads to a black box model, but it is possible to include algorithms that can account for causality and thereby develop at least grey models.
9. Self-organizing maps (SOM), is another mathematical tool that like ANN is able to develop a model from large databases. Like for ANN, it is possible by use of suitable algorithms to include causality into the model.

10. Ecotoxicological models are focusing on ecotoxicological problems. All the other tool boxes can in principle be used to develop ecotoxicological models. The biogeochemical tool box is very frequently applied to develop ecotoxicological models. When it is beneficial to distinguish ecotoxicological models from other model types, it is due to the special properties of ecotoxicological models. They are often simple, because they have often a high uncertainty, which is acceptable because they are applied with a high safety margin. The parameters of the models are often very dependent on the properties of the toxic substance. And as there are a high number of different chemical compounds with different properties, only a relatively limited number of the parameters have been determined by measurements in the laboratory. It has therefore been necessary to introduce estimation methods, which can be used to find approximate values of the parameters.

11. Fugacity models are focusing on the distribution of chemicals (mainly toxic chemicals) in the environment. They are therefore a special type of ecotoxicological models. They have often a high standard deviation, which is, however, acceptable, as they are mainly used to compare the total environmental impacts of the use of two or more chemicals. The comparison makes it possible to select the chemical that should be preferred from an environmental point of view among the possible ones.

12. Fuzzy models are used when the data are fuzzy, meaning only qualitative or semiquantitative. Obviously the model results cannot be presented more accurately than the underlying data and the model results are therefore presented qualitatively or semiquantitatively, for instance a few numbers of levels. Fuzzy models are anyhow important to use in ecology, because a level is important not necessarily the exact number, for instance when the distribution of species is in focus.

13. Stochastic models are considering that many forcing functions are stochastic, which implies that they should be described by using a probability distribution. It is for instance the case with the meteorological variables that often play a significant role for the model results.

14. Climate change models (CCM) encompass models that attempt to predict the climate changes due to increased emissions of greenhouse gases (mainly carbon dioxide and methane) and models that cover the ecological consequences of the climate changes. The latter group models belong to SDM, but it is preferred in this overview to include them in CCM, because the problems associated with the increased emission of greenhouse gases and with the climate changes in general are very central in the environmental debate. There has therefore been an increased number of CCMs and it is therefore beneficial to try to understand the characteristics of these models to learn from previous experience gained by development of CCMs.

1.3 APPLICATION OF THE 14 MODEL TYPES

It is difficult to make a statistical study of the use of different types of models due to the overlap in the application of the types (tool boxes) in practice both when environmental management problems and when scientific ecosystem problems are in focus. It has, however, been attempted to consider 10 different model types as the following types in the overview in Section 1.2 are considered as one type 4+5, 6+14, 8+9, and 11+12. From the statistics of the 10 types, we will able to see the development in the application of models from the mid-1970s, when the journal *Ecological Modelling* started, and up to the past 3 years 2012–2014. To make the statistics more informative, the number of papers of the 11 types in *Ecological Modelling* the first 5 years (1975–1979) and the past 3 years (2012–2014) has been supplemented by the number of theoretical papers covering modeling theory and covering systems ecology. The results of the statistics are summarized in Table 1.2.

The development of *Ecological Modelling* reflects the development of the field ecological modeling, as *Ecological Modelling* is publishing 40–50% of all ecological modeling papers published in international peer-reviewed scientific journals. First of all the field has grown enormously as we today published in *Ecological Modelling* 16 times as many pages per year. The number of pages has been calculated as pages using the format of 1975–1979.

TABLE 1.2 Ecological Model Types, Statistics

Model Type	Number of Papers 2012–2014	Number of Papers 1975–1979
Biogeochemical models	309	62
Population dynamic models	176	13
Spatial distribution models	110	4
Structural dynamic models	50	4
Steady state models	2	0
Ecotoxicological models	16	8
Individual-based models	31	0
ANN and SOM	5	0
Stochastic models	3	0
Fuzzy and fugacity models	6	0
Modeling theory	58	15
Systems ecology	62	14
Socio-economic-ecological models	3	3
Total	830	123
Pages	9690	2323
Pages, format 1975–1979	22,287	2323
Pages/paper format 1975–1979	26.8	18.9
Pages/year format 1975–1979	7429	465
Ratio pages/year	16	1

The number of words per page is 2.3 times as much today as in the years 1975–1979. The number of papers published per year has increased by a factor 12. The papers are therefore about 33% longer today, which is probably due to an increased complexity of the published models. It is also characteristic for the development that the six model types that today cover 7% of the papers were not identified in 1975–1979. The field ecological modeling has therefore increased not only in the number of publications but also in the number of model types applied for model development. Spatial models and IBMs have increased in number more than the number of papers in general, indicating that these model types are clearly more significant than in the 1970s. Biogeochemical models have only increased by a factor about 7–8, but this type is still the most applied model type in ecological modeling. One-third of all models published are biogeochemical models today, while it was about two-thirds in the 1970s.

The theoretical papers—modeling theory and systems ecology—have increased less than the other types of papers. It is understandable as these theoretical questions—How to develop a good model? and How does an ecosystem function?—were core problems in the 1970s. Still, significantly more papers are published today to cover these two important theoretical fields.

1.4 THE ECOSYSTEM AS AN OBJECT FOR MODELING

Ecological models attempt to capture the properties and characteristics of ecosystems. Ecologists generally recognize ecosystems as a specific level of organization, but the open question is the appropriate selection of time and space scales. Any size area could be selected, but in the context of ecological modeling, the following definition presented by Morowitz (1968) will be used: "An ecosystem sustains life under present-day conditions, which is considered a property of ecosystems rather than a single organism or species." This means that a few square meters may seem adequate for microbiologists, while 100 square kilometers may be insufficient if large carnivores are considered (Hutchinson, 1970, 1978). Population–community ecologists tend to view ecosystems as networks of interacting organisms and populations. Tansley (1935) claimed that an ecosystem includes both organisms and chemical–physical components, which of course has to be reflected in the ecological models. It inspired Lindeman (1942) to use the following definition: "An ecosystem composes of physical-chemical-biological processes active within a space-time unit." E.P. Odum (1953, 1959, 1969, 1971) followed these lines and is largely responsible for developing the process–functional approach, which has dominated the last few decades.

This does not mean that different views cannot be a point of entry. Hutchinson (1970, 1978) used a cyclic causal approach, which is often invisible in population–community problems. Measurement of inputs and outputs of total landscape units has been the emphasis in the functional approaches by Likens (1985). O'Neill (1976) has emphasized energy capture, nutrient retention, and rate regulations. H.T. Odum (1957) has underlined the importance of energy transfer rates. Quinlin (1975) has argued that cybernetic views of ecosystems are appropriate and Prigogine (1947), Mauersberger (1983), and Jørgensen (1981, 1982, 1986) have all emphasized the need for a thermodynamic approach for a proper holistic description of ecosystems and to include the important energetic aspects of ecosystems.

For some ecologists, ecosystems are either biotic assemblages or functional systems. The two views are separated. It is, however, important in the context of ecosystem theory to adopt both views and to integrate them. Because an ecosystem cannot be described in detail, it cannot be defined according to Morowitz's definition, before the objectives of our study are presented. Therefore the definition of an ecosystem used in the context of system ecology and ecological modeling, becomes:

An ecosystem is a biotic and functional system or unit, which is able to sustain life and includes all biological and nonbiological variables in that unit. Spatial and temporal scales are not specified a priori but are entirely based upon the objectives of the ecosystem study.

Currently there are several approaches (Likens, 1985) to the study ecosystems:

1. Empirical studies where bits of information are collected, and an attempt is made to integrate and assemble these into a complete picture.
2. Comparative studies where a few structural and a few functional components are compared for a range of ecosystem types.
3. Experimental studies where manipulation of a whole ecosystem is used to identify and elucidate mechanisms.
4. Modeling or computer simulation studies.

The motivation (Likens, 1985) in all of these approaches is to achieve an understanding of the entire ecosystem, giving more insight than the sum of knowledge about its parts relative to the structure, metabolism, and biogeochemistry of the landscape. Likens (1985) has presented an excellent ecosystem approach to Mirror Lake and its environment. The study contains all the above-mentioned studies, although the modeling part is rather weak. The study demonstrates clearly that it is necessary to use all four approaches simultaneously to achieve a good picture of the system properties of an ecosystem. An ecosystem is so complex that you cannot capture all the system properties by one approach!

Ecosystem studies are widely using the notions of order, complexity, randomness, and organization. They are used interchangeably in the literature, which causes much confusion. As the terms are used in relation to ecosystems throughout the volume, it is necessary to give a clear definition of these concepts in this introductory chapter. According to the Third Law of Thermodynamics about entropy at 0 K, see Jørgensen (2008), randomness and order are each other's antithesis and may be considered as relative terms. Randomness measures the amount of information required to describe a system. The more information required to describe the system, the more random it is.

Organized systems are to be carefully distinguished from ordered systems. Neither kind of systems is random, but whereas ordered systems are generated according to simple algorithms and may therefore lack complexity, organized systems must be assembled element by element according to an external wiring diagram with a high level of information. Organization is functional complexity and carries functional information. It is nonrandom by design or by selection, rather than by a priori necessity. Complexity (Jørgensen and Svirezhev, 2005) is a relative concept dependent on the observer. We may distinguish between structural complexity, defined as the number of interconnections between components in the system, and functional complexity, defined as the number of distinct functions carried out by the system.

References

Hutchinson, G.E., 1970. The biosphere. Scientific American 223 (3), 44–53.

Hutchinson, G.E., 1978. An Introduction to Population Ecology. Yale University Press, New Haven.

Jørgensen, S.E., 1981. Application of exergy in ecological models. In: Dubois, D. (Ed.), Progress in Ecological Modelling. Cebedoc, Liege, pp. 39–47.

Jørgensen, S.E., 1982. A holistic approach to ecological modelling by application of thermo-dynamics. In: Mitsch, W., et al. (Eds.), Systems and Energy. Ann Arbor.

Jørgensen, S.E., 1986. Structural dynamic model. Ecological Modelling 31, 1–9.

Jørgensen, S.E., 2008. Evolutionary Essays. Elsevier, Amsterdam, 230 pp.

Jørgensen, S.E., Svirezhev, Y., 2005. Toward a Thermodynamic Theory for Ecological Systems. Elsevier, Amsterdam, Oxford, 366 pp.

Jørgensen, S.E., Fath, B., 2011. Fundamentals of Ecological Modelling, fourth ed. Elsevier, Amsterdam, 400 pp.

Likens, G.E. (Ed.), 1985. An Ecosystem Approach to Aquatic Ecology: Mirror Lake and Its Environment. Springer-Verlag, New York, 516 pp.

Lindeman, R.L., 1942. The trophic dynamic aspect of ecology. Ecology 23, 399–418.

Mauersberger, P., 1983. General principles in deterministic water quality modelling. In: Orlob, G.T. (Ed.), Mathematical Modelling of Water Quality: Streams, Lakes and Reservoirs, International Series on Applied System Analysis, vol. 12. Wiley, New York, pp. 42–115.

Morowitz, H.J., 1968. Energy Flow in Biology. Biological Organisation as a Problem in Thermal Physics. Academic Press, N.Y, 179 pp. (See review by Odum, H.T., 1969. Science 164, 683–684).

O'Neill, R.V., 1976. Ecosystem persistence and heterotrophic regulation. Ecology 57, 1244–1253.

Odum, E.P., 1953. Fundamentals of Ecology. W.B. Saunders, Philadelphia.

Odum, E.P., 1959. Fundamentals of Ecology, second ed. W.B. Saunders, Philadelphia, PA.

Odum, E.P., 1969. The strategy of ecosystem development. Science 164, 262–270.

Odum, E.P., 1971. Fundamentals of Ecology, third ed. W.B. Saunders Co., Philadelphia.

Odum, H.T., 1957. Trophic structure and productivity of silver springs. Ecological Monographs 27, 55–112.

Prigogine, I., 1947. Etude Thermodynamique des Processus Irreversibles. Desoer, Liege.

Quinlin, A.V., 1975. Design and Analysis of Mass Conservative Models of Ecodynamic Systems (Ph.D. dissertation). MIT Press, Cambridge. Massachusetts.

Tansley, A.G., 1935. The use and abuse of vegetational concepts and terms. Ecology 16, 284–307.

2

Biogeochemical Models

S.E. Jørgensen

Emeritus Professor, Copenhagen University, Denmark

E-mail: soerennorsnielsen@gmail.com

OUTLINE

2.1 ADVANTAGES AND PROPERTIES OF BIOGEOCHEMICAL MODELS

Among the modeling types represented in this book, the biogeochemical models are the most applied model type. The journal *Ecological Modelling* started in 1975 and about two-thirds of all papers published in the journal during the 1970s were biogeochemical models. Today, this model type is still the most applied one. Despite the competition with many new model types, about one-third of the papers published in *Ecological Modelling* still belong to this type. In the 1970s, about 10—12 biogeochemical models were annually published in the journal, while today more than 100 of the annually published models in the journal are biogeochemical models.

This dominance of biogeochemical models is easy to understand, as the model type offers the user several advantages. First of all, the model is easy to develop. It is based on the mass conservation principles, which is expressed easily by differential equations. Fig. 2.1 illustrates the idea behind the use of differential equations for the state variables—it means the variables that are chosen to express the state of the modeled system. In mathematics, the

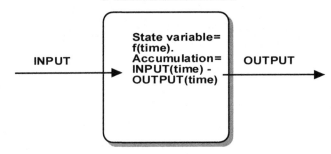

FIGURE 2.1   The idea behind the use of differential equations is illustrated. The differential equation accounts for the increase of a state variable due to what is denoted in the figure accumulation for a selected time step. The accumulation is input − output. Theoretically in mathematics, the time step is infinitely small, but when the equation is solved numerically by use of computers, we are selecting a time step. The shorter the time step, the closer is the solution to the mathematical solution.

differential equations are solved analytically, while the equations are solved numerically by computers by use of the equation accumulation = inputs − outputs. A time step is selected for the model calculations by computers. The shorter the time step, the closer will the computer calculations come to the real time variations of inputs and outputs. It is recommended to test different time steps and use the longest time step that would not give any significant change of the model results by decreasing the time step. Significant changes are of course evaluated relative to the accuracy of the observations that are used as basis for the development of the model. The processes determining the inputs and outputs to the state variables, $C = f(time)$, must, however, be formulated as an algebraic equation. Table 2.1 gives an overview of the most applied process equations:

1. a constant flow rate—also denoted zero-order expression: $dC/dt = k$ ($t = time$)
2. a first-order rate expression, where the rate is proportional to a variable, for instance a concentration of state variables: $rate = dC/dt = k \times C$. This expression corresponds to exponential growth. The following expression is obtained by integration $C(t) = C_o \times e^{kt}$

   First-order decay has the $rate = dC/dt = -k \times t$ and $C(t) = C_o \times e^{-kt}$.

3. a second-order rate expression, where the rate is proportional to two variables simultaneously.
4. a Michaelis-Menten expression or Monod kinetics known from kinetics of enzymatic processes. At small concentrations of the substrate the process rate is proportional to the substrate concentration, while the process rate is at maximum and constant at high substrate concentrations, where the enzymes are fully utilized. The same expression is used when the growth rate of plants are determined by a limiting nutrient according to Liebig's minimum law. The so-called Michaelis–Menten constant or the half saturation constant corresponds to the concentration that gives half the maximum rate. At small concentrations of substrate or nutrients, the rate is very close to a first-order rate expression, while it is close to a zero-order rate expression at high concentrations. Notice furthermore that the rate is regulated from a first-order to a zero-order expression more and more as the concentration increases.

TABLE 2.1 Process Rate Equations

Expression	Mathematical Formulation	Application of the Process Rate Expression
Zero order	Rate = k = constant	Supply is constant f. inst. in film kinetics
First order	Rate = k × variable	Decomposition of organic matter, decay of radioactive components, exponential population growth
Second order	Rate = k × conc1 × conc2	Chemical and biochemical reactions, where the process is based on a reaction of two different components
Monod kinetics	Rate = k × st.var./(st.var. + km)	Plant growth by limiting nutrient (Liebig's law) and growth of organisms or populations limited by the food source, f. inst. grazing
Logistic growth	Rate = k × st.var. × (1 − st.var./carrying capacity)	Growth, where the rate at high values of the st.var. is regulated by another regulation factor
Conc. gradients	Rate = k × dC/dx	Diffusion processes

5. a first-order rate expression with a regulation due to limitation by another factor, for instance the space of the nesting areas. It is expressed by introduction of a carrying capacity or upper limit for the state variables. The general first-order expression is applied regulated by the following factor: (1 − concentration/carrying capacity). When the concentration reaches the carrying capacity, the factor becomes zero and the growth stops. This process of rate expression is denoted logistic growth. These two growth expressions are often applied in population dynamic models, but may also be applied in biogeochemical models for instance to express the growth of an organism.

6. rates governed by diffusion often use a concentration gradient dC/dx to determine the rate, as it is expressed in Fick's laws: rate = k × (dC/dx) (Fick's First Law).

Table 2.1 summarizes the information about the applications of these six expressions: for which processes these equations are most often used. Most processes of ecological relevance are covered with these six expressions, although there of course are a few processes that will require another mathematical formulation. More than 90% of ecologically relevant processes in modeling may presumably be covered by these equations.

This short overview of processes equation used to express the input and outputs to the selected state variables demonstrates another advantage by biogeochemical models: the applied expression build, in most cases, on causality. The equations are rooted in the quantification of real ecological processes.

There are several modifications of the expressions in Table 2.1. For instance, sometimes a threshold concentration tr is used, in the Michaelis−Menten expression. The concentration (state variable) is replaced by the concentration − tr. The concentration has therefore to exceed tr to generate any rate. For grazing and predation is often used to multiply the

Michaelis—Menten expression by (1 − concentration/carrying capacity) similar to what is used in the logistic growth expression. It implies that when the food is abundant (concentration is high) another factor determines the growth for instance the space or the nesting area. These modifications are often used for development of a eutrophication model.

The equations contain, however, also constants as it is called most often in ecological modeling parameters and unfortunately they are frequently not known with low standard deviations, as they are biological constants in contrast to physical—chemical parameters that can be found with low standard deviation in physical—chemical handbooks. The biological parameters are dependent on a number of influential factors that may vary significantly from case to case. For instance the growth rates expressed by the unit 1/24 h for well-defined species vary from ecosystem to ecosystem because the factors that influence the growth rate—for instance the presence of stones, the shelter, the presence of other organisms, the meteorological conditions—are different and the complexity is too high to take all these factors into account. The influence of these factors could be described sometimes but that would inevitably introduce other parameters which would only add to the uncertainty. The biological parameters require in most cases that they are calibrated by a comparison of model results with observations. It implies, however, a disadvantage of the model type, namely that the accuracy of the model is dependent on good observations. To achieve a model with a low standard deviation, it is important to have good observations. Consequently, the cost of developing a good biogeochemical model is mainly determined by the cost of providing quantitatively and qualitatively a good data base of observations to be used for calibration and validation.

Because this model type has been widely applied, applicable software is available to develop the model—relatively easily, which of course should be considered a significant advantage by the application of biogeochemical models.

The next sections will present the modeling elements followed by a section showing how to develop biogeochemical models. The proposed procedure should be considered as general for all model types. If the same steps are not applied in detail for all model types, then the considerations on how to develop the best reliable model are also given for all model types, in general. These two sections are based on the widely applied textbook by Jørgensen and Fath (2011).

2.2 MODELING ELEMENTS

In its mathematical formulation, an ecological model has five components (Jørgensen and Fath, 2011):

1. *Forcing functions, or external variables*, are functions or variables of an external nature that influence the state of the ecosystem. In a management context the problem to be solved can often be reformulated as follows: if certain forcing functions are varied, how will this influence the state of the ecosystem? The model is used to predict what will change in the ecosystem when forcing functions are varied with time. The forcing functions due to the human impact on ecosystems are called *control functions because it is in our hands to change them*. The control functions in ecotoxicological models are, for instance, discharge of toxic

substances to the ecosystems, and in eutrophication models are discharge of nutrients. Other forcing functions of interest could be climatic and natural external variables, which influence the biotic and abiotic components and the process rates. They are in contrast to the control functions not controllable by man. By using models we will be able to address the crucial question: which changes of the control functions are needed to obtain well-defined conditions for a considered ecosystem.

2. *State variables* describe, as the name indicates, the state or the conditions of the ecosystem. The selection of state variables is crucial to the model structure, but often the choice is obvious. If, for instance, we want to model the bioaccumulation of a toxic substance, then the state variables should be the organisms in the most important food chains and concentrations of the toxic substance in the most important organisms. In eutrophication models, the state variables will be the concentrations of nutrients and phytoplankton. When the model is used in a management context, the state variable values simulated by changing the controllable forcing functions provide model results that contain the direct and indirect relations between the forcing functions and the state variables.

3. *Mathematical equations* are used to represent the biological, chemical, and physical processes. They describe the relationship between the forcing functions and state variables and between two or more state variables. The same type of process may be found in many different environmental contexts, which implies that the same equations can be used in different models. This does not imply, however, that the same process is always formulated by use of the same equation. Firstly, the considered process may be better described by another equation because of the influence of other factors. Secondly, the number of details needed or desired to be included in the model may be different from case to case due to a difference in complexity of the system and/or the problem. Some modelers refer to the description and mathematical formulation of processes as submodels. The most applied process formulations are presented by a short overview earlier in this chapter. Most of the applied equations build on causality, which means that they are quantifications of real ecological (physical, chemical, and biological) processes.

4. *Parameters* are coefficients in the mathematical representation of processes. They may be considered constant for a specific ecosystem or part of an ecosystem for a certain time, but they may also be a function of time or vary spatially. In causal models, the parameter will have a scientific definition and a well-defined unit, for instance, the excretion rate of cadmium from a fish—the unit could be mg Cd/(24 h × kg of fish). Many parameters are not indicated in the literature as constants but as ranges, but even that is of great value in the parameter estimation, as will be discussed further in the following text. In Jørgensen et al. (2000), a comprehensive collection of parameters in environmental sciences and ecology can be found. Our limited knowledge of parameters is one of the weakest points in biogeochemical modeling, but also in the applications of other model type—a point that will be touched on often throughout this book. Furthermore, the applications of parameters as constants in our models are unrealistic due to the many feedbacks in real ecosystems. The flexibility and adaptability of ecosystems is inconsistent with the application of constant parameters in the models. A new generation of models that attempts to use parameters varying according to ecological principles seems a possible solution to the problem, but a further development in this direction is absolutely necessary

before we can achieve an improved modeling procedure reflecting the processes in real ecosystems. This topic will be further discussed in Chapter 7.

5. *Universal constants*, such as the gas constant and atomic weights, are also used in many models.

Models can be defined as formal expressions of the essential elements of a problem in mathematical terms. The first recognition of the problem is often verbal. This may be recognized as an essential preliminary step in the modeling procedure, which will be treated in more detail in the next section. The verbal model is, however, difficult to visualize and it is, therefore, more conveniently translated into a *conceptual diagram*, which contains the state variables, the forcing functions, and how these components are interrelated by mathematical formulations of processes. The conceptual diagram shows with other words how the modeling elements 1–3 are related and connected.

Fig. 2.2 illustrates a conceptual diagram of the nitrogen cycle in a lake. The state variables are nitrate, ammonium (which is toxic to fish in the unionized form of ammonia), nitrogen in phytoplankton, nitrogen in zooplankton, nitrogen in fish, nitrogen in sediment, and nitrogen

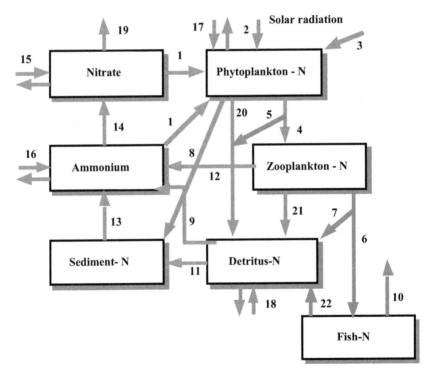

FIGURE 2.2   The conceptual diagram of a nitrogen cycle in an aquatic ecosystem. The processes are (1) uptake of nitrate and ammonium by algae, (2) photosynthesis, (3) nitrogen fixation, (4) grazing with loss of undigested matter, (5)–(7) are predation and loss of undigested matter, (8) settling of algae, (9) mineralization, (10) fishery, (11) settling of detritus, (12) excretion of ammonium from zooplankton, (13) release of nitrogen from the sediment, (14) nitrification, (15)–(18) are inputs/outputs, (19) denitrification, (20)–(22) mortality of phytoplankton, zooplankton, and fish.

in detritus. The state variables are in this conceptual diagram indicated as boxes that are connected by processes (indicated as arrows).

The forcing functions are outflows and inflows, concentrations of nitrogen components in the inflows and outflows, solar radiation, and the temperature, which is not shown in the diagram, but which influences all the process rates. The processes are formulated using quantitative expressions in the mathematical part of the model. Three significant steps in the modeling procedure need to be defined in this section before we go to the modeling procedure in detail. They are verification, calibration, and validation:

Verification is a test of the *internal logic* of the model. Typical questions in the verification phase are: Does the model behave as expected and intended? Is the model long-term stable, as one should expect an ecosystem is? Does the model follow the law of mass conservation, which is often used as the basic for the differential equations of the model? Is the use of units consistent? Verification is, to some extent, a partially subjective assessment of the model behavior and will continue during the model use before the calibration phase.

Calibration is an attempt to find the best agreement between the computed and observed data by variation of some selected parameters. It may be carried out by trial and error, or by using software developed to find the parameters giving the best fit between observed and computed values. In some static models and in some simple models, which contain only a few well-defined, or directly measured, parameters, calibration may not be required, but it is generally recommendable to calibrate the model if observations of a proper quality and quantity are available.

Validation must be distinguished from verification. Validation consists of an objective test on how well the model output fits the data. We distinguish between a structural (qualitative) validity and a predictive (quantitative) validity. A model is said to be structurally valid if the model structure reasonably accurately represents the cause—effect relationship of the real system. The model exhibits predictive validity if its predictions of the system behavior reasonably align with observations of the real system. The selection of possible objective tests will be dependent on the aims of the model, but the standard deviations between model predictions and observations and a comparison of observed and predicted minimum or maximum values of a particularly important state variable are frequently used. If several state variables are included in the validation, then they may be given different weights.

Further detail on these three important steps in modeling will be given in the next section where the entire modeling procedure will be presented.

2.3 THE MODELING PROCEDURE

A tentative modeling procedure is presented in this section. The detailed information is given for the development of biogeochemical models, but also other model types follow the general considerations that are presented. The author has used this procedure successfully numerous times and strongly recommends that all the steps of the procedure are used very carefully. To make shortcuts in modeling is not recommended. Other scientists in the field have published other slightly different procedures, but detailed examinations will reveal that the differences are only very minor. The most important steps of modeling are included in all the recommended modeling procedures for biogeochemical models and to a certain extent for the development of all model types.

Always, the initial focus of research is the definition of the problem. This is the only way in which the limited research resources can be correctly allocated instead of being dispersed into irrelevant activities. The first modeling step is therefore a definition of the problem and the definition will need to be bound by the constituents of space, time, and subsystems. The bounding of the problem in space and time is usually easy, and consequently more explicit, than the identification of the subsystems to be incorporated in the model.

Systems thinking is important in this phase. You must try to grasp the big picture. The focal system behavior must be interpreted as a product of dynamic processes, preferably describable by causal relationships. Fig. 2.3 shows the procedure proposed by the author, but it is important to emphasize that this procedure is unlikely to be correct in the first attempt, so there is no need to aim at perfection in one step. The procedure should be considered as an iterative process and the main requirement is to get started (Jeffers, 1978).

It is difficult, at least in the first instance, to determine the optimum number of subsystems to be included in the model for an acceptable level of accuracy defined by the scope of the model. Due to lack of data, it will often become necessary at a later stage to accept a lower number than intended at the start or to provide additional data for improvement of the model. It has often been argued that a more complex model should account more accurately for the behavior of a real system, but this is not necessarily true. Additional factors are involved, but a more complex model has more parameters and increases the level of uncertainty because parameters have to be estimated either by field observations, by laboratory experiments, or by calibrations, which again are based on field measurements. Parameter estimations are never completely without errors, and the errors are carried through into the model and will thereby contribute to its uncertainty. The problem of selecting the right model complexity will be mentioned several times throughout the book. It is a problem of particular interest for modeling in ecology because ecosystems are very complex, but it does not imply that an ecological model to be used in research or environmental management necessarily should be very complex. It depends on the ecosystem and the problem.

A first approach to the data requirement can be made at this stage, but it is most likely to be changed at a later stage, once experience with the verification, calibration, sensitivity analysis, and validation has been gained. Development of an ecological model should generally be considered an iterative process. In principle, data for all the selected state variables should be available; in only a few cases would it be acceptable to omit measurements of selected state variables, as the success of the calibration and validation is closely linked to the data quality and quantity.

It is helpful at this stage to list the state variables and attempt to get an overview of the most relevant processes by setting up an adjacency matrix. The state variables are listed vertically and horizontally. A 1 is used to indicate that a direct link exists between the two state variables, while 0 indicates that there is no link between the two components. The conceptual diagram Fig. 2.2 can be used to illustrate the application of an adjacency matrix in modeling.

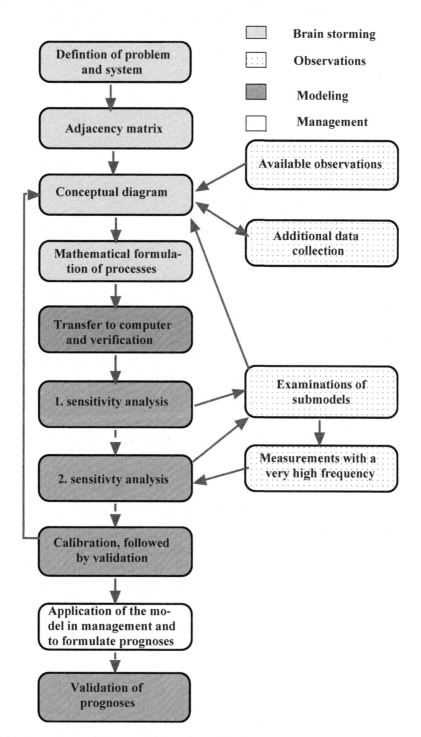

FIGURE 2.3   A tentative modeling procedure is shown. Ideally, as mentioned in the text, one should determine the data collection based on the model, not the other way round. Both possibilities are shown because models in practice have often been developed from available data, supplemented by additional observations. It is shown in the diagram that examinations of submodels and intensive measurements should follow the first sensitivity analysis. Unfortunately, many modelers have not had the resources to do so but have had to bypass these two steps and even the second sensitivity analysis. It must strongly be recommended to follow the sequence of first sensitivity analysis, examinations of submodels and intensive measurements, and second sensitivity analysis. Notice that there are feedback arrows from calibration and validation to the conceptual diagram. It shows that modeling should be considered an iterative process.

Adjacency Matrix for the Model in Fig. 2.2

From / To	Nitrate	Ammonium	Phyt-N	Zoopl-N	Fish N	Detritus-N	Sediment-N
Nitrate	—	0	1	0	0	0	0
Ammonium	0	—	0	1	0	1	1
Phyt-N	1	1	—	0	0	0	0
Zoopl.-N	0	0	1	—	0	0	0
Fish N	0	0	0	1	—	0	0
Detritus-N	0	0	1	1	1	—	0
Sediment-N	0	0	1	0	0	1	—

In this example, the adjacency matrix is made from the conceptual diagram for illustrative purposes, but in practice it is recommended to set up the adjacency matrix before the conceptual diagram. The modeler should ask for each of the possible links: is this link possible? If yes, is it sufficiently significant to be included in the model? If "yes," write 1; if "no," write 0. The adjacency matrix shown above may not be correct for all nitrogen cycles in all aquatic ecosystems. If resuspension is important there should be a link between sediment-N and detritus-N. If the aquatic ecosystem is shallow, resuspension may be significant, while the process is without any effect in deep lakes. This example illustrates clearly the idea behind the application of an adjacency matrix to get the very first overview of the state variables and their interactions. The adjacency matrix can be considered as a checklist to assess which processes of all the possible linkages that actually are realized and should be included in the model.

Once the model complexity, at least as a first attempt, has been selected, it is possible to conceptualize the model as, for instance, in the form of a diagram as shown in Fig. 2.2. It will give information on which state variables, forcing functions and processes are required in the model.

Fig. 2.4 shows a simplification of the procedure illustrated in Fig. 2.3. The more simple procedure in Fig. 2.4 has a more general applicability for all model types.

Ideally, one should determine which data are needed to develop a model according to a conceptual diagram, i.e., to let the conceptual model or even some first more primitive mathematical models determine the data at least within some given economic limitation, but in real life, most models have been developed *after* the data collection as a compromise between model scope and available data. There are developed methods to determine the ideal data set needed for a given model to minimize the uncertainty of the model, but unfortunately the applications of these methods are limited.

The conceptual diagram in Fig. 2.2 indicates the state variables as boxes, for instance nitrate, and the processes as arrows between boxes, for instance process number one. The forcing functions are symbolized by arrows to or from a state variable, for instance 15 and 16. It is possible to use other symbols for the modeling components.

The software STELLA, which will be used to illustrate the development of models in several chapters of the book, applies boxes for state variables (Compartments), thick arrows with a symbol of a valve for the processes (Connections), thick arrows coming or going to a cloud

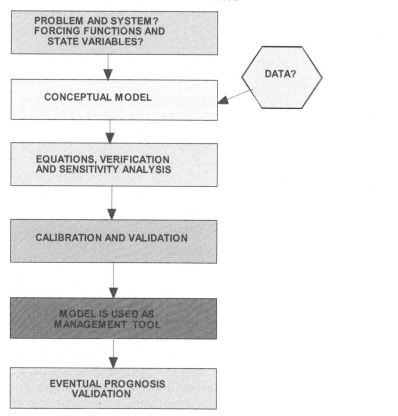

FIGURE 2.4 A slightly simpler model procedure than the one illustrated in Fig. 2.3. It has a more general applicability for all model types.

for the forcing functions (which require a constant, an equation, a table, or a graph), a thin arrow to indicate the transfer of information or variables (Controls—forcing function, parameter, and/or a state variable calculated by an algebraic expression from another state variable and so on); see Fig. 2.5.

There are other symbolic languages for development of conceptual diagrams, for instance Odum's energy circuit language. It has many more symbols than STELLA and is therefore more informative but also more time-consuming to develop. For an overview of the most used symbolic languages including Odum's energy circuit language, see for instance Jørgensen and Bendoricchio (2001).

For each state variable, a differential equation is constructed: accumulation = inputs − outputs. For detritus-N, for instance in Fig. 2.1, the inputs are the processes 20 + 5 + 21 + 7 + 22 + 18 (in) and the outputs are the processes 11 + 9 + 18 (out). The differential equations are attempted to be solved analytically in mathematics, but it is hardly possible with most ecological models because they are too complex. The differential equations are therefore solved numerically within the computer software. A time step is selected for the model calculations. The shorter the time step, the closer will the computer calculations

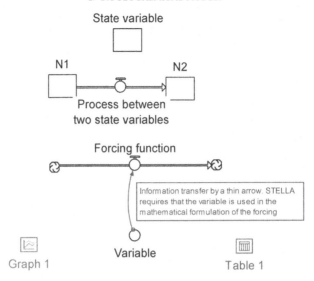

FIGURE 2.5 The symbols applied to erect a conceptual diagram using STELLA. State variables are boxes for which differential equations are erected as accumulation = inputs − outputs. Processes are thick arrows with the symbol of a valve. Forcing functions are thick arrow starting or ending as a cloud. Circles are variables in general. Graph 1 and Table 1 indicate that the results can be presented as graphs or as tables.

come to the real time variations of inputs and outputs; but the shorter the time step, the longer the simulation will take to run. It is recommended as previously mentioned to test different time steps and use the longest time step that would not give any significant change of the model results by decreasing the time step further. The term "significant changes" is of course evaluated relative to the accuracy of the observations that are used as basis for the development of the model.

The software STELLA develops the differential equations directly from the conceptual diagram which is input on the main user interface. The time derivative of the state variables will be equal to all the inputs = all process arrows going into the state variables minus all outputs = all process arrows going out from the state variable. The processes must, however, be formulated as an algebraic equation; see Section 2.1.

The next steps of the modeling procedure include verification and sensitivity analysis. The latter is covered in the next section. Verification is a test of the internal model logic. Crucial questions about the model are asked and answered by the modeler. Verification is to some extent a subjective assessment of the behavior of the model.

Findeisen et al. (1978) give the following definition of verification: A model is said to be verified if it behaves in the way the model builder wanted it to behave. This definition implies that there is a model to be verified, which means that not only the model equations have been set up but also that the parameters have been given reasonable realistic values. The sequence verification, sensitivity analysis, and calibration must consequently not be considered a rigid step-by-step procedure but rather as an iterative operation, which must be repeated a few times. The model is first given realistic parameters from the literature, then it is calibrated coarsely, and finally the model can be verified, followed by a sensitivity analysis and a finer calibration. The model builder will have to go through this procedure several times, before the verification and the model output in the calibration phase will be satisfactory.

It is recommended at this step that answers to the following pertinent questions are provided:

1. **Is the model stable in the long term?** The model is run for a long period with the same annual variations in the forcing functions to observe whether the state variable values remain at approximately the same levels. During the first period, state variables are dependent on the initial values for these and it is recommended that the model is also run with initial values corresponding to the long-term values of the state variables. The procedure also can be recommended for finding the initial values if they are not measured or known by other means. This question presumes that real ecosystems are long-term stable, which is not necessarily the case.

   The model is run for a long period using a certain pattern in the fluctuations of the forcing functions. It should then be expected that the state variables, too, show a certain pattern in their fluctuations. A sufficiently long simulation period should be selected to allow the model to demonstrate any possible instability.

2. **Does the model react as expected?** For example, if the input of toxic substances is increased, then we should expect a higher concentration of the toxic substance in the top carnivores. If this is not so, then it shows that some formulations may be wrong and these should be corrected. This question assumes that we actually know at least some behavior of the ecosystem, which is not always the case. In general, playing with the model is recommended at this phase. Through such exercises the modeler gets acquainted with the model and its reactions to perturbations. Models should generally be considered an experimental tool. The experiments are carried out to compare model results with observations and changes of the model are made according to the modeler's intuition and knowledge of the models behavior. If the modeler is satisfied with the accordance between model and observations, then the model is accepted as a useful description of the real ecosystem at least within the framework of the observations.

   This part of the verification is to a certain extent based upon more subjective criteria. Typically, the model builder formulates several questions about the model behavior and tests the model response by provoking changes in forcing functions or initial conditions. If the responses are not as expected, then the model structure or equations will have to be changed, provided that the parameter space is approved. Examples of typical questions will illustrate this operation: Will increased BOD_5-loading in a stream model imply decreased oxygen concentration? Will increased temperature in the same model imply decreased oxygen concentration? Will the oxygen concentration be at a minimum at sunrise when photosynthesis is included in the model? Will decreased predator concentration in a prey–predator model imply increased prey concentration? Will increased nutrient loadings in a eutrophication model give increased concentration of phytoplankton? etc. Numerous other examples can be given.

3. **Are the applied units correct? It is also recommended to check all the units at this phase of model development.** Check all equations for consistency of units. Are the units the same on both sides of the equation sign? Are the parameters used in the model consistent for the type of equations used and do the units match with the available data?

Almost inevitably it will be necessary at some stage during the verification to make assumptions about the statistical properties of the noise in the model. To conform with the properties of white noise any error sequence should broadly satisfy the following constraints: that its mean value is zero, that it is not correlated with any other error sequence and that it is not correlated with the sequences of measured input forcing functions. Evaluation of the error sequences in this fashion can therefore essentially provide a check on whether the final model invalidates some of the assumptions inherent in the model. Should the error sequences not conform to their desired properties, this suggests that the model does not adequately characterize all of the more deterministic features of the observed dynamic behavior. Consequently, the model structure should be modified to accommodate additional relationships.

To summarize this part of the verification:

1. the errors (comparison model output/observations) must have mean values of approximately zero.
2. the errors are not mutually cross related.
3. the errors are not correlated with the measured input forcing functions.

Results of this kind of analysis are given very illustratively in Beck (1978). Notice that this analysis requires good estimates of standard deviations in sampling and analysis (observations).

Notice finally that during verification, it is possible to perform multiple scenario analyses or "Gedanken Experiments." For example, we can test a eutrophication model by its response to the following test: we rent a chopper and buy 100,000 kg of phosphorus fertilizer and drop it instantly into a lake. The experiment could be made at no cost using the model, while it would be very expensive to rent a chopper and buy 100,000 kg of fertilizer. A major advantage of models is the ease to assess the system behavior under a wide array of scenarios.

Model verification may seem very cumbersome, but it is a very necessary step for the model development process. Through verification, one learns to know the model through its behavior, and the verification is furthermore an important checkpoint in the construction of a workable model. This also emphasizes the importance of good ecological knowledge of the ecosystem, without which the right questions as to the internal logic of the model cannot be posed.

Unfortunately, many models have not been verified properly due to lack of time, but experience shows that what might seem to be a shortcut will lead to an unreliable model, which at a later stage might require more time to compensate for the lack of verification. It must therefore be strongly recommended to invest enough time in the verification and to plan for the necessary allocation of resources in this important phase of the modeling procedure.

Sensitivity analysis follows verification. Through this analysis, the modeler gets a good overview of the most *sensitive components of the model*. Thus, sensitivity analysis attempts to provide a measure of the sensitivity of parameters, forcing functions, or submodels to the state variables of greatest interest in the model. If a modeler wants to simulate a toxic substance concentration in, for instance, carnivorous insects as a result of the use of insecticides, one will obviously choose this state variable as the most important one, maybe besides the concentration of the toxic substance concentration in plants and herbivorous insects. The sensitivity analysis tries to assess the "weak points" of the model: which parameters, forcing functions, and processes should be known better because the model results are very sensitive to the accuracy of this knowledge. In practical modeling, the sensitivity analysis is carried out by changing the parameters, the forcing functions, or the submodels. The corresponding response on the selected state variables is observed. Thus, the sensitivity, S, of a parameter, P, is defined as follows:

$$S = [\partial x/x]/[\partial P/P], \qquad (2.1)$$

where x is the state variable under consideration.

The relative change in the parameter value is chosen based on our knowledge of the certainty of the parameters. If, for instance, the modeler estimates the uncertainty to be about 50%, a change in the parameters at ±10% and ±50% is chosen and the corresponding change in the state variable(s) recorded. It is often necessary to find the sensitivity at two or more levels of parameter changes as the relationship between a parameter and a state variable is rarely linear.

A sensitivity analysis makes it possible to distinguish between high-leverage variables, whose values have a significant impact on the system behavior and low-leverage variables, whose values have minimal impact on the system. Obviously, the modeler must concentrate the effort on improvements of the parameters and the submodels associated with the high-leverage variables. The result of a sensitivity analysis of a eutrophication model with 18 state variables, that is presented in Jørgensen and Fath (2011), is shown in Table 2.2. It is used here to illustrate how to interpret the results of a sensitivity analysis. The sensitivity of the examined parameters by a 10% increase to phytoplankton s-phyt, to zooplankton s-zoo, to soluble nitrogen s-nit, and to soluble phosphorus s-phos is shown. These results indicate clearly that the parameters "maximum growth rate of phytoplankton and zooplankton," "mortality of zooplankton," and the "settling rate of phytoplankton" are very important parameters to determine accurately because they all have a sensitivity to the most important state variable, the phytoplankton, which is more than 0.5 or 50%, meaning that a change of the parameters by 10% would make a change of the phytoplankton concentration of more than 5%. On the other hand, the parameters "maximum denitrification rate," "mortality of fish," and "rate of mineralization" are significantly less important parameters. They all have a sensitivity of less

TABLE 2.2   Results of a Sensitivity Analysis of a 18 State Variable Model

Parameter	s-phyt	s-zoo	s-nit	s-phos
Maximum growth rate of phytoplankton	0.488	0.620	−0.356	−0.392
Maximum growth rate of zooplankton	−2.088	−4.002	2.749	4.052
Denitrification rate	−0.19	−0.010	−0.579	0.013
Fish concentration	0.008	0.012	−0.011	−0.014
Rate of mineralization	0.003	0.010	0.038	0.001
Mortality zooplankton	2.063	1.949	−3.479	−3.350
Settling rate	−1.042	−0.0823	0.321	0.388

than 0.1 or 10%. Therefore, they would change the phytoplankton less than 1% if the parameters are changed by 10%.

The interaction between the sensitivity analysis and the calibration could consequently work along the following lines:

1. A sensitivity analysis is carried out at two or more levels of parameter changes. Relatively great changes are applied at this stage.
2. The most sensitive parameters are determined more accurately either by a calibration or by other means.
3. Under all circumstances, great efforts are made to obtain a relatively good calibrated model.
4. A second sensitivity analysis is then carried out using more narrow intervals for the parameter changes.
5. Still further improvements of the parameter certainty are attempted.
6. A second or third calibration is then carried out focusing mainly on the most sensitive parameters.

A sensitivity analysis on submodels (process equations) can also be carried out. Then the change in a state variable is recorded when the equation of a submodel is deleted from the model or changed to an alternative expression, for instance, with more details built into the submodel. Such results may be used to make structural changes in the model. For example, if the sensitivity shows that it is crucial for the model results to use a more detailed given submodel, then this result should be used to change the model correspondingly.

If it is found that the state variable in focus is very sensitive to a certain submodel, then it should be considered which alternative submodels could be used and they should be tested and/or examined further in detail either in vitro or in the laboratory.

It can generally be stated that those submodels, which contain sensitive parameters are also submodels that are sensitive to the important state variable. But, on the other hand, it is not necessary to have a sensitive parameter included in a submodel to obtain a sensitive submodel. A modeler with a certain experience will find that these statements agree with intuition, but it is also possible to show that they are correct by analytical methods.

A sensitivity analysis of forcing functions gives an impression of the importance of the various forcing functions and tells us what accuracy is required of the forcing functions.

The goal of *calibration* is to improve the parameter estimation. Some parameters in causal ecological models can be found in the literature, not necessarily as constants but as approximate values or intervals. To cover all possible parameters for all possible ecological models including ecotoxicological models, we need, however, to know more than one billion parameters. It is therefore obvious that in modeling there is a particular need for *parameter estimation methods*. This will be discussed further in Chapter 11, where methods to estimate ecotoxicological parameters based upon the chemical structure of the toxic compound are briefly presented. Notice in this context, that ecotoxicological models are biogeochemical models that focus on toxic and ecotoxicological compounds in the environment, but they use entirely the principles of biogeochemical models.

In all circumstances, it is a great advantage to give even approximate values of the parameters before the calibration gets started, as already mentioned above. It is, of course, much easier to search for a value between 1 and 10 than to search between 0 and $+\infty$. Even where all parameters are known within intervals either from the literature or from estimation methods, it is usually necessary to calibrate the model. Several sets of parameters are tested by the calibration and the various model outputs of state variables are compared with measured values of the same state variables. The parameter set that gives the best agreement between model output and measured values is chosen.

The need for the calibration can be explained by use of the following characteristics of biogeochemical ecological models and their parameters:

1. Most parameters in environmental science and ecology are not known as exact values. Therefore, all literature values for parameters (Jørgensen et al., 1991, 2000) have a certain uncertainty. Parameter estimation methods must be used when no literature value can be found, particularly ecotoxicological models, see, for instance, Jørgensen (1991, 1992a) and Chapter 11. In addition, we must accept that unlike many physical parameters, ecological ones are not constant but rather changing in time or situation (Jørgensen, 1986, 1992b, 2002). This point will be discussed further in Chapter 7, where the structurally dynamic model type is presented in detail.

2. All models in ecology and environmental sciences are simplifications of nature. The most important components and processes may be included, but the model structure does not account for every detail. To a certain extent the influence of some unimportant components and processes can be taken into account by the calibration. This will give slightly different values for the parameters from the real, but unknown, values in nature, but the difference may partly account for the influence from the omitted details.

3. Most models in environmental sciences and ecology are "lumped models," which means that one parameter represents the average values of several species. As each species has its own characteristic parameter value, the variation in the species composition with time will inevitably give a corresponding variation in the average parameter used in the model. Adaptation and shifts in species composition will require other approaches as touched on. This will be discussed in more detail in Chapter 7.

A calibration cannot be carried out randomly if more than a couple of parameters have been selected for calibration. If, for instance, 10 parameters have to be calibrated and the uncertainties justify the testing of 10 values for each parameter, the model has to be run 10^{10} times to cover all combinations, which is an impossible task. Therefore, the modeler must learn the behavior of the model by varying one or two parameters at a time and observing the response of the most crucial state variables. In some (few) cases, it is possible to separate the model into several submodels, which can be calibrated approximately independently. Although the calibration described is based to some extent on a systematic approach, it is still a trial-and-error procedure.

However, procedures for automatic calibration are available. This does not mean that the trial-and-error calibration described above is redundant. If the automatic calibration should give satisfactory results within a certain frame of time, then it is necessary to calibrate only six to nine parameters simultaneously. Under any circumstances, it will become easier to find the optimum parameter set, the narrower the ranges of the parameters are before the calibration gets started.

In the trial-and-error calibration, the modeler has to set up, somewhat intuitively, some calibration criteria. For instance, you may want to simulate accurately the minimum oxygen concentration for a stream model and/or the time at which the minimum occurs. When you are satisfied with these model results, you may then want to simulate the shape of the oxygen concentration versus time curve properly, and so on. You calibrate the model step by step to achieve these objectives step by step.

If an automatic calibration procedure is applied, it is necessary to formulate objective criteria for the calibration. A possible function could be based on an equation similar to the calculation of the standard deviation:

$$Y = \left[\left(\sum((x_c - x_m)^2 / x_{m,a})/n\right)\right]^{1/2} \tag{2.2}$$

where x_c is the computed value of a state variable, x_m is the corresponding measured value, $x_{m,a}$ is the average measured value of a state variable, and n is the number of measured or computed values. Y is followed and computed during the automatic calibration and the goal of the calibration is to obtain as low a Y-value as possible. Often, the modeler is, however, more interested in a good agreement between model output and observations for the one or two most important state variables, and less interested in a good agreement with other state variables. Therefore, one may choose weights for the various state variables to account for the emphasis put on each state variable in the model. For a model of the fate and effect of an insecticide, emphasis may be put on the toxic substance concentration of the carnivorous insects and considering the toxic substance concentrations in plants, herbivorous insects, and soil to be of less importance. Therefore, a weight of 10 is applied for the first state variable and only one for the subsequent three.

If it is impossible to calibrate a model properly, then this is not necessarily due to an incorrect model but may be due to poor quality of the data. The quality of the data is crucial for calibration. It is, furthermore, of great importance that the observations reflect the system dynamics. If the objective of the model is to give a good description of one or a few state variables, it is essential that the data can show the dynamics of just these internal variables. The frequency of the data collection should therefore reflect the dynamics of the state variables in focus. This rule has unfortunately often been violated in modeling.

It is strongly recommended that the dynamics of all state variables are considered before the data collection program is determined in detail. Frequently, some state variables have particularly pronounced dynamics in specific periods—often in spring—and it may be of great advantage to have a dense data collection in this period in particular. Jørgensen et al. (1981) show how a dense data collection program in a certain period can be applied to provide additional certainty for the determination of some important parameters. This question will be further discussed later in this chapter; see Section 2.9.

From these considerations, recommendations can now be drawn up about the feasibility of carrying out a calibration of a model in ecology:

1. Find as many parameters as possible from the literature; see in this context Jørgensen et al. (1991, 2000). Even a *wide* range for the parameters should be considered very valuable, as approximate initial guesses for all parameters are needed.
2. If some parameters cannot be found in the literature, which is often the case, the estimation methods mentioned in detail in Jørgensen and Fath (2011) and partially in Chapter 10 may

be used. For some crucial parameters, it may be recommendable to determine them by experiments in situ or in the laboratory.

3. A sensitivity analysis should be carried out to determine which parameters are most important to be known with high certainty. The estimation methods and the determination of the parameters by experiments should focus mainly on the most sensitive parameters.

4. An intensive data collection program for the most important state variables should be used to provide a better estimation for the most crucial parameters. For further details see Section 2.9.

5. First, at this stage, the calibration should be carried out by use of the data not yet applied. The most important parameters are selected and the calibration is limited to these or, at the most, to eight to ten parameters. In the first instance, the calibration is carried out by using the trial-and-error method to get acquainted with the model reaction to changes in the parameters. An automatic calibration procedure is used subsequently to polish the parameter estimation.

6. These results are used in a second sensitivity analysis, which may give results different from the first sensitivity analysis.

7. A second calibration is now used on the parameters that are most important according to the second sensitivity analysis. In this case, too, both the above-mentioned calibration methods may be used. In some cases, the modeler would repeat the steps 6 and 7 one time more and make a third calibration. After this final calibration, the model can be considered calibrated and we can go to the next step: validation.

The calibration should *always* be followed by a validation. By this step the modeler tests the model against an *independent* set of data to observe how well the model simulations fit these data. It may be possible, even in a data rich situation, to force a wrong model by the parameter selection to give outputs that fit well with the data. It must, however, be emphasized that the validation only confirms the model behavior under the range of conditions represented by the available data. So, it is preferable to validate the model using data obtained from a period in which condition other than those of the period of data collection for the calibration prevail. For instance, when a eutrophication model is tested, it should preferably have data sets for the calibration and the validation, which differ by the level of eutrophication. This is often impossible or at least very difficult as it may correspond to a complete validation of a prognosis, which at the best takes place at a later stage of the model development. However, it may be possible and useful to obtain data from a *certain* range of nutrient loadings, for instance from a humid and a dry summer. Alternatively, it may be possible to get data from a similar ecosystem with approximately the same morphology, geology, and water chemistry as the modeled ecosystem. Similarly, a BOD/DO model should be validated under a wide range of BOD-loadings, a toxic substance model under a wide range of concentrations of the considered toxic substances, and a population model by different levels of the populations, etc.

If an ideal validation cannot be obtained, then it is still important to validate the model as best as possible. The method of validation is dependent on the objectives of the model. A comparison between measured and computed data by use of an objective function (2.2) is an obvious test. This is, however, often not sufficient, as it may not focus *on* all the main objectives of the model, but only on the general ability of the model to describe correctly the state variables of the ecosystem. It is necessary, therefore, to translate the main objectives of the model into a few validation criteria. They cannot be formulated generally, but are individual for the model and the modeler. For instance, if we are concerned with the eutrophication in an aquatic ecosystem, it would be useful to compare the measured and computed maximum concentrations of phytoplankton. The discussion of the validation can be summarized by the following issues:

1. Validation is always required to obtain information about the model reliability.
2. Attempts should be made to get data for the validation, which are entirely different from those used in the calibration. It is important to have data from a wide range of forcing functions that are defined by the objectives of the model.
3. The validation criteria are formulated based on the objectives of the model and the quality of the available data. The main purpose of the model may, however, be an exploratory analysis to understand how the system responds to the dominating forcing functions. In this case a structural validation is probably sufficient.

Validation is a very important modeling step because it will give the uncertainty of the model results. It attempts to answer the question: which model uncertainty should we consider when using the model to develop strategies for environmental management? If we use the model as a research tool, the validation will tell us whether the model results can be used to support or reject a hypothesis. The uncertainty determined by the validation relatively to the difference between the hypothesis and the model results will be decisive. In Chapter 8 of Jørgensen and Fath (2011), a eutrophication model with 18 state variables is applied as a case study to demonstrate how the validation results can be used to assess the expected uncertainty of the prognoses developed by the model.

The validation result can also be used to consider the model revisions that would be needed to reduce the uncertainty. In our effort to improve the model, we should ask the following pertinent questions:

1. What is the uncertainty of the observations (measurements)? If the uncertainty of the model is not very different from the uncertainty of the observations, then it will probably be beneficial to get more reliable observations with less uncertainty.
2. Do the observations represent the system dynamics? If not, then more frequent monitoring should be considered for some period to capture the systems dynamics.
3. Are some important processes or components missing or described wrongly in the model? It is in this context, as previously mentioned, important to set up a mass and/or energy balance to reveal the most important processes and sources.
4. It is recommended to give a sufficiently comprehensive answer to question 3 and eventually use the model experimentally to find the best answer; see the examples of using models experimentally presented in Chapter 10. It is quite easy in most cases to replace important equations by other expressions or add new components or processes and so on. Such experiments are very elucidating for the importance of formulations and inclusion of processes. Small changes in process equations that make big changes in the model results uncover the soft points of the model and may inspire additional experiments or observations in situ or in the laboratory, and eventually to further changes of the model.

It should be emphasized that the "ideal" model can never be achieved, but we can, step-by-step, by steadily questioning the model and using the four points again and again, improve the model quality moving asymptotically toward the ideal model. An ideal model is, however, not necessary to have as a useful and powerful tool in environmental management and ecosystem research. A satisfactory calibration and validation with sufficiently low uncertainties to allow application in a defined context would be the general requirement for the pragmatic modeler.

2.4 ECOLOGICAL AND ENVIRONMENTAL BIOGEOCHEMICAL MODELS

This section attempts to present briefly the history of ecological and environmental modeling and to give an overview of how widely biogeochemical models have been applied. From the history we can learn why it is essential to draw upon the previously gained experience and what goes wrong when we do not follow the recommendations that we have been able to set up to avoid previous flaws.

Fig. 2.6 gives an overview of the history of ecological modeling. The nonlinear time axis gives approximate information on the year, when the various development steps took place. The first models of the oxygen balance in a stream (the Streeter-Phelps model) and of the prey—predator relationship (the Lotka—Volterra model) were developed back in the early 1920s. In the 1950s and the 1960s further development of population dynamic models took place. More complex river models were also developed in the 1960s. These developments could be named the second generation of models.

The wide use of ecological models in environmental management and ecological research—mainly biogeochemical models—started around the year 1970, when the first eutrophication models emerged and very complex river models were developed. These models may be named the third generation of models. They are characterized by often being too complex, because it was so easy to write computer programs, which could handle rather complex models. To a certain extent, it was the revolution in computer technology that created this model generation. It became, however, clear in the mid-1970s that the limitations in modeling are not the computer and the mathematics, but the data and our knowledge about ecosystems and ecological processes. So, the modelers became more critical in their acceptance of models. They realized that a profound knowledge about the ecosystem, the problem, and the ecological components were the necessary basis for development of sound ecological models. A result of this period is all the recommendations given in the next chapter:

- follow strictly all the steps of the procedure, i.e., conceptualization, selection of parameters, verification, calibration, examination of sensitivity, validation.
- find a complexity of the model, which considers a balance between data, problem, ecosystem, and knowledge.
- a wide use of sensitivity analyses is recommendable in the selection of model components and model complexity make parameter estimations by using all the methods, i.e., literature review, determination by measurement in laboratory or in situ, use of intensive measurements, calibration of submodels and the entire model, theoretical system

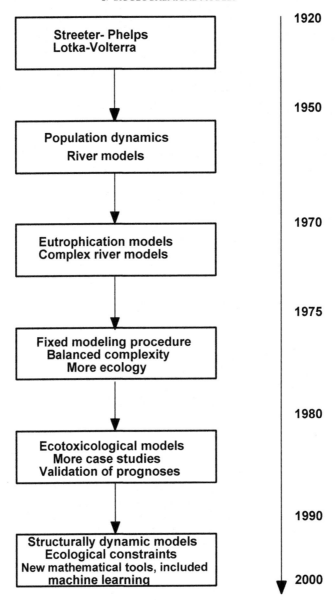

FIGURE 2.6 The development (history) of ecological and environmental models is shown schematically.

ecological considerations and various estimation methods based on allometric principles, and chemical structure of the considered chemical compounds.

Parallel to this development, ecologists became more quantitative in their approach to environmental and ecological problems, probably because of the needs, formulated by environmental management. The quantitative research results by ecology from the late 1960s until today have been of enormous importance for the quality of the ecological models. They are probably just as important as the development in computer technology.

The models from this period, going from the mid-1970s to the mid-1980s, could be called the fourth generation of models. The models from this period are characterized by having a relatively sound ecological basis, and emphasis on realism and simplicity. Many models were validated in this period with an acceptable result and for some (few) it was even possible to validate the prognosis.

The conclusions from this period may be summarized in the following three points:

1. Provided that the recommendations given above were followed and the underlying database was of good quality, it was possible to develop models that could be used as prognostic tool.
2. Models based upon a database of not completely acceptable quality should maybe not be used as a prognostic tool, but they could give an insight into the mechanisms behind the environmental management problem, which is valuable in most cases. Simple models are often of particular value in this context.
3. Ecologically sound models, i.e., models based upon ecological knowledge, are powerful tools in understanding ecosystem behavior and as tools for setting up research priorities. The understanding may be qualitative or semiquantitative, but has in any case proved to be of importance for ecosystem theories and a better environmental management.

The shortcomings of modeling were, however, also revealed clearly in the third and fourth phase of modeling. It became clear that the models were rigid in comparison with the enormous flexibility, which was characteristic of ecosystems. The hierarchy of feedback mechanisms that ecosystems possess was not accounted for in the models, which made the models incapable of predicting adaptation and structural dynamic changes. Since the mid-1980s modelers have proposed many new approaches such as (1) fuzzy modeling, (2) examinations of catastrophic and chaotic behavior of models, and (3) application of goal functions to account for adaptation and structural changes. Application of objective and individual modeling, expert knowledge, and artificial intelligence offers some new additional advantages in modeling. It will in the last chapter of the volume discuss when it is advantageous to apply these approaches and what can be gained by their application. All these recent developments could be named the fifth generation of modeling. The model types from this model generation are covered in other chapters of the book.

The most significant development in ecological modeling is, however, that an enormous wide spectrum of models has been developed for a number of ecosystems and a number of environmental problems.

Table 2.3 reviews types of ecosystems, which have been modeled by biogeochemical models up to year 2000. An attempt has been made to indicate the modeling effort by using a scale from 0 to 5. 5 is very intense modeling effort, more than 50 different modeling approaches can be found in the literature; 4, intense modeling effort, 20—50 different modeling approaches can be found in the literature; 4—5 may be translated to class 4 but on the edge of an upgrading to class 5; 3, some modeling effort, 6—19 different modeling approaches are published; 2, few (2—5) different models that have been fairly well studied have been published; 1, one good study and/or a few not sufficiently well-calibrated and validated models; and 0, almost

TABLE 2.3   Biogeochemical Models of Ecosystems

Ecosystem	Modeling Effort
Rivers	5
Lakes, reservoirs, ponds	5
Estuaries	5
Coastal zone	4
Open sea	3
Wetlands	5
Grassland	4
Desert	1
Forests	5
Agriculture land	5
Savanna	2
Mountain lands	
(Above timberline)	1
Arctic and Antarctic ecosystems	2
Coral reef	3
Waste water systems	5
The ecosphere could maybe be added	5

no modeling effort have been published and even not one well-studied model. Notice that the classification is based on the number of different models not on the number of case studies where these models have been applied. In most cases the same models have been used on several cases studies.

Table 2.4 reviews similarly the environmental problems, which have been modeled up today. The same scale is applied to show the modeling effort as in Table 2.3. Table 2.4 covers besides biogeochemical models also models used for management of population dynamics in national parks and steady state models applied as ecological indicators. It is advantageous to apply goal functions in conjunction with a steady state model to obtain a good ecological indication, as proposed by Christensen (1991) and (1992).

The combinations of ecosystems and environmental problems are $16 \times 19 = 304$ and so far most of the 304 different models combining the two classifications can be found in the ecological modeling literature. Some of these combinations have been modeled even very extensively, for instance the climate change (the greenhouse effect) of the entire ecosphere, eutrophication of lakes, oxygen depletion of rivers, and heavy metal and pesticide pollution of grassland and agricultural land.

TABLE 2.4   Models of Environmental Problems

Problem	Modeling Effort
Oxygen balance	5
Eutrophication	5
Heavy metal pollution, all types of ecosystems	4
Pesticide pollution of terrestrial ecosystems	4–5
Other toxic compounds include ERA	5
Regional distribution of toxic compounds	5
Protection of national parks	3
Management of populations in national parks	3
Endangered species (includes population dynamic models)	3
Ground water pollution	5
Carbon dioxide/greenhouse effect	5
Acid rain	5
Total or regional distribution of air pollutants	5
Change in microclimate	3
As ecological indicator	4
Decomposition of the ozone layer	4
Relationships health–pollution	3
Consequences of climate changes	4
Water stress	5

In addition, numerous biogeochemical models have been developed for important processes in ecosystems—models that could be applied as submodels in ecosystem models associated with various environmental problems and models that are developed to understand better the results of combining important ecological processes or to reveal more details of a core part of the ecosystem that is particularly important for the development of the entire ecosystems.

In most of the other chapters, there are illustrative case studies that demonstrate the characteristic features of the model. There are, however, so many illustrations in the literature of this model type, that it seems redundant. Examples and illustrations can be found in Chapter 8 of Jørgensen and Fath (2011). Details are given in this reference for a medium complex eutrophication model (18 state variables), a relatively simple river model (3 state variables, and a wetland model (removal of nutrients by wetlands) with 25 state variables.

References

Beck, M.B., 1978. Random signal analysis in an environmental sciences problem. Applied Mathematical Modelling 2, 23—29.

Christensen, V., 1991. On ecopath, fishbyte, and fisheries management. Fishbyte 9 (2), 62—66.

Christensen, V., 1992. Network Analysis of Trophic Interactions in Aquatic Ecosystems (Ph. Diss.). Royal Danish School of Pharmacy.

Findeisen, W., Lastebrov, A., Lande, R., Lindsay, J., Pearson, M., Quade, E.S., 1978. A Sample Glossary of Systems Analysis. Working Paper WP-78—12. International Institute for Applied Systems Analysis, Laxenburg, Austria.

Jeffers, N.R.J., 1978. An Introduction to Systems Analysis with Ecological Applications. E. Arnold 220 pp.

Jørgensen, S.E., Fath, B., 2011. Fundamentals of Ecological Modelling, fourth ed. Elsevier, Amsterdam. 400 pp.

Jørgensen, S.E., 1986. Structural dynamic model. Ecological Modelling 31, 1—9.

Jørgensen, S.E., 1991. A model for the distribution of chromium in Abukir Bay. In: Jørgensen, S.E. (Ed.), Modelling in Environmental Chemistry, vol. 17. Elsevier, Amsterdam.

Jørgensen, S.E., 1992a. Parameters, ecological constraints and exergy. Ecological Modelling 62, 163—170.

Jørgensen, S.E., 1992b. Development of models able to account for changes in species composition. Ecological Modelling 62, 195—208.

Jørgensen, S.E., 2002. Integration of Ecosystem Theories: A Pattern, third ed. Kluwer Acad. Publ. Dordrecht, Boston, London. 428 pp.

Jørgensen, S.E., Jørgensen, L.A., Kamp Nielsen, L., Mejer, H.F., 1981. Parameter estimation in eutrophication modelling. Ecological Modelling 13, 111—129.

Jørgensen, S.E., Nors Nielsen, S., Jørgensen, L.A., 1991. Handbook of Ecological Parameters and Ecotoxicology. Elsevier, Amsterdam. Published as CD under the name ECOTOX, with L.A. Jørgensen as first editor in year 2000.

Jørgensen, L.A., Jørgensen, S.E., Nielsen, S.N., 2000. Ecotox. CD.

Jørgensen, S.E., Bendoricchio, G., 2001. Fundamentals of Ecological Modelling, third ed. Elsevier, Amsterdam. 628 pp.

Dynamic Population Models

T. Legović

Ruđer Bošković Institute, Zagreb, Croatia

E-mail: legovic@irb.hr

O U T L I N E

3.1 A DEFINITION OF A POPULATION MODEL

By a population we mean the number of alive organisms in a given area. We are interested in predicting how this number will change in future. For this purpose we create models which include processes that are responsible for a population change. For example, if organisms reproduce in the area of study, the process of reproduction should be included in the population model because it results in population change.

A model which expresses population processes mathematically is called a *mathematical model*. Usually, mathematical models we are interested in result in *population conservation equations*. A common characteristic of these equations is to express mathematically the following:

Population at the time instant $t + \Delta t$ = population at time instant t

+ sum of processes which cause population to increase in the unit time Δt

− sum of processes which cause population to decrease in the unit time Δt.

If we can express all the encompassed processes as a function of the existing population or of a past value of the population which we presently know, we have a *predictive model*.

For example:

$$N(t + \Delta t) = N(t) + \left[\Sigma f_i(N(t), t) - \Sigma g_j(N(t), t) \right] * \Delta t \qquad (3.1)$$

where $N(t)$ stands for the population in a time instant, t (known); $N(t + \Delta t)$ is the population in the time instant $t + \Delta t$ (unknown), f_i is the i-th contribution to population growth and g_j is the j-th contribution to population decline (known) in the time interval Δt (known).

Once, $N(t + \Delta t)$ is computed we substitute this value to the right side instead of $N(t)$ and compute $N(t + 2\Delta t)$. Continuing this process as long as we wish, we are predicting population value into the future.

The above example is of a mathematical model in discrete form, as opposed to a continuous form which will be seen in examples that follow.

3.2 THE FIRST LAW (MODEL) OF POPULATION DYNAMICS: MALTHUS LAW

Let us start with a very simple model and turn it into a mathematical model.

Assume that at the time, t, which we set to be zero, we have $N(t = 0) = N_o$ organisms. The number is known because we can count or estimate the number of organisms.

We ask the following question:

What will be the number of organisms at time $t = 1$ which we denote by $N(t = 1) = N_1$?

Here, the time unit is arbitrary; it may be 1 h, one day, one month, or whatever we find useful. We write:

$N_1 = N_o$ (i.e., the no. of organisms with which we started) + the number of organisms which were born (between $t = 0$ and $t = 1$) − the number of organisms which died (between $t = 0$ and $t = 1$) + the number of immigrated organisms (between $t = 0$ and $t = 1$) − the number of emigrated organisms (between $t = 0$ and $t = 1$).

For simplicity, let us close our area of study so that immigration and emigration are not possible.

Then, we are left with:

$N_1 = N_o$ (i.e., the no. of organisms with which we started) + the number of organisms which were born (between $t = 0$ and $t = 1$) − the number of organisms which died (between $t = 0$ and $t = 1$).

Now we need ecologists to tell us how many organisms are born and how many die per capita in a unit of time. Let us denote by a the number of organisms that are normally born for each existing individual of our population in a unit of time. Similarly, let us denote by b the number of organisms that normally die per existing individual. Hence the number of born individuals in a unit of time will is $a * N_o$. Similarly, the number of died individuals in a unit of time will be $b * N_o$. Now we can put these numbers into the above equation:

$$N_1 = N_o + a * N_o - b * N_o. \tag{3.2}$$

To further simplify this equation, we denote by r the difference between a and b. Hence:

$$N_1 = N_o + r * N_o = (1 + r)N_o. \tag{3.3}$$

Here r is called the *biotic potential*. If the biotic potential is zero (i.e., the number of born is equal to the number of died organisms in a unit of time), then obviously:

$$N_1 = N_o$$

That is, the number in the next time instant will be the same as the number we started with.

Let us now make one key assumption which will enable us to predict the population number far into the future: assume that the biotic potential r is constant for all time instants in the future. Then, after the second time instant we have:

$$N_2 = N_1 + r * N_1 = (1 + r)N_1 \tag{3.4}$$

Now insert N_1 from Eq. (3.3) into Eq. (3.4) and get:

$$N_2 = (1 + r)(1 + r)N_o = (1 + r)^2 N_o.$$

We can generalize the population after any time instant t into the future:

$$N_t = (1 + r)^t N_o \quad where \quad t = 1, 2, 3, \dots \tag{3.5}$$

Furthermore, by using our key assumption, note from Eqs. (3.3) and (3.4) we can generalize:

$$N_{t+1} = (1 + r)N_t. \tag{3.6}$$

Eq. (3.6) is our first mathematical model.

This model is called the *first law of population growth* or the *Malthus law*.

The expression Eq. (3.5) is called the solution to Eq. (3.6) given the initial value N_o.

Namely, to arrive at Eq. (3.5) from Eq. (3.6) we needed to know what was the value of a population at the time $t = 0$.

In mathematics Eq. (3.6) is called a difference equation.

Note also that Eq. (3.6) assumes a jump from t to $t + 1$ and that we do not know anything about the population between t and $t + 1$. For this reason Eq. (3.6) is called discrete as opposed to continuous equation.

To see the difference between discrete and continuous equation let us turn Eq. (3.6) into a continuous equation.

First rewrite Eq. (3.6):

$$N_{t+1} - N_t = rN_t \tag{3.7}$$

Now assume that we are not going to jump from t to $t+1$ but from t to $t + \Delta t$ where Δt is much smaller than 1. Eq. (3.7) becomes:

$$N_{t+\Delta t} - N_t = r^*_{\Delta t} N_t \Delta t \tag{3.8}$$

Now two things must be explained.

1. Why we wrote $r^*_{\Delta t}$ instead of r. This is because r was a number which denoted the difference between born and dead per individual in one unit of time while $r^*_{\Delta t}$ denotes the difference which occurred in Δt which is smaller than 1, and hence $r^*_{\Delta t}$ is smaller than r.
2. Why did we put Δt on the right side? In Eq. (3.7) it did not need to be there because Δt was equal to 1. Eq. (3.8) reads as follows: on the left side is the number of organisms and so it must be on the right side. But the right side reads:

$$[(\text{no. of born} - \text{no. of dead})/(\text{one organism} * \Delta t)] * \text{no. of organisms} * \Delta t$$

Hence, no. of organisms * Δt cancels out and what remains is the number of born $-$ number of dead for the whole population. So, the left side has the same units as the right side.

Now dividing with Δt we have:

$$(N_{t+\Delta t} - N_t)/\Delta t = r^*_{\Delta t} N_t$$

We can further shrink Δt into the infinitesimal increment dt and we have:

$$(N_{t+dt} - N_t)/dt = r_{dt} N_t$$

The left side is the derivative dN/dt which denotes the rate of change of the population at the time instant t. Let us denote r_{dt} by r_c, and we have:

$$dN/dt = r_c N \tag{3.9}$$

where r_c is the instantaneous difference between per capita birth and death. Also, the subscript c stands for the continuous case.

Eq. (3.9) is called the continuous form (as opposed to the discrete form given by Eq. (3.6)) of the Malthus law of population growth. Of course, the word "growth" refers to our expectation that r and r_c are positive because if they are zero the law should be called the first law of population stagnation and if they are negative the law should be called the first law of population decline.

In mathematics, Eq. (3.9) is called a differential equation.

Given $N(t = 0) = N_o$ the solution to Eq. (3.9) is:

$$N(t) = N_o e^{r_c t} \tag{3.10}$$

Hence, the solution to Eq. (3.6) is a geometric growth (Eq. (3.6)) and the solution to Eq. (3.9) is an exponential growth (Eq. (3.10)).

By equating Eq. (3.5) to Eq. (3.10) we see that in the case that $r_c = ln(1 + r)$ the two dynamics coincide in the points of geometric growth.

When do we expect that Eqs. (3.5) and (3.10) will adequately represent population dynamics in nature?

There are two hypotheses that need to be satisfied. First is that population reproduction and death must not be influenced by a finite environment. However, if r and r_c are positive, the population will grow and sooner or later the population will start to deviate from the first law.

Hence, we expect that the first law will apply as long as the population is much smaller than limitations in food supply or space.

The second assumption is that the population must be large enough so that individual differences in organism reproduction and death do not matter. When the population falls below a certain number, individual-based models are more appropriate (Gourney and Nisbet, 1998).

3.3 THE SECOND LAW OF POPULATION DYNAMICS: VERHULST LAW

The first law of population dynamics does not account for the finite environment in which food supply and space are limited. Verhulst (1838) modified Eq. (3.9) by assuming that population growth, besides the Malthus term needs to be multiplied by a linearly decreasing term of N, resulting in the equation:

$$dN/dt = r_c(1 - N/K)N \tag{3.11}$$

where K is the largest population that food sources in the environment can support.

K is called the *carrying capacity* of the environment.

It is clear that Eq. (3.11) is close to Eq. (3.9) when $N \ll K$ (because N/K does not differ much from zero). As N approaches K, the right side of Eq. (3.11) approaches zero and the population becomes constant. If $N > K$ the right side of Eq. (3.11) is negative, which means that the population decreases to K.

Given the initial population $N(t = 0) = N_o$, the solution to Eq. (3.11) is:

$$N(t) = \frac{K}{1 - \left(1 - \frac{K}{N_0}\right)e^{-r_c t}} \tag{3.12}$$

A population whose dynamics is described by Eq. (3.11) and function (3.12) is said to follow the *second law of population dynamics* or *Verhulst law*. The population is called *logistic*.

A graph of the function (3.12) for three initial values is given in Fig. 3.1.

Carrying capacity of the environment is K = 200. For all three populations: $r_c = 0.1$.

The logistic population is often taken when constructing models of populations that interact with each other and thus forming multipopulation models.

Let us consider harvesting the logistic population which is proportional to N so that the dynamics is given by (Schaefer, 1954):

$$dN/dt = rN(1 - N/K) - eN \tag{3.13}$$

where e is a harvesting effort (proportional to the number of fishing days and number of fishing tools). The effort which leads to the maximum sustainable yield (MSY) is given by

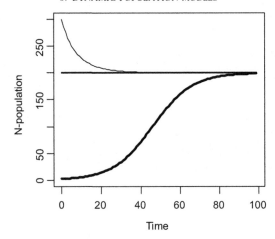

FIGURE 3.1   Dynamics of three logistic populations. $N_1 = 300$, $N_2 = 200$, $N_3 = 2$.

$e_{opt} = r/2$ and the corresponding stationary value of N is $N^* = K/2$. The maximum sustainable yield is given by:

$$MSY = rK/4. \tag{3.14}$$

The resulting value $N^* = K/2$ is stable and the strategy is termed a stable harvesting strategy. This means that as long as e is set approximately equal to e_{opt} no monitoring is needed.

On the other side, if one uses a harvesting quota, i.e., a constant value in Eq. (3.13) instead of eN, the MSY is also given by Eq. (3.14) but now $N^* = K/2$ is unstable in the sense that if the population drops below $K/2$ and due to existing uncertainties in the environment this will certainly happen, the population tends to extinction. Hence, to use this strategy and save the population from extinction, first the adequate monitoring must be put into place.

In case we do have an appropriate monitoring in place, the dynamic control (bang–bang theory) states that the optimum harvesting procedure is:

If $N > K/2$ harvest the population to $K/2$;
If $N = K/2$ harvest $rK/4$;
If $N < K/2$ do not harvest.

The above result is understandable because being at the $N = K/2$ gives the maximum yield. Hence, when N is above that point, it is most advantageous to harvest until $K/2$ is reached. On the other side, when $N < K/2$ the population should recover at the maximum rate, i.e., it should not be harvested at all.

3.4 A LOGISTIC POPULATION IN PERIODIC ENVIRONMENT

In the second law of population dynamics r_c and K are taken to be constant. On the other side, it is clear that food supply changes daily, seasonally, or from one year to the other.

Let us take the carrying capacity to be a periodic function of time. In the simplest case we may assume that:

$$K(t) = K_0 + K_1 sin\omega t \tag{3.15}$$

where K_0 is the average value of the carrying capacity, K_1 is the amplitude of periodic variation, and ω is the circular frequency: $\omega = 2\pi/T$ where T is the period.

Upon inserting Eq. (3.15) into Eq. (3.11) and integrating, one gets a complicated expression which describes the dynamics of logistic population in the periodic environment. The dynamics has roughly two parts: a transient starting with N_0 that continues toward a periodic behavior.

Two limiting cases have been identified (May, 1976):

1. $r_c T \gg 1$. This case means that the population reacts quickly because r_c is large while the carrying capacity changes slowly because the period is large.The population has plenty of time to adapt to changing environment and hence it follows it closely. The asymptotic dynamics is given approximately with:

$$N(t) \approx K_0 + K_1 sin\omega t \tag{3.16}$$

2. $r_c T \ll 1$. In this case the environment changes rapidly and the population is averaging changes in carrying capacity. The asymptotic dynamics is approximately constant:

$$N(t) = \sqrt{K_0^2 - K_1^2} \tag{3.17}$$

It is interesting that in case (1) the average value of the population is K_0 while in case (2) it is smaller. In case (2) as K_1 increases to K_0, the population $N(t)$ decreases to 0.

Fig. 3.2 shows dynamics of a population in cases (1) and (2).

If one were to harvest a logistic population in the periodic environment given by expression (3.16) according to proportional harvesting strategy (Legović and Perić, 1984) so that the dynamics of the population is given by

$$dN/dt = rN[1 - N/K(t)] - eN$$

then the optimum harvesting effort which produces the maximum sustainable average yield (MSAY) is again $e_{opt} = r/2$. In case the population follows the periodic change in $K(t)$ then the MSAY is the same as in the peaceful environment, i.e., $MSAY = rK_0/4$. But if the population is unable to track changes in $K(t)$, then MSAY is smaller and is given by:

$$N(t) \approx r\left(\sqrt{K_0^2 - K_1^2}\right)\Big/4 \tag{3.18}$$

Since populations in nature are all in between the two considered extremes, we conclude that in the periodic environment, MSAY will be smaller than MSY in the peaceful environment.

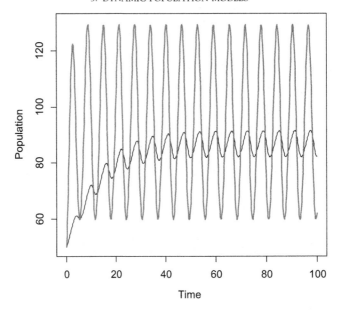

FIGURE 3.2   Dynamics of two logistic populations with periodic carrying capacity $K(t) = 100 + 50 \sin (t)$. $r_c = 1$, $r_c = 0.1$ (black line).

3.5 A LOGISTIC POPULATION IN A RANDOM ENVIRONMENT

Environment may be constant for a while, it may have a periodic component but it may also have a random component. This means that some environmental parameters may vary by chance, hence their extent and timing cannot be predicted. We are interested to find out the population dynamics in such an environment.

If we consider a logistic population immersed in a random environment, it is clear that both r and K may vary in an unpredictable fashion.

Consider a simpler case (Beddington and May, 1977):

$$dN/dt = r(t)N - <r> N^2/K \qquad (3.19)$$

where $r(t) = <r> + g(t)$. $<r>$ is the mean and $g(t)$ is the uncorrelated white noise with the mean equal to 0 and variance s^2. Such a white noise is denoted by:

$$g(t) = [0, s^2]. \qquad (3.20)$$

With the above white noise, the effect of random environment on the biotic potential of the population of organisms is defined (Fig. 3.3). The case is not general because a random variation is included only in the linear term of Eq. (3.19) and not in the nonlinear term. The solution to Eq. (3.19) will be a distribution which will have a transient part and later it will have a stationary phase. We are interested to find out the mean value that N will attain after a long time.

The mean value is:

$$<N> = K(r - s^2/2)/r \qquad (3.21)$$

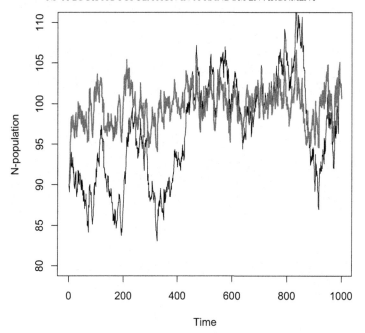

FIGURE 3.3 Dynamics of two simulations are shown: $<r> = 0.01$ (black), 0.1 (red (gray in print versions)) with $g(t) = [0,0.01]$, K = 100.

We see that the mean value of N is not K but smaller by a factor of $Ks^2/2r$ so that in case the variance $s^2 = 2r$, the mean value is zero. Of course, the population will reach extinction for values of s^2 which are smaller than $2r$.

In case we were to harvest the population with a proportional harvesting rate eN, after a long time (asymptotically) the population mean value will reach:

$$< N > = K(r - e - s^2/2)/r \tag{3.22}$$

The optimum harvesting effort which will produce the maximum sustainable average yield (MSAY) is:

$$e_{opt} = r/2 - s^2/4 \tag{3.23}$$

while the corresponding MSAY is:

$$MSAY = (K/4r)(r - s^2/2)^2. \tag{3.24}$$

The mean value of N during the optimum harvesting will reach:

$$< N_{eopt} > = (r - s^2/2)K/2r \tag{3.25}$$

We see that e_{opt}, MSAY, and N_{eopt} will be smaller in the random environment than in peaceful environment.

We see that in all simulations regardless of whether the population reacts faster (with larger r) or slower (with smaller r), the average value $<N>$ is smaller than K.

The message is that the expected average value of N will be smaller in a random environment than in the peaceful environment and that a decrease will grow with the variance of environmental fluctuation.

3.6 PREY–PREDATOR MODELS

The interaction most often found in ecosystems is the one in which predator feeds on prey. The first model of this type was formulated by Lotka (1925). Volterra (1926) independently published the same model in connection to the question by D'Ancona of why the number of predatory fishes rose during the First World War. The model is:

$$dN/dt = rN - bNP \tag{3.26}$$

$$dP/dt = cNP - mP \tag{3.27}$$

In the absence of predator, prey population is assumed to follow Malthus law. Predator population is assumed to feed on prey proportionally to prey population N and predator population P. This interaction is called bilinear collision in an analogy to a collision among two kind of particles.

Given the initial values $N(t = 0) = N_o$ and $P(t = 0) = P_o$ the solution is a pair of periodic functions (Fig. 3.4). A characteristic of these periodic functions is that the peak of prey

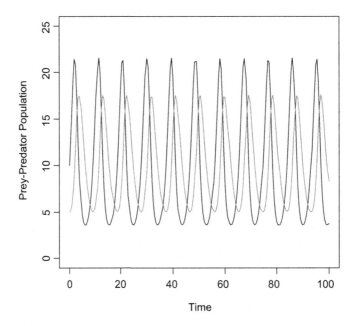

FIGURE 3.4 Prey (black)–predator (red (gray in print versions)) dynamics in the Lotka–Volterra model. $r = 1$, $b = 0.1$, $c = 0.05$, $m = 0.5$.

precedes the peak of predator in such a way that when the prey population peaks, the predator population grows with the maximum rate, and when the predator population is at the maximum, the prey population decreases maximally.

If one uses a proportional fishing eN of prey and eP of predator, where e is a common harvesting effort, one easily shows that the nonextinction equilibrium value (i.e., $N^* \neq 0, P^* \neq 0$ when $dN/dt = 0$ and $dP/dt = 0$) is:

$$N^* = (m + e)/c; \quad P^* = (r - e)/b$$

It turns out that N^* is the same as the average value $<N>$ of the corresponding periodic function and so is $P^* = <P>$.

From the above expressions we see that: if e decreases, N^* decreases and P^* increases.

The above statement constitutes the Volterra principle and represents the answer to the D'Ancona question: Predatory fishes increased and prey fishes decreased because the fishing effort in the Adriatic Sea decreased during the First World War, i.e., most of the fisherman were drafted and sent to battlefields.

Although the above prey–predator model was successful in explaining the D'Ancona question, it is used today mostly as a didactical tool because in the absence of predators we would expect that the prey population is governed by the second law of population growth. When the first term in the right side of Eq. (3.26) is replaced with the right side of Eq. (3.11) we get:

$$dN/dt = rN(1 - N/K) - bNP \tag{3.28}$$

$$dP/dt = cNP - mP \tag{3.29}$$

The dynamics of the model is shown in Fig. 3.5. The dynamics of both prey and predator tends to an equilibrium point in a form of damped oscillations.

It is easy to show that for this model too the Volterra principle is valid.

The above model can be made more realistic by allowing the predator to show saturation in the presence of plenty of prey. One way of taking saturation into account leads to the Rosenzweig and MacArthur (1963) model:

$$dN/dt = rN(1 - N/K) - V_{max}NP/(h + N) \tag{3.30}$$

$$dP/dt = cV_{max}NP/(h + N) - mP \tag{3.31}$$

In this model $V_{max}N/(h + N)$ is the Michaelis and Menten (1913) term, familiar in enzyme kinetics (Wiki, 2016). The term is nearly proportional to N when $h \ll N$ and then the prey–predator interaction does not differ significantly from bNP. But when $N \gg h$ then the term is nearly constant and then the prey-predator interaction is proportional to P.

The model has three equilibrium states:

(0,0)—when both prey and predator disappear. This equilibrium is always unstable.
(K,0)—when predator becomes extinct. This equilibrium will be stable if the predator is too inefficient to catch prey or if its mortality is too high, i.e., if $cV_{max}K/(h + K) < m$.
($N^* = mh/(cV_{max} - m)$, $P^* = (r/V_{max})(h + N^*)(1 - N^*/K)$)—when both prey and predator coexist. For this equilibrium to be stable the following condition must hold:
$m/(cV_{max} - m) \leq K/h \leq (cV_{max} + m)/(cV_{max} - m)$ (Gurney and Nisbet, 1998).

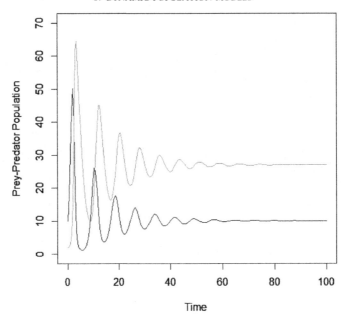

FIGURE 3.5 Prey (black)—predator (red (gray in print versions)) dynamics with $r = 1.5$, $K = 100$, $b = c = 0.05$, $m = 0.5$, and $N_o = 10$, $P_o = 2$.

If K is too small, the predator cannot maintain itself and it goes to extinction, while if K is too high, the predator cannot control the prey sufficiently quickly and the solution tends to a periodic behavior of both prey and predator. In between these two extremes, the equilibrium is stable.

The fact that when K increases the equilibrium is destabilized is termed the paradox of enrichment. When we are dealing with aquatic systems, this phenomenon can also be termed the paradox of eutrophication.

Fig. 3.6 shows a dynamics when the nonextinction equilibrium is stable. The behavior is qualitatively the same as the damped oscillations in Fig. 3.5, except that Fig. 3.6 shows the plot in (N, P) space.

Fig. 3.7 shows a dynamics when the equilibrium is unstable. The prey and predator populations tend to a periodic behavior when the paradox of enrichment occurs.

Proportional harvesting of predator only or prey and predator with a common harvesting effort may stabilize an otherwise unstable equilibrium point.

This means that with harvesting one may control the paradox of enrichment and it is possible to prevent it from occurring. However, too intensive harvesting of prey and predator will result in the extinction of predator.

We conclude with the following laws regarding a prey—predator community:

1. If predator is inefficient in catching the prey, it will tend to extinction and the prey will tend to its carrying capacity.
2. If the prey does not have too high carrying capacity and the predator is efficient in catching the prey, a stable equilibrium will occur in which both prey and predator will coexist.

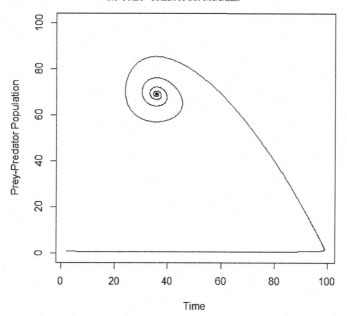

FIGURE 3.6 Dynamics of Rozenzweig–McArthur model for $r = 1.5$, K $= 100$, $V_{max} = 1.2$, $h = 50$, c $= 1$, $m = 0.5$.

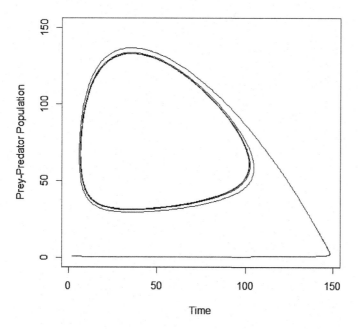

FIGURE 3.7 Dynamics of Rozenzweig–McArthur model for $r = 1.5$, K $= 150$, $V_{max} = 1.2$, $h = 50$, c $= 1$, $m = 0.5$.

3. If prey has a high carrying capacity and the predator is sufficiently efficient to persist, a periodic behavior of both prey and predator will occur and we have the paradox of enrichment.
4. If one applies harvesting of predators only, its population will decrease or it may go to extinction depending on the intensity of harvesting, while the prey population will increase to a stable equilibrium point. The paradox of enrichment will not occur.
5. If both prey and predators are harvested with the same harvesting effort, an otherwise unstable prey−predator system may be stabilized and the paradox of enrichment will not occur. If harvesting is too high, the predator will tend to extinction.

3.7 MODELS OF COMPETITION

In nature, one often observes a competition between populations for nesting space or the source of food (Begon et al., 2006). This competition is called interspecific as opposed to intraspecific which gave the quadratic term in the right hand side of the logistic equation (i.e., rN^2/K).

To consider a simple competition model, assume first the existence of two logistic populations in a finite environment. Each population has its own carrying capacity and biotic potential. Assume further that each population affects the dynamics of the other by drawing on its resources, i.e., both populations are amensals to each other. This can be modeled in such a way that in the nonlinear term of the logistic equation for the first population instead of N_1/K one writes $(N_1 + aN_2)/K$ where a stands for the intensity of amensalism that population N_2 exerts on N_1. In the similar way one can modify the equation for the second population.

The resulting model is the Lotka−Volterra competition model:

$$dN_1/dt = r_1N_1(1 - (N_1 + aN_2)/K_1) \tag{3.32}$$

$$dN_2/dt = r_2N_2(1 - (N_2 + bN_1)/K_2) \tag{3.33}$$

Constants a and b determine the intensity of the corresponding amensalisms.

If *ecological niches* have zero intersection, i.e., if the two populations do not feed on the same source or do not occupy the same space, they do not compete, and then $a = b = 0$. In this case, the above model remains a model of two independent logistic populations N_1 and N_2. If ecological niches intersect at least partially then $a > 0$ and $b > 0$. In case that population N_1 releases toxic material that kills members, or disables development, of the second population, the coefficient b may be large.

There exist four equilibrium solutions. First is the $(N_1^*, N_2^*) = (0,0)$—the total extinction when both populations disappear. Then we have two equilibrium solutions when one of the two populations goes to extinction: $(N_1^*, N_2^*) = (K_1, 0)$ and $(N_1^*, N_2^*) = (0, K_2)$. Finally, one equilibrium solution exists in which both populations are present. Assume $N_1 > 0$ and $N_2 > 0$. From $dN_1/dt = 0$ and $dN_2/dt = 0$ Eqs. (3.32) and (3.33) give:

$$K_1 - N_1^* - aN_2^* = 0 \tag{3.34}$$

$$K_2 - N_2^* - bN_1^* = 0 \tag{3.35}$$

By solving these equations for N_1^* and N_2^* we get:

$$N_1^* = (K_1 - aK_2)/(1 - ab) \tag{3.36}$$

$$N_2^* = (K_2 - bK_1)/(1 - ab). \tag{3.37}$$

where $ab < 1$, $K_1 > aK_2$ and $K_2 > bK_1$.

It turns out that the nonextinction equilibrium is a stable node as long as $N_1^* > 0$ and $N_2^* > 0$. Since $a > 0$ and $b > 0$, $N_1^* < K_1$ and $N_2^* < K_2$.

When a and b are very small we can safely neglect ab because it is of a second-order smallness and we have approximately

$$N_1^* \approx K_1 - aK_2 \tag{3.38}$$

$$N_2^* \approx K_2 - bK_1 \tag{3.39}$$

Suppose $a = b = 1$ and K_1 is different from K_2.
Eqs. (3.34) and (3.35) become:

$$K_1 - N_1^* - N_2^* = 0 \tag{3.40}$$

$$K_2 - N_2^* - N_1^* = 0 \tag{3.41}$$

This system cannot be satisfied for any values $N_1^* > 0$ and $N_2^* > 0$. Therefore the nonextinction equilibrium does not exist and we have competitive exclusion of one population.

Let us summarize the fate of the two competing populations:

$(1/b) < K_1/K_2 > a$	The equilibrium is unstable and the first population wins, so we have competitive exclusion of the second population;
$(1/b) < K_1/K_2 < a$	The equilibrium is unstable, whether the first or second population wins will be determined by initial population values but here too we have competitive exclusion of one population;
$(1/b) > K_1/K_2 < a$	The equilibrium is unstable and the second population wins so here we have competitive exclusion of the second population;
$(1/b) > K_1/K_2 > a$	The equilibrium is stable and we have coexistence of populations.

We see that in three out of four cases, the competition will cause exclusion of one population.

Let us look at another simple model of species (N_1 and N_2) competition for food (F).

$$dF/dt = I - aFN_1 - bFN_2 \tag{3.42}$$

$$dN_1/dt = aFN_1 - m_1N_1 \tag{3.43}$$

$$dN_2/dt = bFN_2 - m_2N_2 \tag{3.44}$$

where I is the inflow of food into the environment occupied by N_1 and N_2 populations; $a(b)$ is the specific efficiency of $N_1(N_2)$ to take the existing food F and $m_1(m_2)$ is the specific mortality rate of $N_1(N_2)$.

Possibly three equilibrium solutions exist. The first is $(I/m_1, m_1/a, 0)$ and represents the extinction of N_2. The second is $(I/m_2, 0, m_2/b)$ and represents the extinction of N_1.

Let us assume that both $N_1^* \neq 0$ and $N_2^* \neq 0$, then from the second equation we have:

$$F^* = m_1/a \tag{3.45}$$

and from the third:

$$F^* = m_2/b \tag{3.46}$$

This is possible only if the two populations are identical in the efficiency of taking food and in mortality. Since in general $m_1 \neq m_2$ and $a \neq b$, the equilibrium will not exist.

In case $m_1/a > m_2/b$, N_1 will reach equilibrium but F will continue to decrease toward m_2/b and hence dN_1/dt will become negative and N_1 will tend to extinction.

The dynamics of the N_1 and N_2 is given in Fig. 3.8.

This model shows that the exclusion of one population due to competition for food will happen always, hence the competitive exclusion principle.

In conclusion, we have the following law: competition among participating populations in a community raises a chance of population extinction. Whether the extinction of less efficient populations will happen faster or slower or even not at all, will depend on spatial characteristics of the environment especially shelter and on fluctuating (seasonal) sources of food, both of which we have not considered here. However, for conservation purposes, where one analyses the chance of survival of a specific population, taking into account natural characteristics of the environment is crucial.

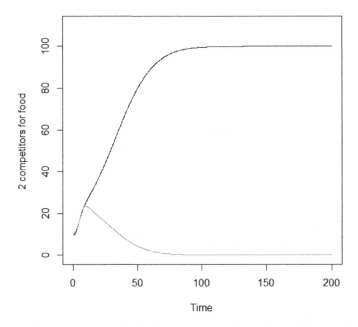

FIGURE 3.8 Two populations competing for food. $I = 10, a = 0.01, b = 0.02, m_1 = 0.1, m_2 = 0.3$. Since $m_1/a < m_2/b$ the first population wins.

3.8 MODELS OF COOPERATION

Recently, cooperation among populations has been found to be much more frequent than thought historically (Bronstein, 2016).

In the simplest case the presence of a commensal population increases the carrying capacity of the other population. The same may be assumed of the other population. Hence the mutualism is a consequence of two commensal interactions: first population helping the second and the second helping the first.

A model of two logistic populations with a mutualism may be written as:

$$dN_1/dt = r_1 N_1 (1 - N_1/(K_1 + aN_2)) \tag{3.47}$$

$$dN_2/dt = r_2 N_2 (1 - N_2/(K_2 + bN_1)) \tag{3.48}$$

There exist four equilibrium solutions: $(0,0)$, $(K_1, 0)$, $(0, K_2)$, and the nonextinction point $(N_1^* \neq 0, N_2^* \neq 0)$:

$$N_1^* = (K_1 + aK_2)/(1 - ab) \tag{3.49}$$

$$N_2^* = (K_2 + bK_1)/(1 - ab) \tag{3.50}$$

$$\text{where } ab < 1. \tag{3.51}$$

We see that $N_1^* > K_1$ and $N_2^* > K_2$.

Given that Eq. (3.51) is satisfied, the nonextinction equilibrium is a stable node, i.e., the populations tend to the equilibrium without oscillations.

If $r_1 = r_2 = r$, the characteristic return time (T_c) to the nonextinction equilibrium is:

$$T_c = 1/r[1 - \sqrt{ab}] \tag{3.52}$$

As ab increases, the return time to the equilibrium will be longer.

We may now formulate the following law:

If two species interact forming a mutualistic community, sooner or later their populations will exceed carrying capacity of the environment.

In case $ab \geq 1$, the above law will also hold simply because some other variable in the finite environment not included in the model will eventually stop the growth of both populations.

Dynamics of two mutualistic populations is shown in Fig. 3.9.

The above model can easily be generalized to a mutualistic community with N_i, $i > 2$ populations. Then the above law will hold for all the participating populations.

In case we harvest all the populations with the same harvesting effort, resulting in the proportional harvesting, we will discover that maximizing the yield (Y):

$$Y = e(\Sigma_i N_i) \tag{3.53}$$

will lead to a maximum sustainable yield (MSY) and none of the participating populations will disappear (Legović and Geček, 2012). This is a consequence of obligatory cooperation. In case some or all populations exist without interaction with the rest of the community some of them may tend to extinction as a consequence of attempting to reach MSY (Legović and Geček, 2010).

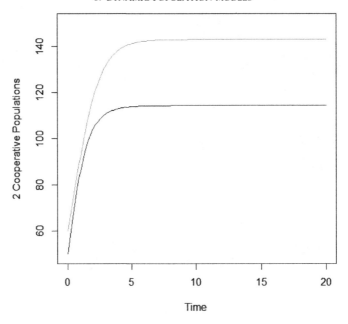

FIGURE 3.9   Dynamics of two populations with mutualism. $R_1 = 1.5$, $K_1 = 100$, $r_2 = 1$, $K_2 = 100$, $a = 0.1$, $b = 0.2$.

In conclusion, we may state the law: mutualism enhances the persistence of participating populations although the return to the equilibrium may take longer.

3.9 FOOD CHAINS

Food webs that we find in ecosystems are composed of many populations and even a greater number of interactions. Since they are essentially nonlinear, it is difficult and time-consuming to construct adequate models and even more difficult to analyze and consequently to understand their dynamics. However, food chains are a class of ecosystems which are relatively easy to construct and analyze.

A food chain can be represented by the following graph of flow of matter:(Fig. 3.10)

Where S is a food source for the first population, N_1, also called the first trophic level. The first trophic level, N_1, is a food source for the second trophic level, N_2, and so on until the last trophic level N_n. N_n is also called the top trophic level or the top predator. Such a food chain is

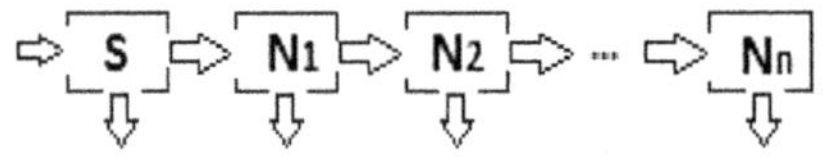

⇨ S ⇨ N_1 ⇨ N_2 ⇨ ⋯ ⇨ N_n

FIGURE 3.10   Food chain with n trophic levels.

said to be composed of n trophic levels. Since the graph in Fig. 3.10 represents a flow of bio-mass, starting from N_1, one has also to include the export of biomass due to mortality, so that each of the trophic levels has a direct outflow from the system.

To avoid the spatial effects which would further complicate the issue, we shall place the food chain into a well mixed biological reactor that will represent a lake, a river, or a coastal sea.

Let us consider the food chain corresponding to the graph in Fig. 3.10. Assume that all that is consumed by one trophic level is turned into biomass of the next trophic level and that we measure biomass of all trophic levels in the units of S. For example if S represents phosphorus in water, than we would measure the amount of phosphorus in each trophic level in a unit of volume. In Eqs. (3.28) and (3.29) this would mean that $b = c$.

Equations of the food chain of length n are:

$$dS/dt = DI - b_1 SN_1 - DS \tag{3.54}$$

$$dN_1/dt = b_1 SN_1 - b_2 N_1 N_2 - DN_1 \tag{3.55}$$

$$dN_2/dt = b_2 N_1 N_2 - b_3 N_2 N_3 - DN_2 \tag{3.56}$$

$$\cdots$$

$$dN_{n-1}/dt = b_{n-1} N_{n-1} N_n - b_n N_{n-1} N_n - DN_{n-1} \tag{3.57}$$

$$dN_n/dt = b_n N_{n-1} N_n - DN_n \tag{3.58}$$

There are n extinction equilibrium points:

$$1. \ \left(S^*, N_1^*, ..., N_n^*\right) = (I, 0, ..., 0) \tag{3.59}$$

$$2. \ \left(S^*, N_1^*, ..., N_n^*\right) = (D/b_1, I - D/b_1, 0, ..., 0) \tag{3.60}$$

$$3. \ \left(S^*, N_1^*, ..., N_n^*\right) = (b_2 I/(b_1 + b_2), D/b_2, b_1 I/(b_1 + b_2) - D/b_2, 0, ..., 0) \tag{3.61}$$

$$\cdots$$

To find the last, (n + 1)-th, equilibrium point (the non-extinction equilibrium) we note that from the last, (n + 1)-th, equation describing the dynamics of the n-th trophic level, one gets:

$$N_{n-1}^* = D/b_n = a_{n-1} D \tag{3.62}$$

Inserting into the (n − 1)-th equation:

$$N_{n-3}^* = D(b_n + b_{n-1})/(b_n b_{n-2}) = a_{n-3} D, \ \text{etc.} \tag{3.63}$$

where $a_{n-1}, a_{n-3}, a_{n-5}, \ldots$ are functions of b_i constants only.

The equilibrium values $N_{n-1}^*, N_{n-3}^*, N_{n-5}^*, \ldots$ are linear functions of the dilution rate, D but not of the inflow concentration I. Starting with N_{n-1}^*, whether the series will end with N_1^* or with N_2^* will depend on whether the number of trophic levels is even or odd.

Assume the food chain has an even number of trophic levels.

Then $n = 2k$ where k is a natural number.

The above series will terminate with $N_1^* = a_1 D$.

Now substitute N_1^* into the first equation and get:

$$S^* = I/(1 + a_1 b_1) = \beta_1 I \tag{3.64}$$

Now substitute S^* into the second equilibrium equation and get:

$$N_2^* = b_2 I/(b_2 + a_{n-2} b_1 b_2) - D/b_2 = \beta_2 I - \gamma_2 D \tag{3.65}$$

Continue in the same manner through all the odd numbers successively until the n-th equation. The obtained equilibrium values of the even trophic levels are:

$$N_{2i}^* = \beta_{2i} I - \gamma_{2i} D \quad \text{where } i = 1, \dots, k. \tag{3.66}$$

Now assume the food chain has an odd number of trophic levels.
Then $n = 2k + 1$. The above series will terminate with $N_2^* = a_2 D$.
Substituting this value into the second equation, one obtains:

$$S^* = (1/b_1 + b_2 a_2/b_1) D. \tag{3.67}$$

Now substitute S^* into the first equation and get N_1^*:

$$N_1^* = I/(1 + b_2 a_2) - D/b_1 = \beta_1 I - \gamma_1 D. \tag{3.68}$$

Substituting into the third, fifth, ... until the last equation successively will lead to the equilibrium value of odd trophic levels. These equilibrium values will be linear functions of I and D only:

$$N_{2i+1}^* = b_{2i+1} I - \gamma_{2i+1} D, \quad i = 1, \dots, k. \tag{3.69}$$

From the above, we have following conclusions:

1. In a food chain with a binary collision among prey and their predator population and an even number of trophic levels, the equilibrium concentration of the first trophic level and all other odd trophic levels is independent of an increase in the concentration of the nutrient in the inflow. They are proportional to the inflow of water.
2. In a food chain with an odd number of trophic levels, the equilibrium concentration of the first trophic level and all other odd trophic levels increases with increasing inflow concentration of the limiting nutrient and decreases with the inflow rate of water.
3. From (1) and (2) it follows that the equilibrium concentration of the top trophic level does not depend on whether the chain is even or odd. The concentration will always increase with increasing nutrient concentration in the inflow and decrease with increasing water flow.
4. Increase of nutrient concentration in the inflow will result in more nutrient being trapped in the food chain which is easily seen from the summation:

$$I = S^* + N_1^* + \dots + N_n^* \tag{3.70}$$

The reader is welcome to attempt the analysis of food chains in which instead of bilinear collision among prey and predator populations, Michaelis–Menten interaction is used.

3.10 CYCLING OF MATTER

Apart from food chains, in food webs of ecosystems one finds that nutrients cycle among various trophic levels.

3.10.1 A Model of Nutrient, Phytoplankton, and Dead Matter

Perhaps one of the simplest ecosystems with a nutrient cycle is the one which contains the source of nutrients, I, nutrient concentration in a lake, S, phytoplankton, N, death of phytoplankton mN, dead matter, M, and decomposition of dead matter, rM. By decomposition, nutrients will be released and will again be available for uptake by phytoplankton. The dilution rate is D.

A graph denoting transport of nutrient in such an ecosystem is shown in Fig. 3.11.

Equations of the ecosystem immersed in a continuous culture reactor are:

$$dS/dt = D(I - S) - aSN + rM \tag{3.71}$$

$$dN/dt = aSN - mN - DN \tag{3.72}$$

$$dM/dt = mN - rM - DM \tag{3.73}$$

where a, D, m, and r are positive.

In the above formulation, it is assumed that phytoplankton, N, takes up the nutrient, S, according to the bilinear collision. As we mentioned earlier, this interaction allows us to state conclusions about a change in parameters if they are not far from original values which lead the system to steady state.

Furthermore, the death of phytoplankton and the remineralization process follow a first-order kinetics.

The system (3.71–3.73) has only two equilibrium points.

The extinction of N and M: $(S^*, N^*, M^*) = (I, 0, 0)$ and the nonextinction steady state.

The nonextinction steady state: $(S^* > 0, N^* > 0, M^* > 0)$ is obtained from Eqs 3.71–3.73 as follows. From (3) in steady state:

$$M^* = mN^*/(r + D) \tag{3.74}$$

From Eq. (3.72) in steady state:

$$S^* = (m + D)/a \tag{3.75}$$

From Eq. (3.71) in steady state, upon substitution of Eq. (3.74) and Eq. (3.75) and by rearranging:

$$N^* = D[I - (m + D)/a]/[m + D - rm/(r + D)] \tag{3.76}$$

In order that S^*, N^*, and M^* be positive, the following condition must be met:

$$I > (m + D)/a \tag{3.77}$$

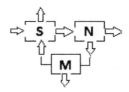

FIGURE 3.11 A graph of nutrient cycling among three components: concentration in water, concentration in the first trophic level, and dead matter.

If incoming nutrient concentration is too small, death rate of phytoplankton, m, or dilution rate, D, are too high, phytoplankton population would be washed out of the reactor. Since organic matter would have no source, it would be washed out too and eventually the extinction state will be reached.

We can now predict what will happen asymptotically to an ecosystem for which the models (3.71) and (3.72) is representative in case characteristics of the environment are changed. For example:

1. An increase in nutrient concentration in the inflow, I, will cause a linear increase in N^* and M^*, but S^* will not be changed.
2. An increase in dilution rate, D, means that the ecosystem is progressing from a relatively stagnant lake to a river. Such a change will increase S^* while N^* and M^* will decrease.
3. An increase in remineralization only, r, will not change S^* while N^* and M^* will decrease.
4. An increase of specific death rate of phytoplankton only, m, will cause S^* and M^* to increase and N^* to decrease.
5. An increase in the efficiency of phytoplankton uptake only, a, will cause S^* to decrease while N^* and M^* will increase.

In case we change the bilinear collision interaction with the more realistic Michaelis–Menten form, the model changes into:

$$dS/dt = D(I - S) - aSN/(h + S) + rM \tag{3.78}$$

$$dN/dt = aSN/(h + S) - mN - DN \tag{3.79}$$

$$dM/dt = mN - rM - DM \tag{3.80}$$

In steady state:

$$S^* = (m + D)h/[a - (m + D)] \tag{3.81}$$

$$N^* = (r + D)M^*/m \tag{3.82}$$

where $a > m + D$ and $I > (m + D)h/[a - (m + D)]$.

In steady state, substitution into Eq. 3.72, gives M^*:

$$M^* = mD[I - S^*]/[(m + D)(r + D) - rm] \tag{3.83}$$

where $(m + D)(r + D) > rm$.

The qualitative conclusion with regard to an increase in I is the same as in the previous model: S^* will not change while M^* and N^* will increase linearly.

A comparison to the case with no cycling, i.e., when m and r are zero, shows that cycling of nutrient increases S^* and decreases N^*.

3.10.2 Food Chain With Two Trophic Levels and Cycling of Nutrients

Consider a cycling of nutrient with an ecosystem composed of nutrient, prey, and predator.(Fig. 3.12)

$$dS/dt = D(I - S) - aSN + rM \tag{3.84}$$

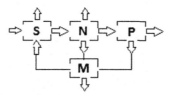

FIGURE 3.12   A graph of nutrient cycling among four components: nutrient in water, first trophic level, second trophic level, and dead matter.

$$dN/dt = aSN - bNP - m_1N - DN \tag{3.85}$$

$$dP/dt = bNP - m_2P - DP \tag{3.86}$$

$$dM/dt = m_1N + m_2P - rM - DM \tag{3.87}$$

In steady state:

$$I = S^* + N^* + P^* + M^* \tag{3.88}$$

$$N^* = (m_2 + D)/b \tag{3.89}$$

$$S^* = (DI + rM^*)/[D + a(m_2 + D)/b] \tag{3.90}$$

$$aS^* - bP^* = m + D \tag{3.91}$$

$$mN^* + m_2P^* = (r + D)M^* \tag{3.92}$$

Concerning an increase in I we have the following conclusions: S^* and M^* will increase, N^* will not change, and P^* will increase because it is the top trophic level. In addition, the total quantity of nutrient captured in the ecosystem will increase.

Using the above model, one can analyze how food chains, with even and odd number of trophic levels and with cycling of nutrients, differ from the respective food chains where cycling is negligible.

3.11 CONCLUSIONS

In this chapter we have gradually progressed from the first law of isolated population dynamics in an infinite environment to the second law, which we expect to hold in a finite peaceful environment. We then examined the dynamics of an isolated population in periodic and random environment including its harvesting. From interactions of two populations we analyzed prey–predator, competition, and mutualism. As examples of simple food webs we briefly considered food chain and food chain with cycling of a nutrient. The treatment serves to see how more complicated systems of populations in nature, being nonlinear, will inevitably lead to unexpected results that could have not been guessed by intuition only. Furthermore, more complicated models inevitably lead to greater variety of effects that are not seen in simpler models. One could have also noticed that for all the populations to persist in progressively more complicated ecosystems, a greater number of conditions must be satisfied. This is particularly easy to see when analyzing food chains. Consequently, more complicated ecosystems are likely to be more vulnerable to at least one population loss due to either natural or man-made perturbations.

3.12 A BRIEF HISTORY OF POPULATION MODELING

We do not have reliable historical data on when the process of mathematical modeling of population dynamics started. The first recorded model of population dynamics that we know of was proposed and solved by Leonardo di Pisa (better known as Fibonacci (1175−1250)) in 1202 in his famous book *Liber abaci*. The problem he was concerned about is:

"If one puts a pair of rabbits in a place surrounded on all sides by a wall. How many pairs of rabbits can be produced from that pair in a year if it is assumed that starting with the second month, each pair becomes productive and every subsequent month, each productive pair brings to life one new pair?" The model produced a series of numbers for the pairs of rabbits: 0 (before rabbits were put into the cage), 1 (an instant after the first pair was introduced), 1 (a month later), 2 (two months later), 3, 5, 8, 13, 21, ..., 233 (after twelfth month). The series extended much further in t is known today as the Fibonacci series. Johannes Kepler (1571−1630) showed that the successive numbers in the series satisfy the following recursive (difference or iterative) equation:

$$N_{t+2} = N_t + N_{t+1} \quad \text{for } t = 0, 1, \ldots$$

The model is a second-order difference equation and it requires two starting values $N_0 = 0$ and $N_1 = 1$ to be solved uniquely. This was done by Daniel Bernoulli who obtained (Bacaër, 2011):

$$N_t = (1/\sqrt{5})((1 + \sqrt{5})/2)^t - (1/\sqrt{5})((1 - \sqrt{5})/2)^t$$

The first law of population dynamics Eq. (3.6) was stated by Leonhard Euler in 1748 (Bacaër, 2011). The law has been unjustly named after Thomas R. Malthus who in 1798 wrote that the human population in England grows according to the geometrical law. In addition, he assumed that quantity of food grows linearly (i.e., along the straight line). Since geometric growth exceeds any linear growth given enough time, he predicted a great starvation and a catastrophe for human population in future. Subsequent data on human population and food production showed that Malthus was wrong about both assumptions: neither has human population grown geometrically until now nor the food production grown linearly. Hence his fear of a catastrophe for human population was unfounded. To get a feeling about the magnitude of his numerical error it suffices to consider present data versus Malthus prediction. According to his prediction, today's human population would be about 256 billion while there would be enough food to feed 9 billion. In fact, human population is about 7.3 billion (as of July 2015) and the present food production is sufficient to feed 57 billion people.

Verhulst (1838) suggested a logistic population model. Given the data on human population during 1961 when human population has grown by 2% per year and assuming $r = 0.039$ one predicts that the carrying capacity for earth is 10 billion. By carefully examining UN data until 2010, Gonzalo et al. (2013) predicted that the human population on Earth will reach 10 billion by 2050 and will start to decrease.

First prey—predator model was suggested by Lotka (1925). Volterra (1926) independently published the same model and used it to determine the cause of predatory fish populations increase in the Adriatic Sea during the First World War. Volterra subsequently analyzed a number of properties of food web models. In fact, germination of many ideas on dynamics of populations of organisms using mathematical models took place between 1923 and 1940 (Scudo and Ziegler, 1978). That development gave rise to expanded interest after the Second World War and an explosion of activity following the advent of computers which enabled simulation and numerical analysis of complicated ecosystem models. The development has been fueled further by a growing need to understand dynamics of populations in ecosystems to protect them better and find ways to reduce human impact.

References

Bacaër, N., 2011. A Short History of Mathematical Population Dynamics. Springer, London.

Beddington, J.R., May, R.M., 1977. Harvesting natural populations in a randomly fluctuating environment. Science 197, 463—465.

Begon, M., Townsend, C.R., Harper, J.L., 2006. Ecology. Blackwell.

Bronstein, J.L. (Ed.), 2016. Mutualism. Oxford Univ. Press, Oxford.

Gonzalo, A.J., Muñoz, F.-F., Santos, D.J., 2013. Using a rate equations approach to model World population trends. Simulation 89, 192—198.

Gurney, W.S.C., Nisbet, R.M., 1998. Ecological Dynamics. Oxford Univ. Press, Oxford.

Legović, T., Geček, S., 2010. Impact of maximum sustainable yield on independent populations. Ecological Modelling 221, 2108—2111.

Legović, T., Geček, S., 2012. Impact of maximum sustainable yield on mutualistic communities. Ecological Modelling 230, 61—72.

Legović, T., Perić, G., 1984. Harvesting population in a periodic environment. Ecological Modelling 24, 221—229.

Lotka, A.J., 1925. Elements of Physical Biology. Williams and Wilkins, Baltimore.

Malthus, T.R., 1798. An Essay on the Principle of Population. Johnson, London.

May, R., 1976. Models for single populations. In: Theoretical Ecology. Blackwell.

Michaelis, L., Menten, M.L., 1913. Die Kinetik der Invertinwirkung. Biochemische Zeitschrift 49, 333—369.

Rosenzweig, M., MacArthur, R., 1963. Graphical representation and stability conditions of predator-prey interaction. American Naturalist 97, 209—223.

Schaefer, M.B., 1954. Some aspects of the dynamics of populations important to the management of commercial marine fisheries. Bulletin Inter-American Tropical Tuna Commission 1, 25—56.

Scudo, F.M., Ziegler, J.R., 1978. The Golden Age of Theoretical Ecology: 1923—1940. Springer.

Verhulst, P.F., 1838. Notice sur la loi que la population suit dans son accroissement. In: Corresp. Math, et Phys, vol. 10, pp. 113—121.

Volterra, V., 1926. Fluctuations in the abundance of a species considered mathematically. Nature 118, 558—560.

Wiki, 2016. https://en.wikipedia.org/wiki/Michaelis%E2%80%93Menten_kinetic.

CHAPTER

4

Steady State Models

X.-Z. Kong, F.-L. Xu[1], W. He, W.-X. Liu, B. Yang

Peking University, Beijing, China

[1]Corresponding author: E-mail: xufl@urban.pku.edu.cn

OUTLINE

4.1 STEADY STATE MODEL: ECOPATH AS AN EXAMPLE

4.1.1 Steady State Model

Steady state ecological models are established to describe conditions in which the modeled components (mass or energy) are stable, i.e., do not change over time (Jørgensen and Fath, 2011). Therefore, outputs from this model type represent the equilibrium of the ecosystems, which would be validated by average values of field observations over a period. In fact, under most conditions, the steady state is essentially dynamic equilibrium, in which the inflow and outflow of the ecosystems are balanced. A classic example of steady state model is to describe ecosystem with alternative stable states, such as the phytoplankton-dominant and vegetation-dominant states in shallow lake ecosystems (Scheffer et al., 1993, 2001). Steady state models were extensively applied in investigating this theory, which provided a deeper insight in understanding the mechanisms of hysteresis shown in many ecosystems. Importantly, the outcomes from steady state models might have critical implications for management.

4.1.2 Ecopath Model

Ecopath model is a well-established modeling tool with a user-friendly interface and a standard modeling procedure to document and analyze food web structure and ecosystem functioning at the steady state (Christensen et al., 2005), particularly focusing on trophic mass—balance analysis (Christensen and Walters, 2004). A typical Ecopath model is composed of multiple groups of state variables, which are usually termed as "functional groups". Each functional group corresponds to one linear equation, all of which could be solved simultaneously under steady state assumption. Ecopath model was firstly developed to evaluate the functioning of marine ecosystem (Polovina, 1984). The model was further modified by the ecological thermodynamic theory and gradually emerged into a powerful tool for analysis of energy flow (Christensen et al., 2005). By importing information for the functional groups such as biomass, production/consumption rate, and diet composition, the model would be able to provide important ecological properties including trophic levels, mass and energy flows, and ecosystem maturity and stability. The model is feasible to be applied for ecosystems with sufficient data. For a detailed review on the methods and capacity of Ecopath model, see Christensen and Walters (2004).

To date, the steady state Ecopath model was used abundantly in aquatic ecosystem including marine and freshwater ecosystems, particularly for gulfs and lakes, considering the fact that these two ecosystems are relatively stable than the others such as rivers and streams. These applications might be simply categorized into the following three groups: Ecopath model for one system at one specific period, for one system at multiple periods, and for multiple systems at one specific period. Different groups of researches were designed with distinct objectives, which would be briefly reviewed as follows.

For the first group, it has been conducted for multiple aquatic ecosystems around the globe (Fetahi and Mengistou, 2007; Fetahi et al., 2011; Hossain et al., 2010, 2012; Jia et al., 2012; Li et al., 2009; Liu et al., 2007b; Pedersen et al., 2008; Shan et al., 2014; Villanueva et al., 2006b, 2008). Outputs from Ecopath model for the studied system at one specific period provide a comprehensive understanding on the basic features of the ecosystems in terms of food web structure, trophic interaction, energy flow, and ecosystem status such as stability and maturity. Moreover, one of the major objectives of Ecopath model approach is to evaluate the impact of fishery on the aquatic ecosystems and to provide suggestions for fishery strategy in a more sustainable way (Coll et al., 2006; Thapanand et al., 2007). For example, it was argued that an ecosystem-level perspective was urgent for fishery management in a fishery-intensive lake, illustrated by an Ecopath approach (Guo et al., 2013). For some other systems, Ecopath model was used to identify important functional groups or key species, such as autotrophs for an estuary (Pavés and González, 2008) and aquaculture fish for a marine ecosystem (López et al., 2008). In addition, the roles of exotic species in one specific ecosystem could be unraveled by Ecopath model (Kao et al., 2014). The introduced exotic fish, in general, exert a negative impact on the target system (Downing et al., 2012), but there are also occasions with positive consequences (Villanueva et al., 2008), where Ecopath played as the essential modeling tool. Overall, in most cases, the results from Ecopath model were considered as the scientific basis for ecological management.

For the second group, this type of research generally aims at investigating the changes in food web structure and ecosystem functioning during a relatively long-term temporary scale. By establishing Ecopath models during different periods, one would be able to compare the outcomes from the model and clearly observed the most significant changes in the properties of ecosystem, thereby obtaining a quantitative understanding of the ecosystem dynamics. For example, two Ecopath steady state models were built for a wetland in China before and after the flow regulation, and the model showed that this activity hampered the natural succession and increased the vulnerability of the wetland ecosystem (Yang and Chen, 2013). In addition, Ecopath models were used to investigate the consequences of the invasive species introduction. For example, Nile perch was introduced to Lake Victoria in East Africa about 30 years ago. Ecopath model showed that both changes in Nile perch's and detritivores' trophic levels were critical in driving the dynamics of the system, and the lake was unlikely to recover even though the main trophic levels would have been restored (Downing et al., 2012). Similar study was also conducted for other lakes to better understand the impact of invasive species at the ecosystem level (Stewart and Sprules, 2011). For lake ecosystems under complex disturbances, Ecopath models were also developed for several distinct periods to reach a quantitative evaluation of changes in ecosystem properties and a deduction of major driving factors (Kong et al., 2016; Li et al., 2010). Overall, to fulfill the research in this group, sufficient data from different periods with equal quality are required, which are, however, difficult to obtain for most of the ecosystems.

For the third group, this type of research is relatively scarce in comparison to the first two groups (Janjua et al., 2014; Mavuti et al., 1996). These studies focused on two or more similar and adjacent ecosystems (usually located in one region) with significant different features such as external conditions. Ecopath model outputs for each ecosystem were comparable, the differences of which would be directly linked to the difference in these features, thereby reaching a better understanding on the causality between driving factors and ecosystem properties. For example, it was revealed by Ecopath that the difference in ecosystem production in two adjacent lakes was attributed to the differences in principle source and pathway of energy flows, i.e., detritus-driven and algae-driven pathways (Villanueva et al., 2006a), which were suggested to play a significant role in determining the ecosystem stability (Moore et al., 2004).

4.1.3 Future Perspectives

Over the two decades of development and application, several hundreds of models based on Ecopath have been documented in literature (Jørgensen and Fath, 2011). As a steady state model, Ecopath has the advantages including (1) the capacity to characterize the food web structure and ecosystem functioning; (2) the small requirements of model input information; and (3) a user-friendly interface to facilitate model development, parameterization, and application. However, there are also disadvantages in Ecopath model, which have been systematically reviewed by Christensen and Walters (2004). Here, we point out that Ecopath is essentially a simplified food web model, in which many processes in nature, such as biogeochemical processes, are not fully considered in the model. This would lead to the deviation of model outputs from observations. Therefore, the model was designed to illustrate the ecosystem on a macroscale, instead of predicting the absolute values of specific ecosystem components. An interesting study attempted to couple a eutrophication model with an Ecopath model, which combined the advantages of both model types to obtain a comprehensive picture of ecosystem behavior in terms of both abiotic and biotic aspects (Cerco et al., 2010). As can be imagined, integration of Ecopath with other process-based ecological models will be a promising but challenging task, which is important for further researches. In addition, Ecopath model might also play a role in comparative limnology studies. As the model can provide multiple indicators representing ecological status for the modeled lake ecosystem, and more importantly, these information were usually provided in relevant literature, it is possible to use these results from steady state Ecopath model for lake classification using cluster analysis. This would be an interesting approach for a synthesis research for Ecopath model and for a promotion of the limnologic studies to categorize different lake types at the ecosystem-level in the future.

4.2 ECOPATH MODEL FOR A LARGE CHINESE LAKE: A CASE STUDY

In the following section, we illustrate a case of Ecopath model as a steady state model study. We managed to establish Ecopath mass—balance models for Lake Chaohu in China during the 1950s, 1980s, and 2000s. Our analysis with Ecopath provides a comprehensive and quantitative evaluation of the changes in the food web structure and ecosystem functioning during these three distinctly different periods. We further analyze the potential driving factors and underlying mechanisms. To our knowledge, this is the first analysis of this type for Lake Chaohu. We expect that the results will provide deeper insight into the changes that occurred in this lake and give essential input for sustainable management strategies by bridging the environmental and ecological perspectives.

4.2.1 Introduction

It is commonly accepted that natural and human systems are strongly coupled and that human activities have currently reached a level that may damage the natural system beyond its adaptive capacity (Liu et al., 2007a; Rockström et al., 2009). Many subsystems on the planet, such as aquatic ecosystems, will exhibit nonlinear behavior as certain thresholds are crossed (Casini et al., 2009; Scheffer et al., 2001). For lake ecosystems, multiple anthropogenic stressors may have synergetic effects that lead to drastic ecological degradation (Yang and Lu, 2014). In addition to fertilization-induced nutrient enrichment and hydrological regulation, fishery in lakes provides important ecological services to local communities, but it is rarely sustainable around the globe (Pauly et al., 2002). Intensive fishery leads to biodiversity loss and simultaneous erosion of the structure and processes that confer stability in the food web (de Ruiter et al., 1995; Rooney et al., 2006), resulting in disastrous consequences, such as "fishing down the food web" (Pauly et al., 1998), food web collapses (Downing et al., 2012), and ultimately catastrophic regime shifts (Casini et al., 2009; Folke et al., 2004).

Over the last several decades, lakes along the Yangtze River floodplain in China have witnessed a strong ecological degradation (Dearing et al., 2012). Lake Chaohu, the fifth-largest freshwater shallow lake in China, is one of the three lakes in China (along with Lake Taihu and Lake Dianchi) that have attracted public concern regarding harmful cyanobacterial blooms (Shan et al., 2014). This lake has suffered from gradual nutrient enrichment from the 1950s onward (Kong et al., 2015), hydrological regulation, which disconnected the lake from the Yangtze River since 1963 (Xu et al., 1999b; Zhang et al., 2014) and intensive fishery since the 1980s (Zhang et al., 2012). Consequently, drastic changes have occurred in Lake Chaohu's ecosystem, resulting in significant changes in the food web structure, a decreasing biodiversity and catastrophic regime shifts toward an unfavorable turbid state dominated by phytoplankton and small-sized planktivorous fish (Liu et al., 2012; Zhang et al., 2015). A significant fourfold to fivefold increase in total biomass was observed, with an increasing dominance of the lower trophic levels (TLs) throughout these years. These changes in the food web structure may strongly influence the functioning of the ecosystem. To date, however, the quantitative changes in ecosystem functioning have not been documented for Lake Chaohu.

The availability of relevant data from the 1950s, 1980s, and 2000s make the present study feasible. Several studies have already used Ecopath for lakes in the Yangtze River floodplain (Guo et al., 2013; Jia et al., 2012; Li et al., 2010) and even a preliminary Ecopath model for Lake Chaohu (Liu et al., 2014), but studies that investigate the changes in the food web structure and ecosystem functioning over a long time span are scarce. The drastic changes in the Lake Chaohu ecosystem, however, urgently call for a quantitative analysis of food web dynamics and ecosystem functioning comparable with studies conducted in, e.g., Lake Taihu (Hu et al., 2011; Li et al., 2010), Lake Chozas (Marchi et al., 2011, 2012), and the Baltic Sea (Casini et al., 2009), to provide deeper insights into the effect of multiple anthropogenic stressors. Furthermore, the development of an integrated modeling tool is becoming imperative to support a sound policy for lake management with both environmental and ecological perspectives (Jørgensen and Nielsen, 2012).

Ecopath requires data of which it is difficult to obtain good empirical estimates (e.g., diet composition). Thus, a validation of the Ecopath model is strongly recommended. We provide a reliable method to validate the diet composition of the Ecopath model for Lake Chaohu using laboratory- and literature-based estimations of TLs from stable isotopes' determination for each functional group in the food web.

Overall, the goals of this study are as follows: (1) to establish three validated Ecopath mass—balance models for Lake Chaohu corresponding to the 1950s, 1980s, and 2000s, (2) to quantify the changes in food web structure in Lake Chaohu during different periods,

(3) to evaluate the changes in the ecosystem functioning of Lake Chaohu during the 1950s, 1980s, and 2000s, (4) to discuss the potential key factors driving the changes in the food web structure and ecosystem functioning, and (5) to provide a sound modeling basis for an integrated management tool with an illustrative example in Lake Chaohu.

4.2.2 Study Site

Lake Chaohu (31°33′59″N, 117°26′40″E) is the fifth-largest shallow lake in China. It covers an area of 760 km^2 and has a depth of 3 m on average (Fig. 4.1). Before the 1950s, the lake was famous for its beautiful scenery with a high water quality, a large amount of vegetation (30% of the surface area), and a high level of biodiversity (Kong et al., 2013; Xu et al., 1999a; Zhang et al., 2012; Zhang et al., 2014). However, in 1963, the connection of the lake with Yangtze River was blocked by the "Chaohu Sluice" on the Yuxi River, and as a result, the water level fluctuation in the lake was largely reduced. Since 1980, the rapid socioeconomic development in the drainage area of the lake led to a gradual elevation in nutrient loading and a deterioration of the water quality (Kong et al., 2015). Intensive fishery from the 1980s onward exacerbated the effect of eutrophication and water level control, resulting in rapid loss of ecological services. Natural riparian areas were reduced to less than 1% of the total area (Ren and Chen, 2011). The west part of the lake could no longer provide drinking water for the city of Hefei, primarily due to frequent cyanobacterial blooms. As the total fishery yield increased rapidly over the past decades (Fig. 4.2A, $p < 0.01$, approximately 2000 t from the 1950s to the 1970s and nearly 20,000 t in 2009), the total number of species in the fish community decreased from 84 (1963) to 62 (1973) to 78 (1981) to 54 (2002), indicating a considerable loss of biodiversity (Lv et al., 2011).

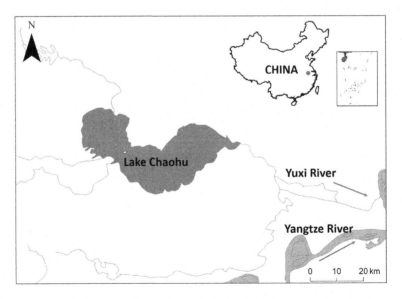

FIGURE 4.1   Location of the Lake Chaohu catchment in China. The blue arrows indicate the direction of water flows in the Yangtze River and in the river that connects Lake Chaohu and the Yangtze River (Yuxi River).

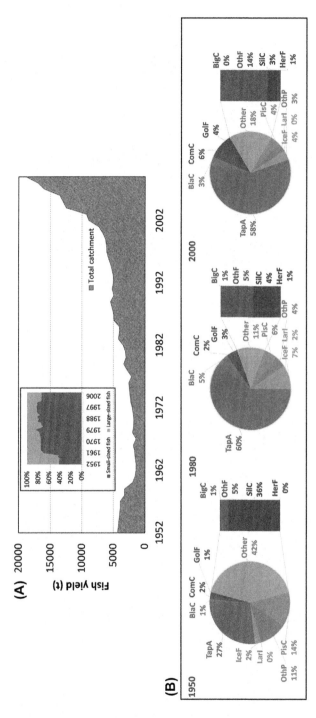

FIGURE 4.2 (A) Total fishery yield (t) from 1952 to 2009 and the fraction of small-sized and large sized fish in the yield. Small-sized fish include, e.g., *Coilia ectenes* and *Neosalanx taihuensis*, whereas large-sized fish include, e.g., *Culter*, silver carps, bighead carps, and common carps. Data collected from Guo (2005), Zhang et al. (2012), and Liu et al. (2014). (B) Composition of fishery yield in the 1950s, 1980s, and 2000s (Details of the abbreviations can be found in Table 4.1).

TABLE 4.1 Basic Input and Estimated Parameters (in Bold) for the Ecopath Models of Lake Chaohu in the 1950s, 1980s, and 2000s

Functional Group	Code	TL	Catchment (t/km²)			Biomass in Habitat Area (t/km²)			P/B			Q/B			Ecotrophic Efficiency			P/Q		
			1950s	1980s	2000s	1950s	1980s	2000s	1950s	1980s	2000s	1950s	1980s	2000s	1950s	1980s	2000s	1950s	1980s	2000s
Piscivorous	PisC	3.8	0.804	0.282	0.612	0.894	0.314	0.700	1.046	1.167	0.974	2.950	3.527	3.200	0.930	0.846	0.983	0.355	0.331	0.304
Other piscivorous	OthP	3.7	0.644	0.226	0.490	0.715	0.251	0.560	1.170	1.423	1.665	4.286	5.212	6.100	0.871	0.734	0.621	0.273	0.273	0.273
Large icefish	LarI	3.9	0.016	0.120	0.057	0.017	0.126	0.060	2.592	1.516	1.983	12.429	11.630	16.650	0.986	0.974	0.947	0.209	0.130	0.119
Other icefish	IceF	3.0	0.142	0.381	0.542	0.150	0.401	0.570	2.639	2.162	2.373	18.862	18.620	27.200	0.944	0.999	0.994	0.140	0.116	0.087
Tapertail anchovy	TapA	3.0	1.618	3.119	8.977	1.692	3.263	9.390	1.987	1.424	1.283	8.664	11.717	11.350	0.925	0.948	0.914	0.229	0.122	0.113
Black carp	BlaC	3.2	0.054	0.261	0.456	0.068	0.326	0.570	0.936	0.859	0.912	3.278	4.937	11.544	0.896	0.935	0.882	0.286	0.174	0.079
Common carp	ComC	2.8	0.095	0.077	0.919	0.123	0.100	1.190	1.069	0.906	0.960	4.767	10.067	10.693	0.906	0.966	0.827	0.224	0.090	0.090
Crucian carp	GolF	2.3	0.063	0.155	0.610	0.082	0.201	0.790	2.825	1.473	1.130	7.211	11.663	12.300	0.974	0.757	0.885	0.392	0.126	0.092
Bighead carp	BigC	2.8	0.068	0.069	0.041	0.076	0.077	0.046	1.436	1.123	1.690	4.693	4.357	6.900	0.943	0.967	0.906	0.306	0.258	0.245
Other fish	OthF	2.8	0.266	0.254	2.073	0.831	0.688	2.303	2.421	1.983	2.155	8.168	9.153	11.000	0.987	0.971	0.799	0.296	0.217	0.196
Silver carp	SilC	2.2	2.099	0.208	0.444	2.332	0.231	0.494	0.926	0.931	1.100	3.565	4.862	8.000	0.978	0.992	0.839	0.260	0.191	0.138
Herbivorous	HerF	2.0	0.014	0.065	0.126	0.025	0.112	0.140	0.697	0.639	0.987	2.424	2.108	7.100	0.928	0.926	0.928	0.288	0.303	0.139
Shrimp	MacS	2.9	0.658	1.118	2.904	0.823	1.398	3.630	4.500	4.500	4.500	21.324	24.400	24.400	0.402	0.279	0.295	0.211	0.184	0.184
Crab	MacC	3.0	0.329	0.559	0.019	0.346	0.588	1.500	2.823	2.120	2.120	8.480	8.480	8.480	0.990	0.743	0.318	0.333	0.250	0.250
Mollusks	Moll	2.1	6.750	11.006	1.010	13.500	22.012	2.020	1.326	1.326	1.326	10.605	10.605	10.000	0.456	0.499	0.750	0.125	0.125	0.133
Other benthos	OthB	2.0	—	—	—	0.775	2.000	0.910	4.030	4.030	4.130	201.500	201.500	50.000	0.589	0.278	0.933	0.020	0.020	0.083
Micro zooplankton	Micz	2.0	—	—	—	0.077	0.684	0.964	67.747	67.747	30.000	981.733	1354.950	600.000	0.950	0.950	0.950	0.069	0.050	0.050
Cladocera	Clad	2.0	—	—	—	0.344	0.379	7.910	21.093	21.093	15.300	826.440	421.858	457.000	0.900	0.900	0.900	0.026	0.050	0.033
Copepoda	Cope	2.0	—	—	—	0.344	0.641	1.820	15.973	15.973	12.165	780.473	319.469	378.000	0.900	0.900	0.900	0.020	0.050	0.032
Cyanobacteria	Cyan	1.0	—	—	—	1.540	16.751	27.386	150.000	150.000	150.000				0.484	0.091	0.070	—	—	—
Chlorophytes	Chlo	1.0	—	—	—	1.391	1.291	16.361	250.000	250.000	250.000				0.512	0.890	0.287	—	—	—
Bacillariophytes	Baci	1.0	—	—	—	1.172	1.956	12.418	200.000	200.000	200.000				0.505	0.422	0.324	—	—	—
Vegetation	SubM	1.0	—	—	—	75.000	5.146	4.457	1.250	1.250	2.253				0.016	0.531	0.977	—	—	—
Detritus	Detr	1.0	—	—	—	1.450	202.500	311.600	-	—	—				0.980	0.471	0.312	—	—	—

4.2.3 Model Development

4.2.3.1 Model Construction and Parameterization

We have built three static mass–balance models for Lake Chaohu, representative of the 1950s, 1980s, and 2000s, using Ecopath with Ecosim, version 6.4.3 (freely available at http://www.ecopath.org). The reasons we focus on these three different periods in the present study are that these three periods (1) represent three distinct stages in the development of the lake ecosystem and (2) correspond to the times when intensive investigations were conducted in Lake Chaohu, with abundant data available in the literature. The basic equation for this model is given in Eq. (4.1):

$$B_i \cdot \left(\frac{P_i}{B_i}\right) \cdot EE_i - \sum_{j=1}^{n} B_j \cdot \left(\frac{Q_j}{B_j}\right) \cdot DC_{ji} - EX_i = 0 \qquad (4.1)$$

where B_i (t/km^2) and B_j (t/km^2) are the biomass of group i and j, respectively, P_i/B_i (per year) is the production/biomass ratio of group i, EE_i (−) is the ecotrophic efficiency of group i, Q_i/B_i (per year) is the consumption/biomass ratio of group j, n is the number of groups, DC_{ji} (−) is the contribution of prey i in the diet of predator j, and EX_i (t/km^2) is the export of group i.

Based on their feeding habits, fish can be categorized into several groups: planktivores, planktivores/benthivores, benthivores, benthivores/piscivores, piscivores, omnivores, detritivores, and herbivores. In Lake Chaohu, small-sized fish are dominated primarily by small pelagic and planktivorous fish, including *Coilia ectenes taihuensis* and *Neosalanx taihuensis*. Large-sized fish include piscivorous fish (e.g., *Erythroculter ilishaeformis*), planktivorous fish (e.g., *Aristichthys nobilis*), benthivorous fish (e.g., *Mylopharyngodon piceus*), herbivorous fish (e.g., *Ctenopharyngodon idella*), and omnivorous fish (e.g., *Hypophthalmichthys molitrix*, *Cyprinus carpio*, and *Carassius auratus*). We defined 24 functional groups in total for the Ecopath model in Lake Chaohu based on a previous study of the food web structure of Lake Chaohu (Zhang et al., 2012). All important biota components are covered by these 24 groups. We separated the phytoplankton group into three subgroups, i.e., Cyanobacteria (Cyan), Chlorophytes (Chlo), and Bacillariophytes (Baci), and added a new group accounting for macrocrustacean shrimp (MacS). For each model, the input data included biomass in certain period of time (*B*), the landings of fishery, diet composition, the parameter values for the production/biomass ratio (*P/B*), the consumption/biomass ratio (*Q/B*), and the ecotrophic efficiency (*EE*) for each functional group. The landing data were collected primarily from peer-reviewed publications, stock assessments, and government reports. Biomass was estimated based on the quote of landing data and estimated fishing mortality (Liu et al., 2014). *P/B* and *Q/B* ratios were primarily estimated according to length-based empirical relations (Palomares and Pauly, 1998; Pauly, 1980), and missing values were assigned based on similar approaches in the same area if data were not available. Most *EE* values were estimated by the model, with the exception of zooplankton. A complete list of data sources could be found in Kong et al. (2016). Diet composition was initially assigned according to the approaches conducted in Lake Taihu (Li et al., 2009), but we involved additional information from Lv et al. (2011) and Guo (2005). The proportion of the predation of different function groups in the three phytoplankton groups was estimated according to food web studies in lakes along the Yangtze River for fish (Guo, 2005), zoobenthos (Liu, 2006), and zooplankton

(Deng, 2004). All input values were adjusted during the model balancing. The values of diet composition were further calibrated based on the comparison between the model calculated and the measurement of TL for each group, and the diet composition with the best fit was applied. The determination of TL based on the nitrogen stable isotope analysis is demonstrated in Section 4.2.3.3.

4.2.3.2 Evaluation of Ecosystem Functioning

The Ecopath model software can calculate multiple indicators for the evaluation of ecosystem functioning (Christensen et al., 2005). The total system throughput (TST) is derived from the sum of all consumption (TC), exports (TEx), respiratory flows (TR), and flows into detritus (TD). It is suggested that TST is positively correlated with the turnover rate of the biomass in the ecosystem (Brando et al., 2004). Moreover, ecosystem maturity is a concept suggesting that ecosystems evolve in succession toward maturity (Odum, 1969). The total net primary production (TPP; $t/km^2/year$), net system production (NSP; $t/km^2/year$), the ratios of TPP with total respiration (TPP/TR), and the total biomass (TPP/TB) are important indicators of ecosystem maturity (Odum et al., 1971); e.g., TPP/TR equals 1 when the system becomes "mature." TPP/TB is positively correlated with a eutrophication state (Barausse et al., 2009). The system omnivory index (SOI) $= (\sum_{i=1}^{N}\sum_{j=1}^{N}(TL_j - (TL_i - 1))^2 \cdot DC_{ij})/N$, where N is the number of living groups, TL_i and TL_j are the trophic level of the predator i and the prey j, respectively, and DC_{ij} is the proportion of prey j that constitutes the diet of predator i. SOI is a weighted measure of food web connectance. Finn's cycling index (FCI) = Tcy/TST, where Tcy is the throughput cycled (including detritus), and TST is the total system throughput. FCI represents the fraction of recycled throughput in the total system throughput (Finn, 1976). Finn's mean path length (FML) = TST/(TEx + TR), where TEx is the sum of all exports and TR is the sum of all respiratory flows. FML indicates the length of the food chain. The connectance index (CI) = $N/2(N - 1)$. CI measures the ratio of the observed links and the possible links in total (Gardner and Ashby, 1970). Ascendancy $= \sum_{i=1}^{N}\sum_{j=1}^{N} T_{ij} \log\left(\frac{T_{ij}T_{..}}{T_{.j}T_{i.}}\right)$, where T_{ij} is the energy flow from j to i, and $T_{j.} = \sum_{k=1}^{N} T_{jk}$, $T_{.i} = \sum_{k=1}^{N} T_{ki}$, and $T_{..} = \sum_{m=1}^{N}\sum_{k=1}^{N} T_{mk}$. Ascendency measures the average mutual information in a system, derived from information theory (Ulanowlcz and Norden, 1990). This indicator is negatively correlated with ecosystem maturity (Christensen, 1995) and positively reflects the gradient of eutrophication (Patrício et al., 2004). Overhead simply equals (1 − Ascendancy). The total transfer efficiencies (TE) is the average of transfer efficiencies between successive discrete TLs, calculated by the ratio between the sum of the exports from a given TL plus the flow that is transferred from one TL to the next and the throughput on the TL (Christensen et al., 2005). Transfer efficiencies from the primary producer (TE p.p.) and transfer efficiencies from detritus (TE det.) are thus the TE for the primary producer (macrophytes and phytoplankton) and detritus, respectively. In addition, mixed trophic impacts (MTI) analysis was utilized to determine trophic interactions, including both the predatory and competitive interactions of a certain functional group on the other groups in an ecosystem (Christensen and Walters, 2004). The element for the matrix, MTI_{ij}, equals $DC_{ij} - FC_{ij}$, where FC_{ij} is the proportion of the predation on j due to i as a predator (Shan et al., 2014). Simply put, FC_{ij} is the proportion of each element in Table 4.2 in the

TABLE 4.2 Diet Composition of the 24 Functional Groups in the Ecopath Model for Lake Chaohu (Details of the Abbreviations can be found in Table 4.1).

No.	Group	PisC	OthP	LarI	IceF	TapA	BlaC	ComC	GolF	BigC	OthF	SilC	HerF	MacS	MacC	Moll	OthB	Micz	Clad	Cope
1	PisC	0.02	0.004																	
2	OthP	0.008	0.021																	
3	LarI	0.007		0.04																
4	IceF	0.047		0.165		0.005														
5	TapA	0.51		0.36		0.005														
6	BlaC	0.001																		
7	ComC	0.001	0.007																	
8	GolF	0.001	0.052																	
9	BigC	0.013																		
10	OthF	0.306	0.281	0.248																
11	SilC	0.005																		
12	HerF	0.001																		
13	MacS	0.01	0.248	0.037			0.15		0.003											
14	MacC	0.003	0.2	0.037			0.025	0.008												
15	Moll		0.092	0.057		0.04	0.8	0.6			0.001									
16	OthB		0.076				0.025	0.157	0.156		0.186						0.001			
17	Micz			0.009	0.007	0.001		0.001	0.01	0.115	0.002	0.003	0.001	0.141	0.141	0.1	0.02		0.016	0.016
18	Clad			0.047	0.433	0.506			0.039	0.375	0.338	0.1		0.34	0.39					
19	Cope			0.057	0.56	0.4			0.05	0.31	0.253	0.097		0.449	0.449					
20	Cyan		0.001						0.05		0.01	0.6				0.5		0.1	0.05	0.05
21	Chlo		0.001			0.02		0.05	0.05	0.09	0.06	0.1				0.1	0.08	0.15	0.25	0.25
22	Baci		0.001			0.023		0.022	0.05	0.09	0.057	0.1				0.1		0.083	0.174	0.174
23	SubM	0.067						0.009	0.415		0.002		0.999	0.05			0.001			
24	Detr		0.016					0.153	0.227	0.02	0.091			0.02	0.02	0.2	0.898	0.667	0.51	0.51
	Sum	1	1	1	1	1	1	1	1	1	1	1	1	1	1	1	1	1	1	1

sum of the corresponding row, whereas DC_{ij} is the proportion in the sum of the corresponding column. We obtained the MTI matrix from the Ecopath software and used R packages (corrplot and cairo) to present the results and obtain high-resolution graphs. More details for these indicators can be found in Christensen et al. (2005).

4.2.3.3 Determination of Trophic Level

The established Ecopath models were validated by comparing calculated and measured TLs for most of the functional groups. Measured data were obtained from both field samples and the literature. For the field data, a total of six dominant fish species were sampled, identified, measured (length), and weighed in March, 2012, from a fishery catchment in Lake Chaohu, including *C. ectenes taihuensis*, *C. carpio*, *A. nobilis*, *Megalobrama amblycephala*, *Hemibarbus maculatus*, and *E. ilishaeformis*. Two benthic invertebrates (*Ballamya purificata* and *Palaemon modestus*) were also collected with a Peterson grab and subsequently kept alive for 24 h in the laboratory to allow for the evacuation of gut contents. Samples of primary producers for a stable isotope analysis were collected with a plankton net. The fish for gut-content analysis were preserved in formalin. Only the dorsal white muscle tissue samples from the adult fish were utilized for gut content analysis. All samples were dried to a constant weight at 60°C and crushed into a fine powder using a mortar and pestle. The TLs of certain functional groups in this study were determined based on nitrogen stable isotope analysis (δ^{15}N), following the method in Jepsen and Winemiller (1980). Nitrogen isotopes (δ^{15}N) were determined at the Institute of Geographic Sciences and Nature Resources Research, Chinese Academy of Sciences in Guangzhou, China, using a Flash EA CN elemental analyzer coupled with a Thermo Finnigan Delta Plus mass spectrometer. The formulation for the TL calculation is shown in Eq. (4.2):

$$TL = \left(\frac{\delta^{15}N_{Fish} - \delta^{15}N_{Reference}}{3.3} \right) + 1 \tag{4.2}$$

where $\delta^{15}N_{Reference}$ is the mean of phytoplankton δ^{15}N, and the denominator value (3.3) is an estimated mean enrichment (fractionation) of δ^{15}N between the fish and food sources (Pauly, 1980). In addition, more TL data for biota in Lake Chaohu were collected from the literature (Xu et al., 2005; Zhang et al., 2012).

4.2.4 Results and Discussion

4.2.4.1 Basic Model Performance

The basic input and the estimated parameters (in bold) for the Ecopath model of Lake Chaohu in the 1950s, 1980s, and 2000s are shown in Table 4.1. The diet composition is shown in Table 4.2. The model outputs of the food web structure and the trophic flows are illustrated in Fig. 4.3. Four TLs are identified by the model for Lake Chaohu's ecosystem, and the majority of the trophic flow occurred between these four TLs (Fig. 4.4). The performance of the Ecopath models was evaluated using the following criteria: calculated EE values for all the functional groups were less than 1, and most of the calculated P/Q ratios (simply the quotient of P/B and Q/B) were between 0.1 and 0.3, representing a mass–balance model (Christensen and Walters, 2004). EE values were generally higher for the fish groups and

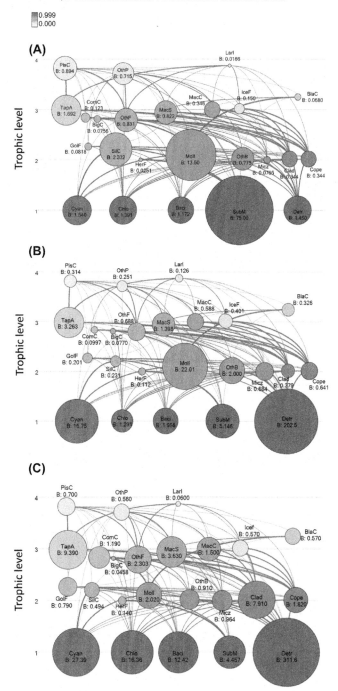

FIGURE 4.3 Food web structure and relative biomasses of Lake Chaohu, China, from the 1950s (A), 1980s (B), and 2000s (C) mass-balanced Ecopath models. The color bar in the left column represents the fraction of the biomass in the total biomass. (Details of the Abbreviations can be found in Table 4.1).

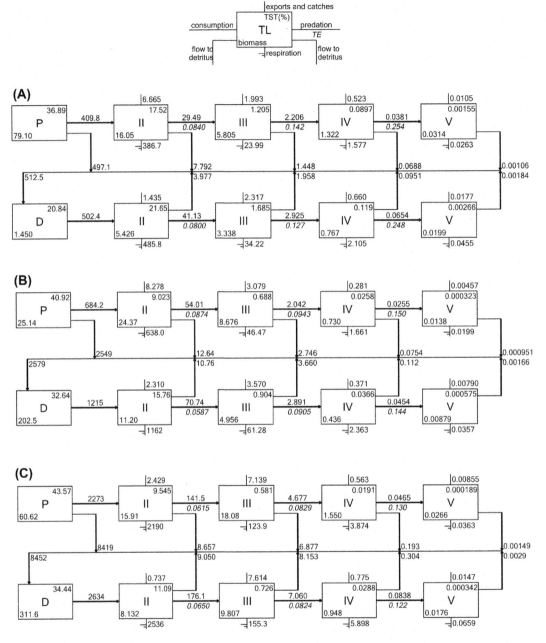

FIGURE 4.4 Lindeman spine representing the trophic flows of Lake Chaohu, China, from the 1950s (A), 1980s (B), and 2000s (C) mass-balanced Ecopath models. 'P' represents phytoplankton and 'D' represents detritus.

lower for zoobenthos and primary producers. In addition, the model-calculated TLs generally agree with the measured values for most functional groups (Fig. 4.5), indicating that the calibrated diet composition is in line with the real situations. For the Ecopath model, the TL for each functional group was largely determined by diet composition, which was, however, mostly arbitrarily assigned. This process has made the diet composition a subjective model input and a source of model uncertainty. An adjustment should be made for model inputs, particularly diet composition (Christensen and Walters, 2004). The calibration in the present study provides a solid basis for our analysis of the changes in the functioning of the ecosystem of Lake Chaohu. Thus, we suggest that the calibration process for diet composition by the comparison of calculated and measured TLs should be a standard procedure in analyses with the Ecopath model in the future.

4.2.4.2 Changes in Ecosystem Functioning

The calculated ecosystem properties of Lake Chaohu in the 1950s, 1980s, and 2000s are shown in Table 4.3. Multiple indicators show the changes in the ecosystem from different angles. TST (t/km^2/year) increased by one order of magnitude from the 1950s (2459.024) through the 1980s (7901.496) to the 2000s (24,541.850), indicating an increasing turnover rate of the biomass in the system related to cyanobacterial blooms (Shan et al., 2014). The profiles of TST were similar in the 1980s and 2000s. However, in the 1950s, the contribution of consumption and respiratory flows were much larger, but exports and flows into detritus were much lower. Similarly, the sum of all production (t/km^2/year) increased by one order of magnitude from the 1950s (960.843) through the 1980s (3351.219) to the 2000s (10,913.150), primarily due to nutrient enrichment from anthropogenic sources. The total catch (t/km^2/year) increased from 13,620 (1950s) to 17,900 (1980s) to 19,280 (2000s). The mean trophic level (MTL) of catch was similar in the 1950s (2.500) and 1980s (2.448) but slightly higher in the

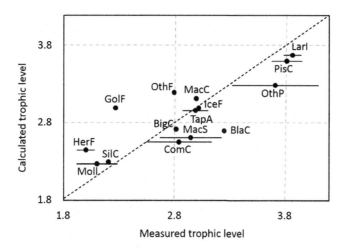

FIGURE 4.5   Comparison of measured and calculated TLs for different functional groups in Lake Chaohu, China. The measured values were based primarily on stable isotopes δ^{13}C and δ^{15}N from our laboratory measurement, also collected from Zhang et al. (2012) and Xu et al. (2005) (Details of the abbreviations can be found in Table 4.1.). The error bars represent one unit of standard deviation.

TABLE 4.3 Ecosystem Properties of Lake Chaohu in the 1950s (CH50′), 1980s (CH80′), and 2000s (CH00′) Based on Outputs From the Ecopath Model

Properties	Abbreviations	Units	CH50′	CH80′	CH00′
Sum of all consumption	TC	t/km^2/year	988.405	2029.502	5236.515
Sum of all exports	TEx	t/km^2/year	23.694	1381.892	5837.949
Sum of all respiratory flows	TR	t/km^2/year	934.462	1911.315	5015.073
Sum of all flows into detritus	TD	t/km^2/year	512.463	2578.786	8452.310
Total system throughput	TST	t/km^2/year	2459.024	7901.496	24,541.850
Sum of all production	TP	t/km^2/year	960.843	3351.219	10,913.150
Mean trophic level of the catch	MTL	—	2.500	2.448	2.912
Total catch	—	t/km^2/year	13.62	17.90	19.28
Gross efficiency	GE	—	0.015	0.006	0.002
Calculated total net primary production	TPP	t/km^2/year	906.900	3233.032	10,691.710
Total primary production/total respiration	TPP/TR	—	0.971	1.692	2.132
Net system production	NSP	t/km^2/year	27.562	1321.717	5676.638
Total primary production/total biomass	TPP/TB	—	8.864	54.858	111.154
Total biomass/total throughput	TB/TST	1/year	0.042	0.007	0.004
Total biomass (excluding detritus)	TB	t/km^2	102.316	58.935	96.188
System omnivory index	SOI	—	0.081	0.066	0.075
Throughput cycled (including detritus)	Tcy	t/km^2/year	836.068	1738.329	2208.767
Finn's cycling index	FCI	%	0.34	0.22	0.09
Finn's mean path length	FML	—	2.566	2.399	2.261
Connectance index	CI	—	0.238	0.238	0.238
Ascendancy	—	%	30.6	39.4	38.6
Overhead	—	%	69.4	60.6	61.4
Total transfer efficiencies	TE	%	14.0	9.8	8.7
Transfer efficiencies from primary producer	TE p.p.	%	14.5	10.7	8.7
Transfer efficiencies from detritus	TE det.	%	13.6	9.1	8.7

2000s (2.912). The gross efficiency (−) gradually decreased from 0.015 (1950s) to 0.006 (1980s) to 0.002 (2000s), suggesting the fraction of primary production that finally transferred into fishery products was decreasing during these years. From the 1950s to the 2000s, TPP, NSP, TPP/TR, and TPP/TB were rapidly increasing. The deviation of TPP/TR from 1 and the increasing TPP/TB in the 1980 and 2000s suggest that the ecosystem was approaching an immature and eutrophic state. SOI was highest in the 1950s (0.081) and lowest in the 1980s (0.066). This result demonstrates a general decrease in the complexity of the food web structure, indicating that the food web did not change from linear to web-like and that the system was not becoming more mature (Odum et al., 1971). Furthermore, the decreasing trend of FCI, FML, and ascendancy also demonstrated a decreasing maturity of system from the 1950s to the 2000s. Regarding to the TE, the geometric means for Lake Chaohu's ecosystem decreased from 14.0% (1950s) to 9.80% (1980s) to 8.70% (2000s), as did the TE p.p. and TE det (Table 4.3), lying in the acceptable range around a theoretical value of 10% (Lindeman, 1942). These results indicated that as the ecosystem of Lake Chaohu was adapting to the changing external conditions, the remaining ability of the system to utilize the available resources was reduced.

The results of the MTI analysis are presented in Fig. 4.6. Only a few differences were observed for the three periods. Piscivorous fish generally had negative effects on other fish groups and on phytoplankton, and they had positive effects on zoobenthos and zooplankton. However, this effect was weakened in the years approaching the 2000s. Planktivorous fish, particularly the tapertail anchovy (TapA), had negative effects on most fish groups and on zooplankton, whereas they positively affected phytoplankton. Most fish groups benefitted from phytoplankton. Fishing (Fleet1) was apparently negatively influencing most fish groups, particularly commercial stocks, such as piscivorous and silver carp. Simultaneously, this anthropogenic activity had a positive effect on phytoplankton, manifesting the intensive fishery in the lake may contribute to the recent blooming of phytoplankton.

4.2.4.3 Collapse in the Food Web: Differences in Structure

Our results clearly illustrate a collapse in the food web structure in Lake Chaohu during the 1950s, 1980s, and 2000s (Fig. 4.3). Here, we demonstrate the structural differences by clusters of functional groups, as follows.

The total fish biomass increased rapidly, but it is clear that the catch composition of the fishery showed an abrupt change in the beginning of 1960s after the onset of the sluice (Fig. 4.2). The dominant group of fish shifted from large-sized piscivorous fish (62.9% in 1952) to small-sized (80% after the 1980s) zooplanktivorous fish, such as pelagic *C. ectenes taihuensis* and *N. taihuensis*. The fraction of young large-sized fish in the fishery catchment was increasing—yet another sign of a growing trend toward small-sized fish in the lake.

Zooplankton and zoobenthos occupied the middle TLs in the lake ecosystem, thereby playing a vital role in the mass and energy cycling. The abundance of zooplankton, including cladocerans (Clad), copepods (Cope), and small-sized zooplankton (Micz) such as rotifers, increased from the 1950s to the 2000s, based on field data (Deng, 2004). The dominant species in the cladocerans changed from the large-sized *Daphnia pulex* and *Daphnia hyalina* to the small-sized *Bosmina coregoni*, whereas in copepods, *Sinocalanus dorii* and *Limnothona sinensis* became the dominant species. Zoobenthos fed primarily on microorganisms such as diatoms, protozoans, rotifers, and detritus and were eaten by carnivorous fish. The dominance of

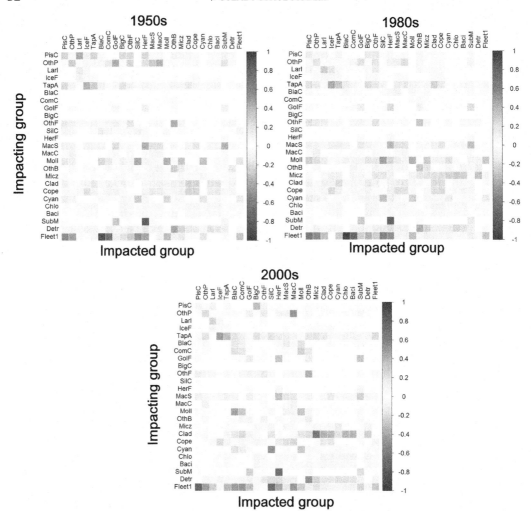

FIGURE 4.6 Mixed trophic impacts of Lake Chaohu's ecosystem in the 1950s, 1980s, and 2000s. Blue values represent a positive impact, whereas red values represent a negative impact, and the absolute values are proportionate to the degree of the impact.

Mollusca (Moll) in the 1960s (*Corbicula fluminea*, *Limnoperna lacustris*, and *Semisulcospira cancellata*) indicates the oligotrophic state of the lake at that time. Later, in the 1980s, the biomass of Moll had increased approximately sixfold (Cai et al., 2012; Hu and Yao, 1981). However, recent studies have shown that the composition of benthic assemblages was largely altered during the 2000s, shifting from a dominance of Moll to oligochaetes and chironomids (OthB; biomass was still primarily attributed to the Moll; Cai et al., 2012; Ning et al., 2012). In addition, the abundance of macrocrustacean shrimp (MacS) and crab (MacC) gradually increased throughout these years, whereas the dominant shrimp species transitioned from *Macrobrachium nipponensis* to a more sedentary *P. modestus* (Hu and Yao, 1981).

At the bottom of the food web, the primary producer shifted from macrophytes in the 1950s to phytoplankton in the 1980s and 2000s. Abundant submerged macrophytes were recorded before the flood in 1954, covering 25–30% of the lake area according to field observation (Xie, 2009; Zhang et al., 2014). The flood led to the degradation but not the extinction of the macrophytes, which recovered to 25% of coverage before 1960, according to field observation, primarily due to the surviving seed banks (Zhang et al., 2014). Thereafter, the macrophytes community gradually shifted toward floating and emergent plants (Xu et al., 1999a), and the observed coverage of macrophytes were as low as 2.54% and 1.54% in the 1980s and 2000s, respectively. Phytoplankton took over and accounted for more than 90% of the lake's primary productivity, with the dominant species of *Microcystis aeruginosa* and *Anabaena spirodies*. The total biomass of phytoplankton in the 2000s became even higher than that in the 1980s with increasing contributions from chlorophytes and bacillariophytes (Guo, 2005).

Overall, we observed a collapse of food web structures toward a simplified structure and decreasing biodiversity in Lake Chaohu's ecosystem. This observation is further supported by the generally decreasing SOI and FML calculated by model (Table 4.3), suggesting a shorter food chain length and a less complex structure. In addition, field data suggest that the MTL of Lake Chaohu declined steadily since 1970, indicating a decrease in the abundance of higher TL species (e.g., large piscivorous bottom fish) relative to lower TL ones (e.g., small pelagic fish; Zhang et al., 2012). This phenomenon, known as "fishing down the food web" (Pauly et al., 1998), has led to a loss of trophic interactions and lower structure complexity in the ecosystem of Lake Chaohu (Zhang et al., 2012). However, in the present study, the MTL of the catch was highest in the 2000s (Table 4.3), which was largely attributed to the relatively high TL of dominant stocks of the TapA (3.0). This value was in line with earlier measurements (2.9–4.1; Xu et al., 2005) and was kept constant due to a fixed diet composition in all three models here, but it could be dynamic in a long time scale. An overestimation of the TL in the 2000s and a bias in the fishery catch data may have led to this result. However, we argue that the TapA could be a special case that occupies high TL but has a specialized diet of zooplankton, thereby simplifying the food web structure and reducing biodiversity and trophic interactions.

4.2.4.4 *Toward an Immature but Stable Ecosystem*

The changes in the food web structure could be the major explanation for the changes in ecosystem functioning, which can be observed in multiple ecosystem property indicators from model outputs (Table 4.3). We have demonstrated that the ecosystem of Lake Chaohu was moving toward an immature state from the 1950s to the 2000s, according to changes in multiple related indicators (TPP/TR, SOI, FCI, FML, and ascendancy). Nonetheless, the stability of the system might have developed in another way such that immature ecosystems were thought to be more stable and vice versa (Pérez-España and Arreguín-Sánchez, 2001). In general, the stability of lake ecosystem decreases as the nutrient loading increases due to the loss of system resilience and maturity (Scheffer et al., 2001). A recent study showed, however, that after the catastrophic shift to the turbid ecosystem state in lakes, there will be a shift toward a higher stability (Kuiper et al., 2015). Because a catastrophic shift in Lake Chaohu was identified around 1980 (Liu et al., 2012), the stability of the ecosystem could have been lower in the 1980s than in the 1950s and 2000s. In addition, the stability of Lake Chaohu ecosystem could be interpreted from the proportion of detritus and primary producer energy channels in the food web such that a stable ecosystem derived energy from both rather than a single pathway to enhance the plasticity and resilience (Fetahi et al., 2011; Moore et al., 2004). What we found was that energy flows were evenly distributed in the 1950s and 2000s, whereas the detritus pathway was more dominant in the 1980s in Lake Chaohu (Fig. 4.4). Additionally, detritus itself increases the stability and persistence of the food web by influencing the composition and dynamics (Moore et al., 2004). The biomass of detritus in the food web in Lake Chaohu kept increasing throughout time (Fig. 4.3), primarily due to increasing production and throughput in the system and decreasing gross efficiency (Table 4.3). Primary production was not utilized efficiently to transfer into fish production but flowed back into detritus through a potential shortcut (Shan et al., 2014). Overall, the ecosystem of Lake Chaohu may develop toward an immature but stable state. The discussion above may contribute to the debate on the relationships among the stability, maturity, and complexity of ecosystems. Further studies may be conducted using thermodynamic ecosystem-level indicators such as exergy and structural exergy (Marchi et al., 2012; Xu et al., 2001) for a more comprehensive evaluation of lake ecosystem status.

4.2.4.5 Potential Driving Factors and Underlying Mechanisms

Unfortunately, as we are using steady state Ecopath model that accounts for the food web components only, the potential external driving factors and underlying mechanisms of the changes in food web structure and ecosystem functioning cannot be directly identified in the present study. We speculated that anthropogenic stressors, including nutrient-loading discharge, intensified fishery, and hydrological regulation, were the major drivers that led to a switch from top-down to bottom-up control in the food web. The onset of the sluice in 1963 might have blocked the migratory route for juvenile large-sized fish, particularly migratory piscivorous, herbivorous, and omnivorous fish, thereby strongly suppressing the survival and growth of these fish in the lake and subsequently favoring the dominance of small-sized, sedentary stocks. The regulated higher water level in spring thereafter led to the abrupt degradation of macrophytes (Zhang et al., 2014), which used to provide refuge for both piscivorous/herbivorous fish and large-sized zooplankton (Carpenter et al., 2008; Schriver et al., 1995). Consequently, the top-down control and trophic cascading were destroyed, resulting in the limitation of piscivorous/herbivorous fish, the domination of zooplanktivorous fish and the enhanced suppression of large-sized zooplankton, all of which favored the dominance of phytoplankton as the primary producer. From the 1980s, the gradual intensification of fishery and eutrophication (Kong et al., 2015; Zhang et al., 2012) might have further overexploited predatory fish and triggered cyanobacterial blooms, which are inedible to zooplankton. All of these factors contributed to the strongly increased productivity and flux cycling in the low TLs of the ecosystem, which might give rise to the bottom-up control in the food web (Oksanen, 1988), stabilize the ecosystem in eutrophic state, enhance the resistance to restoration efforts, and hinder the recovery toward a prior state. As a result, anaerobic conditions caused by cyanobacterial blooms and organic matter accumulation in the benthic environment, along with the toxicity of microcystins produced by cyanobacteria, exerted high pressure on the species of low hypoxia-tolerant Mollusca, such as *C. fluminea*, and favored the dominance of high hypoxia-tolerant oligochaetes and chironomids, thereby largely reducing the biodiversity of benthic communities (Cai et al., 2012; Lv et al., 2011).

Overall, Lake Chaohu might be affected by multiple driving factors, but it remains unclear which factors contribute to the dramatic changes; further evaluation is required. In the Baltic Sea ecosystem, the strong removal of piscivorous fish by fishery may lead to a collapse in the top predators, cascading effects down the food web, and a subsequent shift in ecosystem functioning (Casini et al., 2009). In addition, human-introduced invasive species may alter the food web structure, cause a loss of complexity and in turn trigger a critical transition in ecosystems such as Lake Chozas (Marchi et al., 2011, 2012). However, other than in the Baltic Sea and Lake Chozas, the collapse in the food web structure in Lake Chaohu might initially be triggered by the water level control due to the onset of the sluice. Thus, the major driving factors and the underlying mechanisms for the changes in food web structure and ecosystem function may be unique in each specific case, which requires careful evaluation.

4.2.4.6 Hints for Future Lake Fishery and Restoration

Fishery can be put on a path toward sustainability by limiting fishing effort (Pauly et al., 2002). In fact, there has been a fishery regulation in Lake Chaohu in recent years that releases pressure on economic stocks by prohibiting fishery from February to July. However, this strategy has not led to significant improvements in the lake's state. The alternative stable state theory suggests that the new state after the collapse of the food web may be stabilized by certain mechanisms, which results in difficulties of lake ecosystem restoration even under robust management. There could be a threshold in density of zooplanktivorous fish that separates the ecosystem into two alternative states with different structures and functioning (Casini et al., 2009). It is generally difficult to control small-sized planktivorous fish, but the large-sized fish are much easier to regulate. Thus, we propose that, first, we should evaluate the value of this planktivorous fish density threshold and that the increase in the abundance of large-sized fish should be large enough to keep planktivorous fish density under the threshold, thereby favoring the restoration of the food web. Second, nutrient loading should be controlled; two-thirds of the current loading needs to be cut down to restore the lake to a clear state (Kong et al., 2015). Third, a multiobjective water level control for the hydrological regulation in this lake is strongly suggested (Kong et al., 2013; Zhang et al., 2014). Overall, an integrated strategy incorporating all the strategies above will bring about direct or indirect positive effects on the food web and ecosystem functioning of this lake and together promote the lake toward a mature, stable, and healthy state in the future.

Furthermore, using Ecopath in the case of Lake Chaohu in the present study, we have shown it is possible to develop a new management tool for lakes that considers the changes in the food web and ecosystem functioning as a consequence of the increased anthropogenic effects on the ecosystem. As illustrated in Fig. 4.3 and Table 4.3, the model shows its ability to follow the food web collapse and evaluate the changes in the functioning of an ecosystem, which will undoubtedly bridge environmental and ecological management. In general, there is an increasingly urgent need for integrated ecological–environmental management in lakes (Jørgensen and Nielsen, 2012), where the Ecopath model may play a role.

4.3 CONCLUSIONS

The present study managed to establish Ecopath-based mass–balance models for the 1950s, 1980s, and 2000s to describe systematically the changes in the food web structure and ecosystem functioning in Lake Chaohu, China. We found significant changes in the food web structure throughout these periods and the degradation of the lake ecosystem, which was approaching an immature but stable state. Nutrient enrichment, intensive fishery, and hydrological regulation might be the key drivers of these changes in the food web structure and ecosystem functioning. The present study provided a first comprehensive and quantitative evaluation of the effect of multiple anthropogenic stressors on Lake Chaohu's ecosystem, and it provided hints toward the sustainable management of this important ecosystem. We deem that the Ecopath model can be considered a new management tool for lakes because it incorporates a food web perspective and bridges the strategy of environmental and ecological management, as shown in the concrete case of Lake Chaohu.

References

Barausse, A., Duci, A., Mazzoldi, C., Artioli, Y., Palmeri, L., 2009. Trophic network model of the Northern Adriatic Sea: analysis of an exploited and eutrophic ecosystem. Estuarine, Coastal and Shelf Science 83, 577–590.

Brando, V.E., Ceccarelli, R., Libralato, S., Ravagnan, G., 2004. Assessment of environmental management effects in a shallow water basin using mass-balance models. Ecological Modelling 172, 213–232.

Cai, Y.J., Gong, Z.J., Xie, P., 2012. Community structure and spatiotemporal patterns of macrozoobenthos in Lake Chaohu (China). Aquatic Biology 17, 35–46.

Carpenter, S.R., Brock, W.A., Cole, J.J., Kitchell, J.F., Pace, M.L., 2008. Leading indicators of trophic cascades. Ecology Letters 11, 128–138.

Casini, M., Hjelm, J., Molinero, J.-C., Lövgren, J., Cardinale, M., Bartolino, V., Belgrano, A., Kornilovs, G., 2009. Trophic cascades promote threshold-like shifts in pelagic marine ecosystems. Proceedings of the National Academy of Sciences 106, 197–202.

Cerco, C.F., Tillman, D., Hagy, J.D., 2010. Coupling and comparing a spatially- and temporally-detailed eutrophication model with an ecosystem network model: an initial application to Chesapeake Bay. Environmental Modelling & Software 25, 562–572.

Christensen, V., 1995. Ecosystem maturity-towards quantification. Ecological Modelling 77, 3–32.

Christensen, V., Walters, C.J., 2004. Ecopath with Ecosim: methods, capabilities and limitations. Ecological Modelling 172, 109–139.

Christensen, V., Walters, C.J., Pauly, D., 2005. Ecopath with Ecosim: A User's Guide. Fisheries Centre, University of British Columbia. Vancouver 154.

Coll, M., Palomera, I., Tudela, S., Sardà, F., 2006. Trophic flows, ecosystem structure and fishing impacts in the South Catalan Sea, Northwestern Mediterranean. Journal of Marine Systems 59, 63–96.

de Ruiter, P.C., Neutel, A.-M., Moore, J.C., 1995. Energetics, patterns of interaction strengths, and stability in real ecosystems. Science 1257–1260.

Dearing, J.A., Yang, X., Dong, X., Zhang, E., Chen, X., Langdon, P.G., Zhang, K., Zhang, W., Dawson, T.P., 2012. Extending the timescale and range of ecosystem services through paleoenvironmental analyses, exemplified in the lower Yangtze basin. Proceedings of the National Academy of Sciences 109, E1111–E1120.

Deng, D.G., 2004. Ecological Studies on the Effects of Eutrophication on Plankton Communities in a Large Shallow Lake, Lake Chaohu (Doctoral dissertation). Institute of Hydrobiology, Chinese Academy of Sciences, Wuhan, China.

Downing, A.S., van Nes, E.H., Janse, J.H., Witte, F., Cornelissen, I.J.M., Scheffer, M., Mooij, W.M., 2012. Collapse and reorganization of a food web of Mwanza Gulf, Lake Victoria. Ecological Applications 22, 229–239.

Fetahi, T., Mengistou, S., 2007. Trophic analysis of Lake Awassa (Ethiopia) using mass-balance Ecopath model. Ecological Modelling 201, 398–408.

Fetahi, T., Schagerl, M., Mengistou, S., Libralato, S., 2011. Food web structure and trophic interactions of the tropical highland Lake Hayq, Ethiopia. Ecological Modelling 222, 804–813.

Finn, J.T., 1976. Measures of ecosystem structure and function derived from analysis of flows. Journal of Theoretical Biology 56, 363–380.

Folke, C., Carpenter, S., Walker, B., Scheffer, M., Elmqvist, T., Gunderson, L., Holling, C., 2004. Regime shifts, resilience, and biodiversity in ecosystem management. Annual Review of Ecology, Evolution, and Systematics 557–581.

Gardner, M.R., Ashby, W.R., 1970. Connectance of large dynamic (cybernetic) systems: critical values for stability. Nature 228, 784.

Guo, C.B., Ye, S.W., Lek, S., Liu, J.S., Zhang, T.L., Yuan, J., Li, Z.J., 2013. The need for improved fishery management in a shallow macrophytic lake in the Yangtze River basin: evidence from the food web structure and ecosystem analysis. Ecological Modelling 267, 138–147.

Guo, L.G., 2005. Studies on Fisheries Ecology in a Large Eutrophic Shallow Lake, Lake Chaohu. Institute of Hydrobiology, Chinese Academy of Sciences, Wuhan, China (Doctoral Dissertation).

Hossain, M., Arhonditsis, G.B., Koops, M.A., Minns, C.K., 2012. Towards the development of an ecosystem model for the Hamilton Harbour, Ontario, Canada. Journal of Great Lakes Research 38, 628–642.

Hossain, M.M., Matsuishi, T., Arhonditsis, G., 2010. Elucidation of ecosystem attributes of an oligotrophic lake in Hokkaido, Japan, using Ecopath with Ecosim (EwE). Ecological Modelling 221, 1717–1730.

Hu, J., Yao, W., 1981. Investigation of macrozoobenthos in lake Chaohu, China. Journal of Anhui University and Technology (Natural Science) 2, 159–173 (in Chinese).

Hu, W., Jørgensen, S.E., Zhang, F., Chen, Y., Hu, Z., Yang, L., 2011. A model on the carbon cycling in Lake Taihu, China. Ecological Modelling 222, 2973–2991.

Janjua, M.Y., Tallman, R., Howland, K., 2014. Elucidation of ecosystem attributes of two Mackenzie great lakes with trophic network analysis. Aquatic Ecosystem Health and Management 17, 151–160.

Jepsen, D.B., Winemiller, K.O., 2002. Structure of tropical river food webs revealedby stable isotope ratios. Oikos 96, 46–55.

Jia, P.Q., Hu, M.H., Hu, Z.J., Liu, Q.G., Wu, Z., 2012. Modeling trophic structure and energy flows in a typical macrophyte dominated shallow lake using the mass balanced model. Ecological Modelling 233, 26–30.

Jørgensen, S.E., Fath, B.D., 2011. Fundamentals of Ecological Modelling, fourth ed. Elsevier, Amsterdam.

Jørgensen, S.E., Nielsen, S.N., 2012. Tool boxes for an integrated ecological and environmental management. Ecological Indicators 21, 104–109.

Kao, Y.C., Adlerstein, S., Rutherford, E., 2014. The relative impacts of nutrient loads and invasive species on a Great Lakes food web: an Ecopath with Ecosim analysis. Journal of Great Lakes Research 40, 35–52.

Kong, X., Dong, L., He, W., Wang, Q., Mooij, W.M., Xu, F., 2015. Estimation of the long-term nutrient budget and thresholds of regime shift for a large shallow lake in China. Ecological Indicators 52, 231–244.

Kong, X., He, W., Liu, W., Yang, B., Xu, F., Jørgensen, S.E., Mooij, W.M., 2016. Changes in food web structure and ecosystem functioning of a large shallow Chinese lake during 1950s, 1980s and 2000s. Ecological Modelling 319, 31–41.

Kong, X., Jørgensen, S.E., He, W., Qin, N., Xu, F., 2013. Predicting the restoration effects by a structural dynamic approach in Lake Chaohu, China. Ecological Modelling 266, 73–85.

Kuiper, J.J., van Altena, C., de Ruiter, P.C., van Gerven, L.P.A., Janse, J.H., Mooij, W.M., 2015. Food-web stability signals critical transitions in temperate shallow lakes. Nature Communications 6, 7727.

Li, Y.K., Chen, Y., Song, B., Olson, D., Yu, N., Chen, L., 2009. Ecosystem structure and functioning of Lake Taihu (China) and the impacts of fishing. Fisheries Research 95, 309–324.

Li, Y.K., Song, B., Chen, Y., Chen, L.Q., Yu, N., Olson, D., 2010. Changes in the trophic interactions and the community structure of Lake Taihu (China) ecosystem from the 1960s to 1990s. Aquatic Ecology 44, 337–348.

Lindeman, R.L., 1942. The trophic-dynamic aspect of ecology. Ecology 23, 399–417.

Liu, E., Li, Y., Zang, R., Wang, H., 2014. A preliminary analysis of the ecosystem structure and functioning of Lake Chaohu based on Ecopath model. Journal of Fisheries in China 38, 417–425 (in Chinese with English abstract).

Liu, J.G., Dietz, T., Carpenter, S.R., Alberti, M., Folke, C., Moran, E., Pell, A.N., Deadman, P., Kratz, T., Lubchenco, J., Ostrom, E., Ouyang, Z., Provencher, W., Redman, C.L., Schneider, S.H., Taylor, W.W., 2007a. Complexity of coupled human and natural systems. Science 317, 1513–1516.

Liu, Q., Yang, X., Anderson, N.J., Liu, E., Dong, X., 2012. Diatom ecological response to altered hydrological forcing of a shallow lake on the Yangtze floodplain, SE China. Ecohydrology 5, 316–325.

Liu, Q.G., Chen, Y., Li, J.L., Chen, L.Q., 2007b. The food web structure and ecosystem properties of a filter-feeding carps dominated deep reservoir ecosystem. Ecological Modelling 203, 279–289.

Liu, X.Q., 2006. Food Composition and Food Webs of Zoobenthos in Yangtze Lakes (Doctoral Dissertation). Institute of Hydrobiology, Chinese Academy of Sciences, Wuhan, China.

López, B.D., Bunke, M., Shirai, J.A.B., 2008. Marine aquaculture off Sardinia Island (Italy): ecosystem effects evaluated through a trophic mass-balance model. Ecological Modelling 212, 292–303.

Lv, Y., Wang, H., Tang, H., 2011. Review of the fishery development in lake Chaohu since the founding of the People's Republic of China. Journal of Anhui Agricultural Sciences 39, 4018–4020 (in Chinese with English abstract).

Marchi, M., Jørgensen, S.E., Bécares, E., Corsi, I., Marchettini, N., Bastianoni, S., 2011. Dynamic model of Lake Chozas (León, NW Spain)—decrease in eco-exergy from clear to turbid phase due to introduction of exotic crayfish. Ecological Modelling 222, 3002–3010.

Marchi, M., Jørgensen, S.E., Bécares, E., Fernández-Aláez, C., Rodríguez, C., Fernández-Aláez, M., Pulselli, F.M., Marchettini, N., Bastianoni, S., 2012. Effects of eutrophication and exotic crayfish on health status of two Spanish lakes: a joint application of ecological indicators. Ecological Indicators 20, 92–100.

Mavuti, K., Moreau, J., Munyandorero, J., Plisnier, P.D., 1996. Analysis of trophic relationships in two shallow equatorial lakes Lake Naivasha (Kenya) and Lake Ihema (Rwanda) using a multispecifics trophic model. Hydrobiologia 321, 89–100.

Moore, J.C., Berlow, E.L., Coleman, D.C., Ruiter, P.C., Dong, Q., Hastings, A., Johnson, N.C., McCann, K.S., Melville, K., Morin, P.J., 2004. Detritus, trophic dynamics and biodiversity. Ecology Letters 7, 584–600.

Ning, Y., Gao, F., Deng, J., Hu, W., Gao, J., Zhao, Z., 2012. Biological assessment of water quality in Chaohu Lake watershed: a case study of benthic macroinvertebrate. Chinese Journal of Ecology 31, 916–922 (in Chinese with English abstract).

Odum, E.P., 1969. The strategy of ecosystem development. Science 164, 262–270.

Odum, E.P., Odum, H.T., Andrews, J., 1971. Fundamentals of Ecology. Saunders, Philadelphia.

Oksanen, L., 1988. Ecosystem organization: mutualism and cybernetics or plain Darwinian struggle for existence? American Naturalist 131, 424–444.

Palomares, M.L.D., Pauly, D., 1998. Predicting food consumption of fish populations as functions of mortality, food type, morphometrics, temperature and salinity. Marine and Freshwater Research 49, 447–453.

Patrício, J., Ulanowicz, R., Pardal, M., Marques, J., 2004. Ascendency as an ecological indicator: a case study of estuarine pulse eutrophication. Estuarine, Coastal and Shelf Science 60, 23–35.

Pauly, D., 1980. On the interrelationships between natural mortality, growth parameters, and mean environmental temperature in 175 fish stocks. Journal du Conseil 39, 175–192.

Pauly, D., Christensen, V., Dalsgaard, J., Froese, R., Torres, F., 1998. Fishing down marine food webs. Science 279, 860–863.

Pauly, D., Christensen, V., Guénette, S., Pitcher, T.J., Sumaila, U.R., Walters, C.J., Watson, R., Zeller, D., 2002. Towards sustainability in world fisheries. Nature 418, 689–695.

Pavés, H.J., González, H.E., 2008. Carbon fluxes within the pelagic food web in the coastal area off Antofagasta (23 S), Chile: the significance of the microbial versus classical food webs. Ecological Modelling 212, 218–232.

Pedersen, T., Nilsen, M., Nilssen, E.M., Berg, E., Reigstad, M., 2008. Trophic model of a lightly exploited cod-dominated ecosystem. Ecological Modelling 214, 95–111.

Pérez-España, H., Arreguín-Sánchez, F., 2001. An inverse relationship between stability and maturity in models of aquatic ecosystems. Ecological Modelling 145, 189–196.

Polovina, J.J., 1984. Model of a coral reef ecosystem. Coral Reefs 3, 1–11.

Ren, Y.Q., Chen, K.N., 2011. Status of submerged macrophytes and its relationship with environmental factors in Lake Chaohu, 2010. Journal of Lake Sciences 23, 409–416 (in Chinese with English abstract).

Rockström, J., Steffen, W., Noone, K., Persson, Å., Chapin, F.S., Lambin, E.F., Lenton, T.M., Scheffer, M., Folke, C., Schellnhuber, H.J., 2009. A safe operating space for humanity. Nature 461, 472–475.

Rooney, N., McCann, K., Gellner, G., Moore, J.C., 2006. Structural asymmetry and the stability of diverse food webs. Nature 442, 265–269.

Scheffer, M., Carpenter, S., Foley, J.A., Folke, C., Walker, B., 2001. Catastrophic shifts in ecosystems. Nature 413, 591–596.

Scheffer, M., Hosper, S.H., Meijer, M.L., Moss, B., Jeppesen, E., 1993. Alternative equilibria in shallow lakes. Trends in Ecology and Evolution 8, 275–279.

Schriver, P., Bøgestrand, J., Jeppesen, E., Søndergaard, M., 1995. Impact of submerged macrophytes on fish-zooplankton-phytoplankton interactions: large-scale enclosure experiments in a shallow eutrophic lake. Freshwater Biology 33, 255–270.

Shan, K., Li, L., Wang, X.X., Wu, Y.L., Hu, L.L., Yu, G.L., Song, L.R., 2014. Modelling ecosystem structure and trophic interactions in a typical cyanobacterial bloom-dominated shallow Lake Dianchi, China. Ecological Modelling 291, 82–95.

Stewart, T.J., Sprules, W.G., 2011. Carbon-based balanced trophic structure and flows in the offshore Lake Ontario food web before (1987–1991) and after (2001–2005) invasion-induced ecosystem change. Ecological Modelling 222, 692–708.

Thapanand, T., Moreau, J., Jutagate, T., Wongrat, P., Lekchonlayut, T., Meksumpun, C., Janekitkarn, S., Rodloi, A., Dulyapruk, V., Wongrat, L., 2007. Towards possible fishery management strategies in a newly impounded man-made lake in Thailand. Ecological Modelling 204, 143–155.

Ulanowlcz, R.E., Norden, J.S., 1990. Symmetrical overhead in flow networks. International Journal of Systems Science 21, 429–437.

Villanueva, M.C., Isumbisho, M., Kaningini, B., Moreau, J., Micha, J.C., 2008. Modeling trophic interactions in Lake Kivu: what roles do exotics play? Ecological Modelling 212, 422–438.

Villanueva, M.C., Laleye, P., Albaret, J.J., Lae, R., de Morais, L.T., Moreau, J., 2006a. Comparative analysis of trophic structure and interactions of two tropical lagoons. Ecological Modelling 197, 461–477.

Villanueva, M.C., Ouedraogo, M., Moreau, J., 2006b. Trophic relationships in the recently impounded Bagre reservoir in Burkina Faso. Ecological Modelling 191, 243–259.

Xie, P., 2009. Reading About the Histories of Cyanobacteria, Eutrophication and Geological Evolution in Lake Chaohu. Science Press, Beijing (in Chinese).

Xu, F.L., Jorgensen, S.E., Tao, S., Li, B.G., 1999a. Modeling the effects of ecological engineering on ecosystem health of a shallow eutrophic Chinese lake (Lake Chao). Ecological Modelling 117, 239–260.

Xu, F.L., Tao, S., Dawson, R.W., Li, B.G., Cao, J., 2001. Lake ecosystem health assessment: indicators and methods. Water Research 35, 3157–3167.

Xu, F.L., Tao, S., Xu, Z.R., 1999b. The restoration of riparian wetlands and macrophytes in Lake Chao, an eutrophic Chinese lake: possibilities and effects. Hydrobiologia 405, 169–178.

Xu, J., Xie, P., Zhang, M., Yang, H., 2005. Variation in stable isotope signatures of seston and a zooplanktivorous fish in a eutrophic Chinese lake. Hydrobiologia 541, 215–220.

Yang, X., Lu, X., 2014. Drastic change in China's lakes and reservoirs over the past decades. Scientific Reports 4, 6041. http://dx.doi.org/10.1038/srep06041.

Yang, Y., Chen, H., 2013. Assessing impacts of flow regulation on trophic interactions in a wetland ecosystem. Journal of Environmental Informatics 21, 63–71.

Zhang, K., Dearing, J.A., Dawson, T.P., Dong, X., Yang, X., Zhang, W., 2015. Poverty alleviation strategies in eastern China lead to critical ecological dynamics. Science of the Total Environment 506, 164–181.

Zhang, M., Xie, C., Hansson, L.-A., Hu, W., Che, J., 2012. Trophic level changes of fishery catches in lake Chaohu, Anhui Province, China: trends and causes. Fisheries Research 131, 15–20.

Zhang, X., Liu, X., Wang, H., 2014. Developing water level regulation strategies for macrophytes restoration of a large river-disconnected lake, China. Ecological Engineering 68, 25–31.

Earth's Surface Modeling

T.X. Yue*,§,1, N. Zhao*,§, Z.P. Du*

*Institute of Geographical Science and Natural Resources Research, Beijing, China and
§University of Chinese Academy of Sciences, Beijing, China
1Corresponding author: E-mail: yue@lreis.ac.cn

5.1 INTRODUCTION

The Earth's surface environment is an active and complex place, at the interface of the lithosphere, the hydrosphere, the atmosphere, and the biosphere (Phillips, 1999). An earth surface system is a set of interconnected components of the earth surface environment that function together as a complex whole. Earth's surface modeling is generally defined as a spatially explicitly digital description of an earth surface system or a component of the earth surface environment (Yue, 2011).

Surface modeling began to be used in the 1960s, with the general availability of computers (Lo and Yeung, 2002), but because it requires powerful software and a large amount of spatially explicit data, its development was limited before the 1990s. The major advances that enabled the use of surface modeling included trend surface analysis (Ahlberg et al., 1967; Schroeder and Sjoquist, 1976; Legendre and Legendre, 1983), the digital terrain model (Stott, 1977), surface approximation (Long, 1980), spatial simulation of wetland habitats (Sklar et al., 1985), spatial pattern matching (Costanza, 1989), spatial prediction (Turner et al., 1989), and modeling costal landscape dynamics (Costanza et al., 1990). Surface modeling has greatly progressed since the early 1990s, with rapid development of remote sensing (RS) and a geographical information system (GIS), as well as the accumulation of spatially explicit data.

It was learnt that slope and curvature are significant variables of Earth's surface analysis in the early 1980s (Evans, 1980). In fact, a plane curve is uniquely determined by its curvature and a space curve is uniquely determined by its curvature and torsion if a translation followed by a rotation is allowed in terms of curve theorems in the plane and in the space (Spivak, 1979). Following this consideration, two equivalent indexes (EQIs) of curves were developed respectively for plane curves and space curves to simulate surfaces (Yue and Ai, 1990) and detect surface changes (Yue et al., 2002) by fitting section lines of a surface and combining them together.

In the early 2000s, it was found that Earth's surface systems are controlled by a combination of global factors and local factors, which cannot be understood without accounting for both the local and global components. The system dynamic cannot be recovered from the global or local controls alone (Phillips, 2002). In fact, in terms of the fundamental theorem of surfaces (FTS), a surface is uniquely defined by the first and the second fundamental coefficients (Somasundaram, 2005). The first fundamental coefficients express the information about the details of the surface, which are observed when we stay on the surface. The second fundamental coefficients express the change of the surface observed from outside the surface (Yue et al., 2015a).

To significantly reduce uncertainty of simulating Earth's surfaces, we suggest an alternative method, high accuracy surface modeling (HASM), which takes global approximate information (e.g., RS images or model simulation results) as its driving field and local accurate information (e.g., ground observation data and/or sampling data) as its optimum control constraints (Yue et al., 2007). HASM completes its operation when its output satisfies the iteration stopping criterion which is determined by application requirement for accuracy (Zhao and Yue, 2014a). A fundamental theorem of earth surface modeling (FTESM) is abstracted on the basis of applying HASM to simulating surfaces of elevation, soil properties, changes of ecosystem services, and driving forces of ecosystem changes on multiscales for about 20 years (Yue et al., 2016).

5.2 EQUIVALENT INDEXES

For curves in space, the EQI of curves can be formulated as follows in terms of curve theorem if a translation and a rotation are not allowed (Yue and Ai, 1990):

$$Eq = \frac{1}{S - S_0} \int_{S_0}^{S} \left((l_1(S_0) - l_2(S_0))^2 + (\tau_1(s) - \tau_2(s))^2 + (k_1(s) - k_2(s))^2 + |\overrightarrow{n}_1(s) - \overrightarrow{n}_2(s)|^2 \right)^{\frac{1}{2}} ds$$

(5.1)

where $k_i(s)$, $\tau_i(s)$, and $\overrightarrow{n}_i(s)$ are respectively curvature, torsion, direction of curve l_i ($i = 1,2$); s is arc length; $l_i(S_0)$ is the initial of curve l_i; $L = S - S_0$.

If a Euclidean motion of a translation and a rotation is allowed, the EQI can be simplified as,

$$Eq = \frac{1}{L} \int_{S_0}^{S} \left((k_1(s) - k_2(s))^2 + (\tau_1(s) - \tau_2(s))^2 \right)^{\frac{1}{2}} ds$$

(5.2)

For curves in plane, if the Euclidean motion is not allowed, Eq. (5.1) can be expressed as,

$$Eq = \frac{1}{S - S_0} \int_{S_0}^{S} \left((l_1(S_0) - l_2(S_0))^2 + (\alpha_1(s) - \alpha_2(s))^2 + (k_1(s) - k_2(s))^2 \right)^{\frac{1}{2}} ds \quad (5.3)$$

where $k_i(s)$ and $\alpha_i(s)$ are, respectively, the slope and curvature of curve l_i ($i = 1,2$) at an arc length of s; $l_i(S_0)$ is the initial value.

If the Euclidean motion is allowed, Eq. (5.3) can be simplified as,

$$Eq = \frac{1}{L} \int_{S_0}^{S} |k_1(s) - k_2(s)| ds \quad (5.4)$$

It can be proven (Yue et al., 1999; Yue and Zhou 1999) that $Eq(L_1,L_2)$ has the following three properties: (a) $Eq(L_1,L_2) \geq 0$; $Eq(L_1,L_2) = 0$ if and only if $L_1 = L_2$; (b) $Eq(L_1,L_2) = Eq(L_2,L_1)$; (c) $Eq(L_1,L_3) \leq Eq(L_1,L_2) + Eq(L_2,L_3)$. In functional analysis, $Eq(L_1,L_2)$ is a kind of distance in metric space of curves (Taylor, 1958).

If curves L_i could be simulated as

$$y = f_i(x) \quad (5.5)$$

then, α_i and k_i can be respectively formulated as

$$\alpha_i(x) = \frac{df_i(x)}{dx} \quad (5.6)$$

$$k_i(x) = \frac{d\alpha_i(x)}{dx} \cdot \left(1 + \alpha_i^2(x)\right)^{-\frac{3}{2}} \quad (5.7)$$

$$ds = \left(1 + \alpha^2(x)\right)^{\frac{1}{2}} dx \quad (5.8)$$

where x is abscissa and s is arc length.

Although the EQIs of curves are very useful for curve fitting and comparison, it is incomplete to use them to simulate surfaces. In fact, a surface is uniquely defined by the first fundamental coefficients, about the details of the surface observed when we stay on the surface, and the second fundamental coefficients, the change of the surface observed from outside the surface, in terms of FTS (Somasundaram, 2005; Yue et al., 2015a).

5.3 HIGH ACCURACY SURFACE MODELING

If $\{(x_i,y_j)\}$ is an orthogonal division of a computational domain and h represents the simulation step length, the central point of lattice (x_i,y_j) could be expressed as $(0.5h + (i-1)h, 0.5h + (j-1)h)$, in which $i = 0,1,2, \ldots,I,I+1$ and $j = 0,1,2, \ldots,J,J+1$. If $f_{i,j}^{(n)}$ ($n \geq 0$) represents the iterants of $f(x,y)$ at (x_i,y_j) in the nth iterative step, in which $\{f_{i,j}^{(0)}\}$ are interpolations based on sampling values $\{\bar{f}_{i,j}\}$, in terms of numerical mathematics (Quarteroni et al., 2000), the iterative formulation of the HASM master equation set can be expressed as (Yue et al., 2013a,b; Zhao and Yue, 2014b),

$$
\left\{
\begin{aligned}
&\frac{f_{i+1,j}^{(n+1)} - 2f_{i,j}^{(n+1)} + f_{i-1,j}^{(n+1)}}{h^2} = \left(\Gamma_{11}^1\right)_{i,j}^{(n)} \frac{f_{i+1,j}^{(n)} - f_{i-1,j}^{(n)}}{2h} + \left(\Gamma_{11}^2\right)_{i,j}^{(n)} \frac{f_{i,j+1}^{(n)} - f_{i,j-1}^{(n)}}{2h} + \frac{L_{i,j}^{(n)}}{\sqrt{E_{i,j}^{(n)} + G_{i,j}^{(n)} - 1}} \\[2mm]
&\frac{f_{i,j+1}^{(n+1)} - 2f_{i,j}^{(n+1)} + f_{i,j-1}^{(n+1)}}{h^2} = \left(\Gamma_{22}^1\right)_{i,j}^{(n)} \frac{f_{i+1,j}^{(n)} - f_{i-1,j}^{(n)}}{2h} + \left(\Gamma_{22}^2\right)_{i,j}^{(n)} \frac{f_{i,j+1}^{(n)} - f_{i,j-1}^{(n)}}{2h} + \frac{N_{i,j}^{(n)}}{\sqrt{E_{i,j}^{(n)} + G_{i,j}^{(n)} - 1}} \\[2mm]
&\frac{f_{i+1,j+1}^{(n+1)} - f_{i+1,j}^{(n+1)} - f_{i,j+1}^{(n+1)} + 2f_{i,j}^{(n+1)} - f_{i-1,j}^{(n+1)} - f_{i,j-1}^{(n+1)} + f_{i-1,j-1}^{(n+1)}}{2h^2} \\[2mm]
&\quad = \left(\Gamma_{12}^1\right)_{i,j}^{(n)} \frac{f_{i+1,j}^{(n)} - f_{i-1,j}^{(n)}}{2h} + \left(\Gamma_{12}^2\right)_{i,j}^{(n)} \frac{f_{i,j+1}^{(n)} - f_{i,j-1}^{(n)}}{2h} + \frac{M_{i,j}^{(n)}}{\sqrt{E_{i,j}^{(n)} + G_{i,j}^{(n)} - 1}}
\end{aligned}
\right.
$$

$$(5.9)$$

where,

$$
E_{i,j}^{(n)} = 1 + \left(\frac{f_{i+1,j}^{(n)} - f_{i-1,j}^{(n)}}{2h}\right)^2
$$

$$
F_{i,j}^{(n)} = \left(\frac{f_{i+1,j}^{(n)} - f_{i-1,j}^{(n)}}{2h}\right)\left(\frac{f_{i,j+1}^{(n)} - f_{i,j-1}^{(n)}}{2h}\right)
$$

$$
G_{i,j}^{(n)} = 1 + \left(\frac{f_{i,j+1}^{(n)} - f_{i,j-1}^{(n)}}{2h}\right)^2
$$

$$
L_{i,j}^{(n)} = \frac{\dfrac{f_{i+1,j}^{(n)} - 2f_{i,j}^{(n)} + f_{i-1,j}^{(n)}}{h^2}}{\sqrt{1 + \left(\dfrac{f_{i+1,j}^{(n)} - f_{i-1,j}^{(n)}}{2h}\right)^2 + \left(\dfrac{f_{i,j+1}^{(n)} - f_{i,j-1}^{(n)}}{2h}\right)^2}}
$$

$$
M_{i,j}^{(n)} = \frac{\left(\dfrac{f_{i+1,j+1}^{(n)} - f_{i+1,j-1}^{(n)}}{4h^2}\right) - \left(\dfrac{f_{i-1,j+1}^{(n)} - f_{i-1,j-1}^{(n)}}{4h^2}\right)}{\sqrt{1 + \left(\dfrac{f_{i+1,j}^{(n)} - f_{i-1,j}^{(n)}}{2h^2}\right)^2 + \left(\dfrac{f_{i,j+1}^{(n)} - f_{i,j-1}^{(n)}}{2h^2}\right)^2}}
$$

$$
N_{i,j}^{(n)} = \frac{\dfrac{f_{i,j+1}^{(n)} - 2f_{i,j}^{(n)} + f_{i,j-1}^{(n)}}{h^2}}{\sqrt{1 + \left(\dfrac{f_{i+1,j}^{(n)} - f_{i-1,j}^{(n)}}{2h}\right)^2 + \left(\dfrac{f_{i,j+1}^{(n)} - f_{i,j-1}^{(n)}}{2h}\right)^2}}
$$

$$\left(\Gamma_{11}^1\right)_{i,j}^{(n)} = \frac{G_{i,j}^{(n)}\left(E_{i+1,j}^{(n)} - E_{i-1,j}^{(n)}\right) - 2F_{i,j}^{(n)}\left(F_{i+1,j}^{(n)} - F_{i-1,j}^{(n)}\right) + F_{i,j}^{(n)}\left(E_{i,j+1}^{(n)} - E_{i,j-1}^{(n)}\right)}{4\left(E_{i,j}^{(n)}G_{i,j}^{(n)} - \left(F_{i,j}^{(n)}\right)^2\right)h}$$

$$\left(\Gamma_{12}^1\right)_{i,j}^{(n)} = \frac{G_{i,j}^{(n)}\left(E_{i,j+1}^{(n)} - E_{i,j-1}^{(n)}\right) - 2F_{i,j}^{(n)}\left(G_{i+1,j}^{(n)} - G_{i-1,j}^{(n)}\right)}{4\left(E_{i,j}^{(n)}G_{i,j}^{(n)} - \left(F_{i,j}^{(n)}\right)^2\right)h}$$

$$\left(\Gamma_{22}^1\right)_{i,j}^{(n)} = \frac{2G_{i,j}^{(n)}\left(F_{i,j+1}^{(n)} - F_{i,j-1}^{(n)}\right) - G_{i,j}^{(n)}\left(G_{i+1,j}^{(n)} - G_{i-1,j}^{(n)}\right) - F_{i,j}^{(n)}\left(G_{i,j+1}^{(n)} - G_{i,j-1}^{(n)}\right)}{4\left(E_{i,j}^{(n)}G_{i,j}^{(n)} - \left(F_{i,j}^{(n)}\right)^2\right)h}$$

$$\left(\Gamma_{11}^2\right)_{i,j}^{(n)} = \frac{2E_{i,j}^{(n)}\left(F_{i+1,j}^{(n)} - F_{i-1,j}^{(n)}\right) - E_{i,j}^{(n)}\left(E_{i,j+1}^{(n)} - E_{i,j-1}^{(n)}\right) - F_{i,j}^{(n)}\left(E_{i,j+1}^{(n)} - E_{i,j-1}^{(n)}\right)}{4\left(E_{i,j}^{(n)}G_{i,j}^{(n)} - \left(F_{i,j}^{(n)}\right)^2\right)h}$$

$$\left(\Gamma_{12}^2\right)_{i,j}^{(n)} = \frac{E_{i,j}^{(n)}\left(G_{i+1,j}^{(n)} - G_{i-1,j}^{(n)}\right) - F_{i,j}^{(n)}\left(E_{i,j+1}^{(n)} - E_{i,j-1}^{(n)}\right)}{4\left(E_{i,j}^{(n)}G_{i,j}^{(n)} - \left(F_{i,j}^{(n)}\right)^2\right)h}$$

$$\left(\Gamma_{22}^2\right)_{i,j}^{(n)} = \frac{E_{i,j}^{(n)}\left(G_{i,j+1}^{(n)} - G_{i,j-1}^{(n)}\right) - 2F_{i,j}^{(n)}\left(F_{i,j+1}^{(n)} - F_{i,j-1}^{(n)}\right) + F_{i,j}^{(n)}\left(G_{i+1,j}^{(n)} - G_{i-1,j}^{(n)}\right)}{4\left(E_{i,j}^{(n)}G_{i,j}^{(n)} - \left(F_{i,j}^{(n)}\right)^2\right)h}$$

where $E_{i,j}^{(n)}$, $F_{i,j}^{(n)}$, and $G_{i,j}^{(n)}$ are the iterants of the first fundamental coefficients at the nth iterative step; $L_{i,j}^{(n)}$, $M_{i,j}^{(n)}$, and $N_{i,j}^{(n)}$ represent the iterants of the second fundamental coefficients at the nth iterative step; $(\Gamma_{11}^1)_{i,j}^{(n)}$, $(\Gamma_{11}^2)_{i,j}^{(n)}$, $(\Gamma_{22}^1)_{i,j}^{(n)}$, and $(\Gamma_{22}^2)_{i,j}^{(n)}$ the iterants of the Christoffel symbols of the second kind at the nth iterative step, which depend only upon the first fundamental coefficients and their derivatives.

The matrix formulation of HASM master equations can be respectively expressed as,

$$\mathbf{A} \cdot \mathbf{z}^{(n+1)} = \mathbf{d}^{(n)} \tag{5.10}$$

$$\mathbf{B} \cdot \mathbf{z}^{(n+1)} = \mathbf{q}^{(n)} \tag{5.11}$$

$$\mathbf{C} \cdot \mathbf{z}^{(n+1)} = \mathbf{p}^{(n)} \tag{5.12}$$

where $\mathbf{A}$, $\mathbf{B}$, and $\mathbf{C}$ represent coefficient matrixes of the first equation, the second equation, and the third equation; $\mathbf{d}^{(n)}$, $\mathbf{q}^{(n)}$, and $\mathbf{p}^{(n)}$ are right-hand side vectors of the three equations respectively; $\mathbf{z}^{(n+1)} = (f_{1,1}^{(n+1)}, ..., f_{1,J}^{(n+1)}, ..., f_{I,1}^{(n+1)}, ..., f_{I,J}^{(n+1)})^T = (z_1^{(n+1)}, ..., z_J^{(n+1)}, ..., z_{(I-1)\cdot J+1}^{(n+1)}, ..., z_{I\cdot J}^{(n+1)})^T$, $f_{i,j}^{(n)}$ is the value of the nth iteration of $f(x,y)$ at grid cell (x_i,y_i), and $z_{(i-1)\cdot J+j}^{(n+1)} = f_{i,j}^{(n+1)}$ for $1 \le i \le I, 1 \le j \le J$.

If $\bar{f}_{i,j}$ is the value of $z = f(x,y)$ at the pth sampled point (x_i,y_j), $s_{p,(i-1) \times J + j} = 1$, and $k_p = \bar{f}_{i,j}$. There is only one nonzero element, 1, in every row of the coefficient matrix, $\mathbf{S}$, making it a sparse matrix. The solution procedure of HASMabc, taking the sampled points as its constraints, can be transformed into solving the following linear equation set in terms of the least squares principle:

$$[\mathbf{A}^T \quad \mathbf{B}^T \quad \mathbf{C}^T \quad \lambda \cdot \mathbf{S}^T] \begin{bmatrix} \mathbf{A} \\ \mathbf{B} \\ \mathbf{C} \\ \lambda \cdot \mathbf{S} \end{bmatrix} \mathbf{z}^{(n+1)} = [\mathbf{A}^T \quad \mathbf{B}^T \quad \mathbf{C}^T \quad \lambda \cdot \mathbf{S}^T] \begin{bmatrix} \mathbf{d}^{(n)} \\ \mathbf{q}^{(n)} \\ \mathbf{p}^{(n)} \\ \lambda \cdot \mathbf{k} \end{bmatrix} \tag{5.13}$$

The parameter λ is the weight of the sampling points and determines the contribution of the sampling points to the simulated surface. λ could be a real number, which means all sampling points have the same weight, or a sector, which means every sampling point has its own weight. An area affected by a sampling point in a complex region is smaller than in a flat region. Therefore, a smaller value of λ is selected in a complex region and a larger value of λ is selected in a flat region.

$$\text{Let } \mathbf{W} = [\mathbf{A}^T \quad \mathbf{B}^T \quad \mathbf{C}^T \quad \lambda \cdot \mathbf{S}^T] \begin{bmatrix} \mathbf{A} \\ \mathbf{B} \\ \mathbf{C} \\ \lambda \cdot \mathbf{S} \end{bmatrix} \text{ and } \mathbf{v}^{(n)} = [\mathbf{A}^T \quad \mathbf{B}^T \quad \mathbf{C}^T \quad \lambda \cdot \mathbf{S}^T] \begin{bmatrix} \mathbf{d}^{(n)} \\ \mathbf{q}^{(n)} \\ \mathbf{p}^{(n)} \\ \lambda \cdot \mathbf{k} \end{bmatrix}, \text{ then}$$

HASMabc has the following formulation:

$$\mathbf{W} \cdot \mathbf{z}^{(n+1)} = \mathbf{v}^{(n)} \tag{5.14}$$

If the third equation is deleted in the Eq. (5.9), we can get an equation set as follows:

$$\begin{cases} \dfrac{f_{i+1,j}^{(n+1)} - 2f_{i,j}^{(n+1)} + f_{i-1,j}^{(n+1)}}{h^2} = (\Gamma_{11}^1)_{i,j}^{(n)} \dfrac{f_{i+1,j}^{(n)} - f_{i-1,j}^{(n)}}{2h} + (\Gamma_{11}^2)_{i,j}^{(n)} \dfrac{f_{i,j+1}^{(n)} - f_{i,j-1}^{(n)}}{2h} + \dfrac{L_{i,j}^{(n)}}{\sqrt{E_{i,j}^n + G_{i,j}^n - 1}} \\[4ex] \dfrac{f_{i,j+1}^{(n+1)} - 2f_{i,j}^{(n+1)} + f_{i,j-1}^{(n+1)}}{h^2} = (\Gamma_{22}^1)_{i,j}^{(n)} \dfrac{f_{i+1,j}^{(n)} - f_{i-1,j}^{(n)}}{2h} + (\Gamma_{22}^2)_{i,j}^{(n)} \dfrac{f_{i,j+1}^{(n)} - f_{i,j-1}^{(n)}}{2h} + \dfrac{N_{i,j}^{(n)}}{\sqrt{E_{i,j}^{(n)} + G_{i,j}^{(n)} - 1}} \end{cases} \tag{5.15}$$

Eq. (5.15) has the following matrix formulation:

$$\begin{cases} \mathbf{A} \cdot \mathbf{z}^{(n+1)} = \mathbf{d}^{(n)} \\ \mathbf{B} \cdot \mathbf{z}^{(n+1)} = \mathbf{q}^{(n)} \end{cases} \tag{5.16}$$

If optimum control constraints are considered, a method for high accuracy surface modeling, termed HASMab, can be formulated as (Yue, 2011),

$$\begin{cases} \min \left\| \begin{bmatrix} \mathbf{A} \\ \mathbf{B} \end{bmatrix} \cdot z^{(n+1)} - \begin{bmatrix} \mathbf{d}^{(n)} \\ \mathbf{q}^{(n)} \end{bmatrix} \right\| \\ s.t. \quad \mathbf{S} \cdot \mathbf{z}^{(n+1)} = \mathbf{k} \end{cases} \tag{5.17}$$

According to the Method of Lagrange Multipliers (Kolman and Trend, 1971), Eq. (5.17) can be transferred into

$$\mathbf{z}^{(n+1)} = \left(\mathbf{A}^T \cdot \mathbf{A} + \mathbf{B}^T \cdot \mathbf{B} + \lambda^2 \cdot \mathbf{S}^T \cdot \mathbf{S} \right)^{-1} \left(\mathbf{A}^T \cdot \mathbf{d}^{(n)} + \mathbf{B}^T \cdot \mathbf{q}^{(n)} + \lambda^2 \cdot \mathbf{S}^T \cdot \mathbf{k} \right) \tag{5.18}$$

Accuracy of HASMabc is much higher than the one of HASMab when the driving field has a bigger error, while HASMabc is little more accurate than HASMab when the driving field is relatively accurate (Yue, 2016). HASMabc has a much more complex computation because it includes the third equation, which makes it need about two times more memory and has much slower computational-speed compared to HASMab, especially with increase in computational size. In other words, Eq. (5.12) can be ignored to reduce the calculated amount and improve computational-speed.

5.4 THE FUNDAMENTAL THEOREM OF EARTH'S SURFACE MODELING AND ITS COROLLARIES

HASM has been successfully applied to constructing digital elevation models (Yue et al., 2007; Yue and Wang, 2010; Yue et al., 2010a, b; Chen and Yue, 2010; Chen et al., 2012, 2013a, b), modeling surface soil properties (Shi et al., 2011) and soil pollution (Shi et al., 2009) as well as soil antibiotics (Shi et al., 2016), filling voids in the Shuttle Radar Topography Mission dataset (Yue et al., 2012), simulating climate change (Yue et al., 2013a, b; Zhao and Yue, 2014a, b), filling voids on remotely sensed XCO_2 surfaces (Yue et al., 2015b), analyzing ecosystem responses to climatic change (Yue et al., 2015c), and mapping superresolution land cover (Chen et al., 2015). In all these applications, HASM produced more accurate results than the classical methods.

A FTESM and its Corollaries have been developed on the basis of summarizing the successful applications of HASM (Yue et al., 2016): "an Earth's surface system or a component surface of the Earth's surface environment can be simulated with HASM when its spatial resolution is fine enough, which is uniquely defined by both extrinsic and intrinsic invariants of the surface."

The approaches to Earth surface modeling can be classified into five categories: (1) spatial interpolation, (2) data fusion, (3) data assimilation, (4) upscaling, and (5) downscaling. Spatial interpolation is defined as predicting the values of a primary variable at points within the same region of sampled locations in terms of spatial data in the form of discrete points or in the form of data partition (Wang and Wang, 2012; Li and Heap, 2014). Data fusion is the process of integration of multiple data and knowledge streams representing the same real-world object into a consistent, accurate, and useful representation (Mitchell, 2012). Data assimilation is the process by which measured observations are incorporated into a system model (Nichols, 2010).

The transfer of knowledge from a finer resolution to a coarser resolution is referred to as upscaling to mostly reduce computational costs (Schlummer et al., 2014). However, spatial resolutions of many models or data are sometimes too coarse to be used for analyses on regional or local scales. To overcome this problem, downscaling approaches are developed to obtain information at finer spatial resolution from the coarser-spatial-resolution models and data (Zhang et al., 2004).

In terms of the **FTESM**, the following seven corollaries can be derived:

Corollary 1 (Interpolation): If only an intrinsic invariant is available, HASM can be used to create the Earth's surface or a component surface of the Earth's surface environment with higher accuracy after the necessary extrinsic invariants have been extracted from the intrinsic invariant by geostatistics.

Corollary 2 (Upscaling): If a surface on a finer spatial resolution is transferred to the one on a coarser-spatial-resolution, ground-based observations are necessary to operate HASM for obtaining higher accuracy.

Corollary 3 (Downscaling): If a surface at a coarse spatial resolution is available, it is necessary to supplement ground-based observations to obtain a corresponding surface at finer spatial resolution with higher accuracy by means of HASM.

Corollary 4 (Data fusion): When remotely sensed data from satellites are available, ground measurements have to be obtained and incorporated before HASM can be used to generate a more accurate surface.

Corollary 5 (Data fusion): When both remotely sensed data from satellites and ground measurements are available, HASM can be used to generate a surface that is more accurate than the one from either the satellite observations or the ground measurements.

Corollary 6 (Data assimilation): When a system model is available, a more accurate surface can be produced when ground observations are incorporated into the system model.

Corollary 7 (Data assimilation): When both a system model and ground observations are available, a surface can be produced by using HASM to incorporate the ground observations into the system model, which is more accurate than the one either from system model or from ground observations.

5.5 CONCLUSIONS

Error problems and slow-computational-speed problems are the two critical challenges currently faced by GIS and Computer-Aided Design Systems (CADS). The method for HASM provides solutions to these problems that have long troubled GIS and CADS (Jorgensen, 2011).

HASM can take advantage of limited observation data to construct a continuous surface by filling missing data, with higher accuracy comparing with the classical methods such as triangulated irregular network (TIN), inverse distance weighting (IDW), ordinary Kriging (OK), and Spline (Yang et al., 2015). TIN calculates the value of each point within a triangle by means of a linear function based on its location, while it ignores nonlinear information and is unable to represent cliffs, caves, or holes. IDW uses an IDW function to determine the interpolation value for any given point within the calculated area, but it fails to incorporate the spatial structure and ignores information beyond the neighborhood. OK tries to have the mean residual or error equal to zero and aims at minimizing the variance of the errors, but the goals are practically unattainable since the mean error and the variance of the errors are always unknown. Spline is approximately used to simulate surfaces, while few types of surfaces fit the formulation of Splines.

HASM can improve the quality of the information so that it is more accurate than would be possible if the data sources were used individually, by means of its function of data fusion. Since data from different sources have varying accuracy and coverage, the benefits of this data fusion include improved system reliability, extended coverage, and reduced uncertainty. HASM can operate data assimilation to use measured observations in combination with a system model to derive accurate estimates of the current and future states of the system, together with estimates of the uncertainty in the estimates.

HASM is able to transfer information from one spatial scale to another with improved errors, which are mainly caused by the spatial heterogeneity of objects and process nonlinearities, the scale dependency of the characteristics of objects and processes, feedbacks associated with process interactions at small and large scales, emergent properties that arise at larger scales through the interaction of small-scale processes, and the time lags of system response to external perturbation.

The further research focuses of HASM include (1) theoretical analyses of convergence and stability of numerical solution procedure of HASM equation set, (2) clarification of physical significance of HASM parameters and variables as well as their effects on solution accuracy and speed, (3) construction of finite element method of HASM under spherical coordinates, (4) development of a faster numerical solver by selecting an optimal preconditioning operator, and (5) parallelization of HASM to find a solution for the problems of large memory requirements and slow computing speed.

References

Ahlberg, J.H., Nilson, E.N., Walsh, J.L., 1967. The Theory of Splines and Their Application. Academic Press, New York.

Chen, C.F., Yue, T.X., 2010. A method of DEM construction and related error analysis. Computers & Geosciences 36, 717−725.

Chen, C.F., Yue, T.X., Li, Y.Y., 2012. A high speed method of SMTS. Computers & Geosciences 41, 64−71.

Chen, C.F., Yue, T.X., Dai, H.L., Tian, M.Y., 2013a. The smoothness of HASM. International Journal of Geographical Information Science 27, 1651−1667.

Chen, C.F., Li, Y.Y., Yue, T.X., 2013b. Surface modeling of DEMs based on a sequential adjustment method. International Journal of Geographical Information Science 27, 1272−1291.

Chen, Y.H., Ge, Y., Song, D.J., 2015. Superresolution land-cover mapping based on high-accuracy surface modeling. IEEE Geoscience and Remote Sensing Letters 12 (12), 2516−2520.

Costanza, R., 1989. Model goodness of fit: a multiple resolution procedure. Ecological Modelling 47, 199−215.

Costanza, R., Sklar, F.H., White, M.L., 1990. Modeling costal landscape dynamics. Bioscience 40 (2), 91−107.

Evans, I.S., 1980. An integrated system of terrain analysis and slope mapping. Zeitschrift für Geomorphologie, Supplementband 36, 274−295.

Jorgensen, S.E., 2011. Book review. Ecological Modelling 222, 3300.

Kolman, B., Trend, W.F., 1971. Elementary Multivariable Calculus. Academic Press, New York.

Legendre, L., Legendre, P., 1983. Numerical Ecology. Elsevier Scientific Pul. Co., Amsterdam.

Li, J., Heap, A.D., 2014. Spatial interpolation methods applied in the environmental sciences: a review. Environmental Modelling & Software 53, 173—189.

Lo, C.P., Yeung, A.K.W., 2002. Concepts and Techniques of Geographic Information Systems. Prentice Hall, Upper Saddle River, New Jersey.

Long, G.E., 1980. Surface approximation: a deterministic approach to modeling partially variable systems. Ecological Modelling 8, 333—343.

Mitchell, H.B., 2012. Data Fusion: Concepts and Ideas. Springer-Verlag, Berlin.

Nichols, N.K., 2010. Mathematical concepts of data assimilation. In: Lahoz, W., Khattatov, B., Menard, R. (Eds.), Data Assimilation: Making Sense of Observations. Springer-Verlag, Berlin, pp. 13—39.

Phillips, J.D., 1999. Earth Surface Systems. Blackwell Publishers, Oxford.

Phillips, J.D., 2002. Global and local factors in earth surface systems. Ecological Modelling 149, 257—272.

Quarteroni, A., Sacco, R., Saleri, F., 2000. Numerical Mathematics. Springer, New York.

Schlummer, M., Thomas, H., Dikau, R., Eickmeier, M., Fischer, P., Gerlach, R., Holzkämper, J., Kalis, A.J., Kretschmer, I., Lauer, F., Maier, A., Meesenburg, J., Meurers-Balke, J., Münc, U., Pätzoldg, S., Steininger, F., Stobbe, A., Zimmermann, A., 2014. From point to area: upscaling approaches for late quaternary archaeological and environmental data. Earth-Science Reviews 131, 22—48.

Schroeder, L.D., Sjoquist, D.L., 1976. Investigation of population density gradients using trend surface analysis. Land Economics 52, 382—392.

Shi, W.J., Liu, J.Y., Song, Y.J., Du, Z.P., Chen, C.F., Yue, T.X., 2009. Surface modeling of soil pH. Geoderma 150, 113—119.

Shi, W.J., Liu, J.Y., Du, Z.P., Stein, A., Yue, T.X., 2011. Surface modeling of soil properties based on land use information. Geoderma 162, 347—357.

Shi, W.J., Yue, T.X., Du, Z.P., Wang, Z., Li, X.W., 2016. Surface modeling of soil antibiotics. Science of the Total Environment 543, 609—619.

Sklar, F.H., Costanza, R., Day Jr., J.W., 1985. Dynamic spatial simulation modeling of coastal wetland habitat succession. Ecological Modelling 29 (1—4), 261—281.

Somasundaram, D., 2005. Differential Geometry. Alpha Science International Ltd., Harrow.

Spivak, M., 1979. A Comprehensive Introduction to Differential Geometry, vol. 3. Publish or Peril, Inc., Boston.

Stott, J.P., 1977. Review of surface modeling. In: Proceedings of Surface Modeling by Computer, a Conference Jointly Sponsored by the Royal Institution of Chartered Surveyors and the Institution of Civil Engineers, Held in London on 6 October, 1976, pp. 1—8.

Taylor, A.E., 1958. Introduction to Functional Analysis. John Wiley & Sons, Inc., New York.

Turner, M.G., Costanza, R., Sklar, F.H., 1989. Methods to evaluate the performance of spatial simulation models. Ecological Modelling 48 (1—2), 1—18.

Wang, H.W., Wang, Y.H., 2012. Analysis of spatial interpolation methods. Lecture Notes in Electrical Engineering 129, 507—512.

Yang, H., Wang, C.H., Ma, T.F., Guo, W.J., 2015. Accuracy assessment of interpolation methods in grid DEMs based on a variance-scale relation. Environmental Earth Sciences 74, 6525—6539.

Yue, T.X., Ai, N.S., 1990. The mathematic model of cirque morphology. Journal of Glaciology and Geocryology 12, 227—234 (in Chinese With English Abstract).

Yue, T.X., Haber, W., Grossmann, W., Kasperidus, H., 1999. A method for strategic management of land. In: Pykh, Y.A., Hyatt, D.E., Lenz, R.J.M. (Eds.), Environmental Indices: Systems Analysis Approaches. London: EOLSS Publishers Co Ltd., pp. 181—201.

Yue, T.X., Zhou, C.H., 1999. An approach of differential geometry to data mining. ECOMOD 10, 1—6.

Yue, T.X., Chen, S.P., Xu, B., Liu, Q.S., Li, H.G., Liu, G.H., Ye, Q.H., 2002. A curve theorem based approach for change detection and its application to Yellow River Delta. International Journal of Remote Sensing 23, 2283—2292.

Yue, T.X., Du, Z.P., Song, D.J., Gong, Y., 2007. A new method of surface modeling and its application to DEM construction. Geomorphology 91, 161—172.

Yue, T.X., Wang, S.H., 2010. Adjustment computation of HASM: a high-accuracy and high-speed method. International Journal of Geographical Information Science 24, 1725—1743.

Yue, T.X., Chen, C.F., Li, B.L., 2010a. An adaptive method of high accuracy surface modeling and its application to simulating elevation surfaces. Transactions in GIS 14, 615—630.

Yue, T.X., Song, D.J., Du, Z.P., Wang, W., 2010b. High-accuracy surface modelling and its application to DEM generation. International Journal of Remote Sensing 31, 2205–2226.

Yue, T.X., 2011. Surface Modelling: High Accuracy and High Speed Methods. CRC Press, New York.

Yue, T.X., Chen, C.F., Li, B.L., 2012. A high accuracy method for filling SRTM voids and its verification. International Journal of Remote Sensing 33, 2815–2830.

Yue, T.X., Zhao, N., Yang, H., Song, Y.J., Du, Z.P., Fan, Z.M., Song, D.J., 2013a. The multi-grid method of high accuracy surface modelling and its validation. Transactions in GIS 17 (6), 943–952.

Yue, T.X., Zhao, N., Ramsey, R.D., Wang, C.L., Fan, Z.M., Chen, C.F., Lu, Y.M., Li, B.L., 2013b. Climate change trend in China, with improved accuracy. Climatic Change 120, 137–151.

Yue, T.X., Zhang, L.L., Zhao, N., Zhao, M.W., Chen, C.F., Du, Z.P., Song, D.J., Fan, Z.M., Shi, W.J., Wang, S.H., Yan, C.Q., Li, Q.Q., Sun, X.F., Yang, H., Wang, C.L., Wang, Y.F., Wilson, J., Xu, B., 2015a. A Review of recent developments in HASM. Environmental Earth Sciences 74 (8), 6541–6549.

Yue, T.X., Zhao, M.W., Zhang, X.Y., 2015b. A high-accuracy method for filling voids on remotely sensed XCO_2 surfaces and its verification. Journal of Cleaner Production 103, 819–827.

Yue, T.X., Du, Z.P., Lu, M., Fan, Z.M., Wang, C.L., Tian, Y.Z., Xu, B., 2015c. Surface modelling of ecosystem responses to climatic change. Ecological Modelling 306, 16–23.

Yue, T.X., Liu, Y., Zhao, M.W., Du, Z.P., Zhao, N., 2016. A fundamental theorem of Earth's surface modelling. Environmental Earth Sciences 75 (9), 1–12. Article 751.

Yue, T.X., 2016. Principles and Methods for Earth's Surface Modelling. Science Press, Beijing (in Chinese).

Zhang, X.Y., Drake, N.A., Wainwright, J., 2004. Scaling Issues in Environmental Modelling, Environmental Modelling. Finding Simplicity in Complexity. John Wiley & Sons, Ltd., pp. 319–334

Zhao, N., Yue, T.X., 2014a. A modification of HASM for interpolating precipitation in China. Theoretical and Applied Climatology 116, 273–285.

Zhao, N., Yue, T.X., 2014b. Sensitivity studies of a high accuracy surface modelling method. Science China Earth Sciences 57, 1–11.

Application of Structurally Dynamic Models (SDMs)

S.E. Jørgensen

Emeritus Professor, Copenhagen University, Denmark
E-mail: soerennorsnielsen@gmail.com

OUTLINE

6.1 INTRODUCTION

Ecological models attempt to capture the characteristics of ecosystems. However, ecosystems differ from most other systems by being extremely adaptive, having the ability of self-organization, and having a large number of feedback mechanisms. The real challenge of modeling ecosystems is: How can we construct models that are able to reflect these extremely dynamic characteristics? They are of particular importance when we want to develop ecological models that considered the ecological changes due to the impacts of radical changes in the forcing functions (impacts). Some recent development in ecological modeling has attempted to meet this challenge by the use of a new model type denoted structurally dynamic models (SDMs). Section 6.2 describes i.e., the properties of ecosystems that make it compulsory at least in some situations to use SDMs. Section 6.3 will focus on how to develop this model type and how it can be applied to consider the ecological changes that are a result of impact changes on ecosystems. It means that the changes of the state variables due to the changes of the forcing functions are considered.

SDMs can be developed by two methods: either by the use of expert knowledge or by the use of a goal function. Expert knowledge can be used to change the parameter of crucial species according to what is known about which species are characteristic for the focal ecosystem at various impacts or forcing functions or expressed differently by the prevailing conditions. The use of a goal function implies that changes of state variables due to changing forcing functions can be described by a function that is able to capture the regulating feedbacks of the ecosystem caused by changes. The most applied goal function for development of SDMs is the thermodynamic variable eco-exergy [it is exergy or work energy capacity (abbreviated as WE) defined for ecosystems; the definition and presentation is given in Section 6.3].

Section 6.4 gives an overview of SDMs developed by use of WE as goal function and Section 6.5 presents one SDM example to illustrate the model type. As it is expected that this model type will be used more generally in the future to assess the consequences of global warming, which is a very massive impact change, Section 6.6 is devoted to population dynamic examples of how the WE will change when temperature changes are realized. It means that this section will illustrate the applicability of SDMs to describe the consequences of global climate changes.

6.2 ECOSYSTEM PROPERTIES

Many researchers have advocated for a holistic approach to ecosystem science (e.g., Odum, 1953; Ulanowicz, 1986, 1995). Holism is taken to mean a description of the system level properties of an ensemble, rather than simply an exhaustive detailed description of all the components. It is thought that by adopting a holistic viewpoint, certain properties become apparent and other behaviors are made visible that otherwise would be undetected. It is, however, clear that the complexity of ecosystems has set the limitations for our understanding and for the possibilities of proper management. We cannot capture the complexity as such with all its details, but we can understand how ecosystems are complex and we can set up a realistic strategy for how to get sufficient knowledge about the system—not knowing all the details, but still understanding and knowing the average behavior and the important reactions of the system, particularly to specified impacts. It means that we can only try to reveal the basic properties behind the complexity. We have therefore no other choice than to go holistic. The results from the more reductionistic ecology are essential in our effort "to go to the root" of the system properties of ecosystems, but we need systems ecology, which consists of many new ideas, approaches, and concepts, to follow the path to the roots of the basic system properties of ecosystems. The idea may also be expressed in another way: we cannot find the properties of ecosystems by analyzing all the details because they are simply too many, but only by trying to reveal the system properties of ecosystems by examination of the entire systems. A brief overview of the most important system properties of ecosystems are given below.

The number of feedbacks and regulations is extremely high and makes it possible for the living organisms and populations to survive and reproduce despite changes in external conditions. The feedbacks are furthermore constantly changing, i.e., the adaptation itself is adaptable in the sense that if a regulation is not sufficient another regulation process higher in the hierarchy of feedbacks will take over. The change by adaptation within the same species is limited. When this limitation has been reached, other species will take over. It implies that not only the processes and the components, but also the feedbacks can be replaced, if it is needed to achieve a better utilization of the available resources. All these regulation mechanisms are rooted in the enormous amount of information that ecosystems possess.

Ecosystems show a high degree of heterogeneity in space and in time.

An ecosystem is a very dynamic system. All its components and particularly the biological ones are steadily changing and their properties are steadily modified, which is why an ecosystem will never return to the same situation again. Every point is furthermore different from any other point and therefore offering different conditions for the various life forms. This enormous heterogeneity explains why biodiversity is so high on earth. There is, so to say, an ecological niche for "everyone" and "everyone" may be able to find a niche where he is best fitted to utilize the resources. Ecotones, the transition zones between two ecosystems, offer a particular variability in life conditions, which often results in a particular richness of species diversity. Studies of ecotones have recently drawn much attention from ecologists because ecotones have pronounced gradients in the external and internal variables, which give a clearer picture of the relation between external and internal variables.

Margalef (1991) claims that ecosystems are anisotropic, meaning that they exhibit properties with different values when measured along axes in different directions. It means that the ecosystem is not homogeneous in relation to properties concerning matter, energy, and information and that the entire dynamics of the ecosystem works toward increasing the differences. These variations in time and space make it particularly difficult to model ecosystems and to capture the essential features of ecosystems. However, the hierarchy theory applies these variations to develop a natural hierarchy as framework for ecosystem descriptions and theory. The strength of the hierarchy theory is that it facilitates the studies and modeling of ecosystems.

Ecosystems and their biological components, the species, evolve steadily and over the long time toward higher complexity. Darwin's theory describes the competition among species and states that those species best fitted to the prevailing conditions in the ecosystem will survive. Darwin's theory can, in other words, describe the changes in ecological structure and the species composition, but cannot directly be applied quantitatively, e.g., in ecological modeling; see, however the next section.

All species in an ecosystem are confronted with the question: how is it possible to survive or even grow under the prevailing conditions? The prevailing conditions are considered as all factors influencing the species, i.e., all external and internal factors including those originating from other species. This explains coevolution, as any change in the properties of one species will influence the evolution of the other species. The environmental stage on which the selection plays out is comprised of all the interacting species, each influencing another.

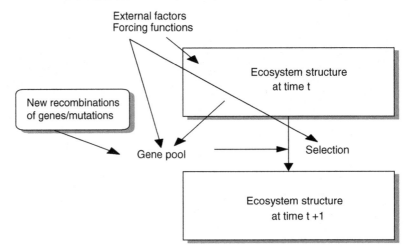

FIGURE 6.1 Conceptualization of how the external factors steadily change the species composition. The possible shifts in species composition are determined by the gene pool, which is steadily changed due to mutations and new sexual recombinations of genes. The development is, however, more complex. This is indicated by (1) arrows from "structure" to "external factors" and "selection" to account for the possibility that the species can modify their own environment (see text) and thereby their own selection pressure; (2) an arrow from "structure" to "gene pool" to account for the possibilities that the species can to a certain extent change their own gene pool.

All natural external and internal factors of ecosystems are dynamic—the conditions are steadily changing, and there are always many species waiting in the wings, ready to take over, if they are better fitted to the emerging conditions than the species dominating under the present conditions. There is a wide spectrum of species representing different combinations of properties available for the ecosystem. The question is which of these species are best able to survive and grow under the present conditions and which species are best able to survive and grow under the conditions one time step further and two time steps further and so on? The necessity in Monod's sense is given by the prevailing conditions—the species must have genes or maybe rather phenotypes (meaning properties) that match these conditions, to be able to survive. But the natural external factors and the genetic pool available for the test may change randomly or by "chance."

Steadily, new mutations (misprints are produced accidentally) and sexual recombinations (the genes are mixed and shuffled) emerge and give steadily new material to be tested by the question: which species are best fitted under the conditions prevailing just now? These ideas are illustrated in Fig. 6.1. The external factors are steadily changed and some even relatively fast—partly at random, such as the meteorological or climatic factors. The species within the system are selected among the species available and represented by the genetic pool, which again is slowly, but surely changed at random or by "chance." The selection in Fig. 6.1 includes a selection of the organisms that possess the properties best fitted to the prevailing conditions.

Species are continuously tested against the prevailing conditions (external as well as internal factors) and the better they are fitted, the better they are able to maintain and even increase their biomass. The specific rate of population growth may even be used as a measure for the fitness (see, e.g., Stenseth, 1986). But the property of fitness must be heritable to have any effect on the species composition and the ecological structure of the ecosystem in the long run. Natural selection has been criticized for being a tautology: fitness is measured by survival, and survival of the fittest therefore means survival of the survivors. However, the entire Darwinian theory, including the above-mentioned three assumptions, cannot be conceived as a tautology but may be interpreted as follows: species offer different solutions to survive under the given prevailing conditions and the species that have the best combinations of properties to match the conditions have also the highest probability of survival and growth.

If we follow the modeling procedure (see for instance Jørgensen and Fath, 2011), we will attain a model that describes the processes in the focal ecosystem, but the parameters will represent the properties of the state variables as they are in the ecosystem during the examination period. They are not necessarily valid for another period because we know that an ecosystem can regulate, modify, and change them, if needed as response to changes in the existing conditions, determined by the forcing functions and the interrelations between the state variables—see Fig. 6.1. Our present models have rigid structures and a fixed set of parameters meaning that no changes or replacements of the components are possible. We need, however, to introduce parameters (properties) that can change according to changing forcing functions and general conditions for the state variables (components) to optimize continuously the ability of the system to move away from thermodynamic equilibrium (Jørgensen et al., 2000) and described by Odum (1971). The model type that can account for the change in species composition as well as for the ability of the species, i.e., the biological components of our models, to change their properties, i.e., to adapt to the existing conditions imposed on the species, is sometimes called SDM, to indicate that they are able to capture structural changes.

It could of course be argued that the ability of ecosystems to replace present species with other, better fitted species, can be considered by constructing models that encompass all actual species for the entire period that the model attempts to cover. This approach has, however, two essential disadvantages. This will introduce a high uncertainty to the model and will render the application of the model very case specific (Nielsen, 1992a,b). In addition, the model will still be rigid and not allow the model to have continuously changing parameters even without changing the species composition (Fontaine, 1981). Straskraba (1979) uses a maximization of biomass as the governing principle. The model computes the biomass and adjusts one or more selected parameters to achieve the maximum biomass at every instance. The model has a routine which computes the biomass for all possible combinations of parameters within a given realistic range. The combination that gives the maximum biomass is selected for the next time step and so on.

Eco-exergy [work energy capacity (WE) including the work energy of information] has been used widely as a goal function in ecological models, and a few of the available case studies will be presented and discussed below in this section. Eco-exergy or WE has two pronounced advantages as goal function. It is defined far from thermodynamic equilibrium and it is related to the state variables, which are easily determined or measured, opposite for instance maximum power that is related to the flows. As exergy is not a generally used thermodynamic function, we need, however, first to present this concept properly. Let us translate Darwin's theory into thermodynamics, applying eco-exergy (WE) as the basic concept. Survival implies biomass maintenance, and growth means biomass increase. It costs WE to construct biomass and biomass therefore possesses WE, which is transferable to support other work-energy requiring processes. Survival and growth can therefore be measured by use of the thermodynamic concept eco-exergy (WE). Darwin's theory may therefore be reformulated in thermodynamic terms as follows: *The prevailing conditions of an ecosystem steadily change and the system will continuously select the species and thereby the processes that can contribute most to the maintenance or even growth of the work energy of the system. This hypothesis is often denoted ELT—the Ecological Law of Thermodynamics.*

The hypothesis describes how ecosystems react when the forcing functions (impacts) are changed. These properties of ecosystems are theoretically very important for the development of SDMs.

Notice that the thermodynamic translation of Darwin's theory requires that populations have the properties of reproduction, inheritance, and variation. The selection of the species that contributes most to the WE of the system under the existing conditions requires that there are enough individuals with different properties so that a selection can take place—it means that the reproduction and the variation must be high and that once a change has taken place due to better fitness, it can be conveyed to the next generation.

6.3 DEVELOPMENT OF STRUCTURALLY DYNAMIC MODELS

Notice also the change in WE (eco-exergy) of ecosystems according to ELT is not necessarily ≥ 0; it depends on the changes of the resources of the ecosystem. The proposition claims, however, that the ecosystem attempts to reach the highest possible WE level under the given circumstances or prevailing conditions and with the available genetic pool ready for this attempt (Jørgensen and Mejer, 1977, 1979); see Fig. 6.2. Jørgensen and Mejer (1979) have shown by the use of thermodynamics that the following equation is valid for the components of an ecosystem:

$$Ex = RT \sum_{i=1}^{i=n} (C_i \times \ln(C_i/C_{eq,i}) - (C_i - C_{eq,i})), \qquad (6.1)$$

where R is the gas constant, T the temperature of the environment (Kelvin), while C_i represents the ith component expressed in a suitable unit, e.g., for phytoplankton in a lake C_i could be milligrams of a focal nutrient in the phytoplankton per liter of lake water, $C_{eq,i}$ is the concentration of the ith component at thermodynamic equilibrium. The quantity $c_{eq,i}$ represents a very small, but nonzero, concentration (except for $i = 0$, which is considered to cover the inorganic compounds), corresponding to the very low probability of forming complex organic compounds spontaneously in an inorganic soup at thermodynamic equilibrium.

The idea of the new type of models presented here is to find continuously a new set of parameters (limited for practical reasons to the most crucial, i.e., sensitive parameters) that is better fitted for the prevailing conditions of the ecosystem. "Fitted" is defined in the Darwinian sense by the ability of the species to survive and grow, which may be measured by the use of exergy (see Jørgensen, 1986, 1988, 1990, 1992; Jørgensen and Mejer, 1977, 1979).

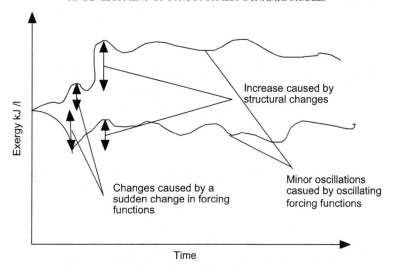

FIGURE 6.2   Exergy (eco-exergy, WE) response to increased and decreased nutrient concentration. The level of work energy can thereby increase or decrease, but when these changes have been introduced, the system will attempt to gain as much work energy as possible under the prevailing conditions.

Fig. 6.3 shows the proposed modeling procedure, which has been applied for all the cases studies listed in Section 6.4.

Eco-exergy or work energy capacity (WE) is defined as the work the system can perform when it is brought into equilibrium with the environment or another well-defined reference state. If we presume a reference environment for a system at thermodynamic equilibrium, meaning that all the components are (1) inorganic, (2) at the highest possible oxidation state signifying that all free energy has been utilized to do work, and (3) homogeneously distributed in the system, meaning no gradients, then the situation illustrated in Fig. 6.4 is valid.

It is possible to distinguish in Eq. (5.1) between the contribution to the eco-exergy from the information and from the biomass. We define p_i as c_i/A, where

$$A = \sum_{i=1}^{n} c_i \qquad (6.2)$$

is the total amount of matter density in the system. With introduction of this new variable, we get:

$$Ex = ART \sum_{i=1}^{n} p_i \, \ln p_i/p_{io} + A \, \ln A/A_o \qquad (6.3)$$

As $A \approx A_o$, eco-exergy becomes a product of the total biomass A (multiplied by RT) and Kullback measure:

$$K = \sum_{i=1}^{n} p_i \, \ln(p_i/p_{io}) \qquad (6.4)$$

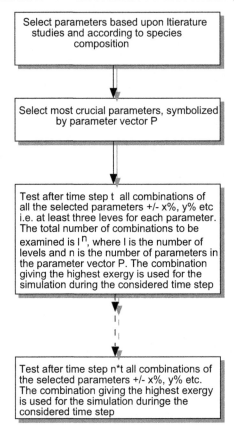

FIGURE 6.3   The procedure used for the development of structurally dynamic models.

where p_i and p_{io} are probability distributions, a posteriori and a priori to an observation of the molecular detail of the system. It means that K expresses the amount of information that is gained as a result of the observations. For different organisms that contribute to the eco-exergy of the ecosystem, the eco-exergy density becomes $cRT \ln(p_i/p_{io})$, where c is the concentration of the considered organism. $RT \ln(p_i/p_{io})$, denoted β, is found by calculating the probability to form the considered organism at thermodynamic equilibrium, which would require that organic matter is formed and that the proteins (enzymes) controlling the life processes in the considered organism have the right amino acid sequence. These calculations can be seen in Jørgensen and Svirezhev (2005). In the latter reference, the latest information about the β values for various organisms is presented; see Table 5.1. For human, the β value is 2173, when the eco-exergy is expressed in detritus equivalent or 18.7 times as much or 40,635 kJ/g if the eco-exergy should be expressed as kJ and the concentration unit g per unit of volume or area. One hypothesis, apparently confirmed by observation is that the β values increase as a result of evolution. To mention a few β values from Table 8.2: bacteria 8.5, protozoa 39, flatworms 120, ants 167, crustaceans 232, mollusks 310, fish 499, reptiles 833,

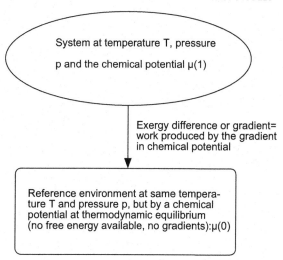

FIGURE 6.4 Illustration of the eco-exergy, work energy capacity, concept used to compute the exergy index for an ecological model. Temperature and pressure are the same for both the system and the reference state which implies that only the difference in chemical potential can contribute to the exergy.

birds 980, and mammals 2127. Evolution resulted in a more and more effective transfer of what we could call the classical work capacity to the work capacity of the information. A β value of 2.0 means that the eco-exergy embodied in the organic matter and the information are equal. As the β values, see above, are much bigger than 2.0 (except for virus, where the β value is 1.01—slightly more than 1.0), the information eco-exergy is the most significant part of the eco-exergy of organisms.

In accordance with the above-presented interpretation of Eqs. (6.3) and (6.4), it is now possible to find the eco-exergy density for a model as:

$$\text{Eco-exergy density} = \sum_{i=1}^{i=n} \beta_i c_i \qquad (6.5)$$

The eco-exergy due to the "fuel" value of organic matter (chemical energy) is about 18.7 kJ/g (compare with coal: about 30 kJ/g and crude oil: 42 kJ/g). It can be transferred to other energy forms for instance mechanical work directly and be measured by bomb calorimetry, which requires destruction of the sample (organism), however. The information eco-exergy = (β − 1) × biomass or density of information eco-exergy = (β − 1) × concentration. The information eco-exergy controls the function of the many biochemical processes. The ability of a living system to do work is contingent upon its functioning as a living dissipative system. Without the information eco-exergy, the organic matter could only be used as fuel similar to fossil fuel. But due to the information eco-exergy, organisms are able to make a network of the sophisticated biochemical processes that characterize life. The eco-exergy (of which the major part is embodied in the information) is a measure of the organization (Jørgensen and Svirezhev, 2005). This is the intimate relationship between energy and organization that Schrödinger (1944) was struggling to find.

The eco-exergy is a result of the evolution and of copying again and again in a long chain of copies where only minor changes are introduced for each new copy. The energy required for the copying process is very small, but it required a lot of energy to come to the "mother" copy through evolution from prokaryotes to human cells. To cite Margalef (1991) in this context: the evolution provides for cheap—unfortunately often "erroneous," i.e., not exact—copies of messages or pieces of information. The information concerns the degree of uniqueness of entities that exhibit one characteristic complexion that may be described.

Eco-exergy has successfully been used to develop SDMs in 25 case studies so far. The eco-exergy goal function is found using Eq. (6.5), while the β values are found using Table 6.1.

The application is based on what may be considered thermodynamic translation of survival of the fittest. Biological systems have many possibilities for moving away from thermodynamic equilibrium, and it is important to know along which pathways among the possible ones a system will develop. This leads to the following hypothesis sometimes denoted the ecological law of thermodynamics (ELT) (Jørgensen et al., 2007): If a system receives an input of exergy, then it will utilize this exergy to perform work. The work performed is first applied to maintain the system (far) away from thermodynamic equilibrium whereby exergy is lost by transformation into heat at the temperature of the environment. If more exergy is available, then the system is moved further away from thermodynamic equilibrium, reflected in growth of gradients. If there is offered more than one pathway to depart from equilibrium, then the one yielding the highest eco-exergy storage (denoted Ex) will tend to be selected. Or expressed differently: Among the many ways for ecosystems to move away from thermodynamic equilibrium, the one maximizing dEx/dt under the prevailing conditions will have a propensity to be selected.

This hypothesis is supported by several ecological observations and case studies (Jørgensen and Svirezhev, 2005; Jørgensen et al., 2007; Jørgensen, 2008a,b, 2012). Survival implies maintenance of the biomass, and growth means increase of biomass and information. It costs exergy to construct biomass and gain information and biomass and information possess exergy. Survival and growth can therefore be measured using the thermodynamic concept eco-exergy, which may be understood as the work capacity the ecosystem possesses.

6.4 OVERVIEW OF STRUCTURALLY DYNAMIC MODELS USING WORK ENERGY AS GOAL FUNCTION

WE, the work energy of biomass and the information that the organisms carry, also called eco-exergy, has been applied to develop SDMs in 25 cases; several of these 25 cases are presented by Zhang et al. (2010) and in the textbook Fundamentals of Ecological Modelling by Jørgensen and Fath (2011). The last four cases are published in the following papers: Cho et al. (2011), Marchi et al. (2011), Kong et al. (2013), and Jørgensen and Nielsen (2015). The 25 case studies are:

1–8. Eight eutrophication models of six different lakes,
   9. A model to explain the success and failure of biomanipulation based on removal of planktivorous fish,
  10. A model to explain under which circumstances submerged vegetation and phytoplankton are dominant in shallow lakes,

TABLE 6.1   β Values = Eco-exergy Content Relative to the Eco-exergy of Detritus (Jørgensen et al., 2005)

Early organisms	Plants		Animals
Detritus		1.00	
Viroids		1.0004	
Virus		1.01	
Minimal cell		5	
Bacteria		8.5	
Archaea		13.8	
Protists	Algae	20	
	Yeast	17.8	
		33	Mesozoa, Placozoa
		39	Protozoa, amoeba
		43	Phasmida (stick insects)
	Fungi, molds	61	
		76	Nemertina
		91	Cnidaria (corals, sea anemones, jellyfish)
	Rhodophyta	92	
		97	Gastrotricha
		98	Porifera
		109	Brachiopoda
		120	Platyhelminthes
		133	Nematoda
		133	Hirudinea
		143	Gnathostomulida
	Mustard weed	143	
		165	Kinorhyncha
	Seedless angiosperms	158	
		163	Rotifera
		164	Entoprocta
	Moss	174	
		167	Insecta
		191	Coleodiea (sea squirt)

(*Continued*)

TABLE 6.1   β Values = Eco-exergy Content Relative to the Eco-exergy of Detritus (Jørgensen et al., 2005)—cont'd

Early organisms	Plants		Animals
		221	Lepidoptera
		232	Crustacea
		246	Chordata
	Rice	275	
	Gymnosperms	314	
		310	Mollusca
		322	Mosquito
	Angiosperms	393	
		499	Fish
		688	Amphibia
		833	Reptilia
		980	Aves
		2127	Mammalia
		2138	Monkeys
		2145	Anthropoid apes
		2173	*Homo sapiens*

11. A model of Lake Balaton which was used to support the intermediate disturbance hypothesis,
12–15. For small population dynamic models, a eutrophication model of
16. The Lagoon of Venice and
17. The Mondego Estuary and
18. An ecotoxicological model focusing on the influence of copper on zooplankton growth rates.
19. A model of Darwin's finches,
20. A model of the interaction between parasites and birds and
21. The SDM included in Pamolare 1 applied on Lake Fure in Denmark,
22. The role of conjugation in the gene–individual population relationship,
23. SDM of Lake Chozas,
24. Prediction the restoration effects by a structural dynamic approach in Lake Chaoku,
25. Landscape modeling.

It is not yet possible to present case studies where the structural changes have been successfully modeled in the case of climatic changes, but Section 6.5 will illustrate some first model approaches of ecosystem changes due to climatic changes. It is shown that the use of WE as goal function seems in principle to work also in the case of climatic changes. In the last section, it will be attempted to make some first-hand conclusions to the extent that it is possible at this stage.

6.5 DEVELOPMENT OF STRUCTURALLY DYNAMIC MODEL FOR DARWIN'S FINCHES

The development of an SDM for Darwin's finches (see Jørgensen and Fath, 2011) illustrates the advantages of SDMs very clearly; see further details in Jørgensen and Fath (2004). This illustration of SDM has therefore been chosen as an example to demonstrate the applicability of SDM. The model reflects—as all models—the available knowledge which in this case is comprehensive and sufficient to validate even the ability of the model to describe the changes in the beak size as a result of climatic changes, causing changes in the amount, availability, and quality of the seeds that make up the main food item for the finches. The Medium ground finches, *Geospiza fortis*, on the island Daphne Major were selected for these modeling case due to very detailed case specific information found in Grant (1986). The model has three state variables: seed, Darwin's finches adult, and Darwin's finches juvenile. The juvenile finches are promoted to adult finches 120 days after birth. The mortality of the adult finches is expressed as a normal mortality rate plus an additional mortality rate due to food shortage and an additional mortality rate caused by a disagreement between bill depth and the size and hardness of seeds. Due to a particular low precipitation in 1977–79, the population of the Medium ground finches declined significantly and the beak size increased at the same time about 6%. An SDM was developed to be able to describe this adaptation of the beak size due to bigger and harder seeds as a result of the low precipitation.

The beak depth can vary between 3.5 and 10.3 cm according to Grant. The beak size is furthermore equal to square root of D×H, where D is the diameter and H the hardness of the seeds. Both D and H are dependent on the precipitation, particularly from January to April. The coordination or fitness of the beak size with D and H is a survival factor for the finches. The fitness function is based on the seed handling time and it influences the mortality as mentioned above, but has also an impact on the number of eggs laid and the mortality of the juveniles. The growth rate and mortality rate of the seeds is dependent on the precipitation and the temperature, which are forcing functions known as f(time). The food shortage is calculated from the food required by the finches which is known according to Grant and the actual available food according to the state function seed. How the food shortage influences the mortality of the adults and juveniles can be found in Grant (1986). The seed biomass and the number of finches are known as a function of time for the period 1975–82; see Grant (1986). The observations of the state variables from 1975 to 1977 were applied for calibration of the model, focusing on the following parameters:

1. the influence of the fitness function on (a) the mortality of adult finches, (b) the mortality of juvenile finches, and (c) the number of eggs laid,
2. the influence of food shortage on the mortality of adult and juvenile finches is known (Grant, 1986). The influence is therefore calibrated within a narrow range of values,
3. the influence of precipitation on the seed biomass (growth and mortality).

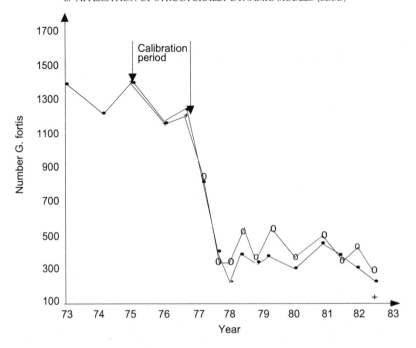

FIGURE 6.5   The observed number of finches (·) from 1973 to 1983, compared with the simulated result (0). 75 and 76 were used for calibration and 77/78 for the validation.

All other parameters are known from the literature (see Grant, 1986).

The eco-exergy density is calculated (estimated) as $275 \times$ the concentration of seed $+ 980 \times$ the concentration of finches (see Table 6.1). Every 15 days, it is found if a feasible change in the beak size taken the generation time and the variations in the beak size into consideration will give a higher exergy. If it is the case, then the beak size is changed accordingly. The modeled changes in the beak size were confirmed by the observations. The model results of the number of Darwin's finches are compared with the observations in Fig. 6.5. The standard deviation between modeled and observed values was 11.6%. For the validation and the correlation coefficient, r^2, for modeled versus observed values is 0.977. The results of a non-SDM would not be able to predict the changes in the beak size and would therefore give too low values for the number of Darwin's finches because their beak would not adapt to the lower precipitation yielding harder and bigger seeds. The calibrated model not using the eco-exergy optimization for the SDMs in the validation period 1977–82 resulted in complete extinction of the finches. A non-SDM—a normal biogeochemical model—could not describe the impact of the low precipitation, while the SDM gave an approximately correct number of finches and could describe the increase of the beak at the same time.

6.6 APPLICATION OF STRUCTURALLY DYNAMIC MODEL FOR THE ASSESSMENT OF ECOLOGICAL CHANGES DUE TO CLIMATE CHANGES

There is a scientific accordance that major climate changes can be expected in the coming decades due to the increasing emission of greenhouse gases. It is therefore very understandable that we ask the question: which ecological changes can we expect as a consequence of the foreseen global warming? How will the different types of ecosystems react to the impact of the climatic changes? Obviously, SDM is a model type that should be able to give the answer to these crucial questions. No SDMs focusing on climatic changes has, however, been developed so far, because it is of course necessary to have some observations of the influence of climate changes on the ecosystems and the ecological processes, before a model can be published and applied more generally. The calibration and validation of an SDM will inevitably require that the foreseen adaptation and/or shifts in species composition by the model are observed with a reasonable and acceptable standard deviation. Although a global temperature increase of 0.8°C has been observed, it would therefore be beneficial to provide observations over a longer period, for instance the coming decade, and furthermore preferably after a slightly higher temperature increase.

It is, however, possible to examine whether an increased temperature will give a decrease in the WE (eco-exergy = WE of biomass and information) and whether adaptation to the increased temperature afterward will yield an increase of the WE and thereby compensate for the previous drop in WE. Such an examination would with a positive result maybe not ensure that SDMs could be used to answer all the relevant questions associated with ecological changes due to climatic changes, but it would indicate that there is a high probability that SDMs could be good tools to model the ecological consequences of climate changes, applying WE (eco-exergy) as goal function.

The examination has been carried out by two population dynamic models, presented in detail in the reference Jørgensen (2015). The models applied in these two examples are shown in Figs. 6.6 and 6.7. The first model has only one state variable—a population, while the second model covers a food chain: plants, herbivores, and carnivores populations. In both models, the relatively simple Arrhenius expression for the influence of the temperature on the ecological processes has been applied: rate as f(temp) = k^|temperature − optimum temperature|. An optimum temperature of 20°C was applied, except when an adaptation was presumed. The absolute/numeric values were applied to account for the influence of the temperature difference from the optimum temperature of 20°C. A characteristic temperature pattern as f(time) for a temperate latitude was applied as a table or graph function. k was for all growth rates 1.05 and for the other processes 1.1. The result of the first model—see Fig. 6.6—is shown in Table 6.2. The biomass is indicated and as there is only one state variable, the focal population, the WE is proportional to the biomass, as the work energy is equal to biomass × the β value for the considered population × 18.7 kJ (if the biomass is in grams).

It can be seen from the results in Table 6.2, that the maximum biomass value and the final biomass value at the end of the year (the model is running 12 months) decrease when the

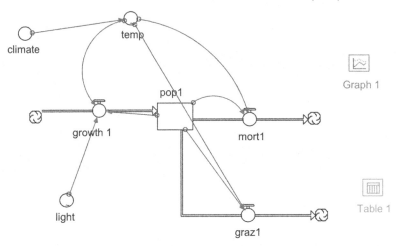

FIGURE 6.6   Population dynamic model applied for the examination.

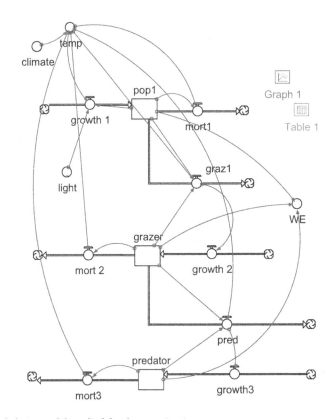

FIGURE 6.7   Food chain model applied for the examinations.

TABLE 6.2 Biomass (g per Unit of Area) for the Population in Fig. 6.6 as f(Increase of Temperature, °C) and Optimum Temperature, °C

Optimum Temperature	Increase of Temperature	Maximum Biomass	Final Biomass
20	0	1018	999
20	1	1015	1000
20	2	1011	999
20	3	1004	992
20	4	993	992
20	5	984	973
23	3	1018	1000
25	5	1018	1000

temperature increases, while a change in the optimum temperature by adaptation is able to eliminate completely the drop in biomass or WE.

The results obtained by the food chain model, Fig. 6.7 are shown in Figs. 6.8 and 6.9. Fig. 6.8 shows the results of biomass for the three populations by an optimum temperature of 20°C and with no change in the temperature, while Fig. 6.9 shows the results obtained with the same optimum temperature but with 3°C increase of the temperature as f(time). Fig. 6.10

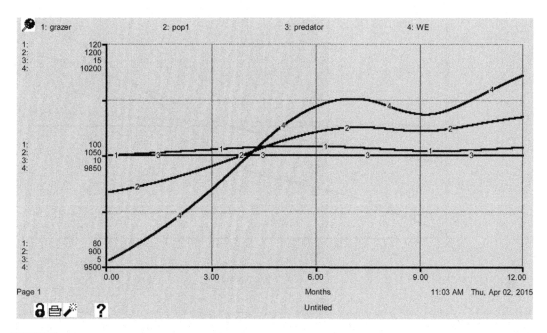

FIGURE 6.8 Biomass and work energy as f(time) for 20°C as optimum temperature and present temperature pattern.

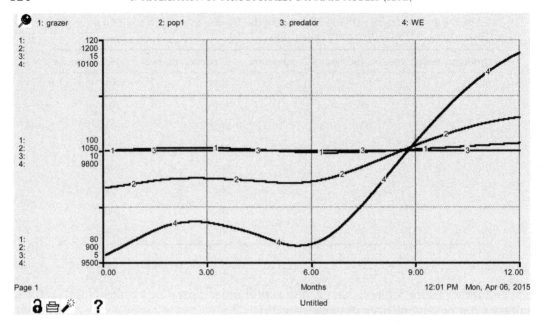

FIGURE 6.9 Biomass and work energy as f(time) for 20°C as optimum temperature and present temperature pattern +3°C. Compare with Fig. 5.7. The work energy has decreased due to the higher temperature.

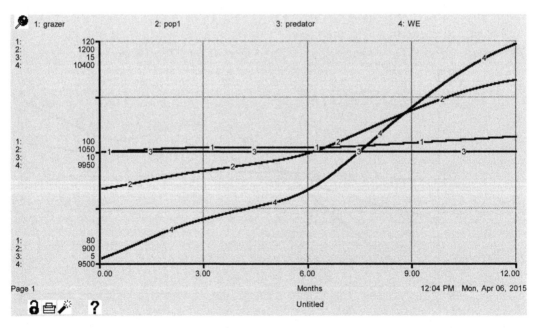

FIGURE 6.10 Biomass and work energy as f(time) for 23°C as optimum temperature and present temperature pattern +3°C. Compare with Figs. 5.7 and 5.8. The work energy has increased compared with no adaptation of the optimum temperature (Fig. 5.8) and is even slightly higher than the work energy in Fig. 5.7. The grazers have clearly benefitted by the higher temperature, but adaptation of the optimum temperature has also to take place.

shows the results achieved by an optimum temperature of 23°C and the general temperature pattern—temperature as f(time). The WE is calculated by the assumption that the food chain is grass, deer, and wolves with β values, respectively, 285, 2027, and 2027. A comparison of the results in Figs. 6.8–6.10 illustrates that the 3°C increased temperature gives a decrease in the WE, but if the optimum temperature is increased correspondingly by adaptation, the WE is regained.

6.7 CONCLUSIONS

SDM seems to be a suitable model tool to describe the expected changes of ecosystems due to impact changes, including the changes expected from global warming. It is therefore recommendable to attempt to use SDMs in our effort in the coming years to develop ecological models that are able to describe the ecological consequences of the climate changes.

References

Cho, W.S., Jørgensen, S.E., Chon, T.-S., 2011. The role of conjugation in the gene-individual population relationship in increasing eco-exergy. Ecological Modelling 222, 407–418.

Fontaine, T.D., 1981. A self-designing model for testing hypotheses of ecosystem development. In: Dubois, D. (Ed.), Progress in Ecological Engineering and Management by Mathematical Modelling. Proc. 2nd Int. Conf. State-of-the-Art Ecological Modelling, 18–24 April, 1980, Liège, Belgium, pp. 281–291.

Grant, P.R., 1986. Ecology and Evolution of Darwin's Finches. Reprinted in 1999. Princeton University Press, New Jersey, 492 pp.

Jørgensen, S.E., 1986. Structural dynamic model. Ecological Modelling 31, 1–9.

Jørgensen, S.E., 1988. Use of models as experimental tools to show that structural changes are accompanied by increased exergy. Ecological Modelling 41, 117–126.

Jørgensen, S.E., 1990. Ecosystem theory, ecological buffer capacity, uncertainty and complexity. Ecological Modelling 52, 125–133.

Jørgensen, S.E., 1992. Development of models able to account for changes in species composition. Ecological Modelling 62, 195–208.

Jørgensen, S.E., 2008a. An overview of the model types available for development of ecological models. Ecological Modelling 215, 3–9.

Jørgensen, S.E., 2008b. Evolutionary Essays. Elsevier, Amsterdam, 230 pp.

Jørgensen, S.E., 2012. Introduction to Systems Ecology. CRC, Boca Raton. Chinese edition 2013. 320 pp.

Jørgensen, S.E., 2015. Application of structurally dynamic models (SDMs) to determine impacts of climate changes. In: Park, Y.-S., Lek, S., Baehr, C., Jørgensen, S.E. (Eds.), Advanced Modelling Techniques Studying Global Changes in Environmental Sciences. Elsevier, Amsterdam, Oxford, pp. 69–86 (Chapter 4) 361 pp.

Jørgensen, S.E., Mejer, J.F., 1977. Ecological buffer capacity. Ecological Modelling 3, 39–61.

Jørgensen, S.E., Mejer, H.F., 1979. A holistic approach to ecological modelling. Ecological Modelling 7, 169–189.

Jørgensen, S.E., Patten, B.C., Straskraba, M., 2000. Ecosystem emerging: 4. Growth. Ecological Modelling 126, 249–284.

Jørgensen, S.E., Fath, B.D., 2004. Modelling the selective adaptation of Darwin's Finches. Ecological Modelling 176, 409–418.

Jørgensen, S.E., Svirezhev, Y., 2005. Toward a Thermodynamic Theory for Ecological Systems. Elsevier, Amsterdam, Oxford, 366 pp.

Jørgensen, S.E., Ladegaard, N., Debeljak, M., Marques, J.C., 2005. Calculations of exergy for organisms. Ecological Modelling 185, 165–175.

Jørgensen, S.E., Fath, B., Bastiononi, S., Marques, M., Müller, F., Nielsen, S.N., Patten, B.C., Tiezzi, E., Ulanowicz, R., 2007. A New Ecology. Systems Perspectives. Elsevier, Amsterdam, 288 pp.

Jørgensen, S.E., Fath, B., 2011. Fundamentals of ecological modelling. In: Application in Environmental Management and Research, fourth ed. Elsevier. 390 pp.

Jørgensen, S.E., Nielsen, S.N., 2015. Hierarchical networks. Ecological Modelling 295, 59–66.

Kong, X.-Z., Jørgensen, S.E., He, W., Qin, N., Xu, F., 2013. Prediction the restoration effects by a structural dynamic approach in Lake Chaoku, China. Ecological Modelling 266, 73–85.

Marchi, M., Jørgensen, S.E., Bacares, E., Corsi, I., Marchettini, N., Bastiononi, S., 2011. Dynamic model of Lake Chozas (Ledo, NW Spain). Ecological Modelling 222, 3002–3010.

Margalef, R., 1991. Networks in ecology. In: Higashi, M., Burns, T.P. (Eds.), Theoretical Studies of Ecosystems: The Network Perspective. Cambridge University Press, pp. 41–57.

Nielsen, S.N., 1992a. Application of Maximum Exergy in Structural Dynamic Models (Ph.D. thesis). National Environmental Research Institute, Denmark, 51 pp.

Nielsen, S.N., 1992b. Strategies for structural-dynamical modelling. Ecological Modelling 63, 91–102.

Odum, E.P., 1953. Fundamentals of Ecology. W.B. Saunders, Philadelphia.

Odum, E.P., 1971. Fundamentals of Ecology, third ed. W.B. Saunders, Philadelphia.

Schrödinger, E., 1944. What Is Life? The Physical Aspect of the Living Cell. Cambridge at the University Press.

Stenseth, N.C., 1986. Darwinian evolution in ecosystems: a survey of some ideas and difficulties together with some possible solutions. In: Casti, J.L., Karlqvist, A. (Eds.), Complexity, Language, and Life: Mathematical Approaches. Springer-Verlag, Berlin, pp. 105–129.

Straskraba, M., 1979. Natural control mechanisms in models of aquatic ecosystems. Ecological Modelling 6, 305–322.

Ulanowicz, R.E., 1986. Growth and Development: Ecosystem Phenomenology. Springer-Verlag, New York.

Ulanowicz, R.E., 1995. Ecosystem trophic foundations: Lindeman exonerata. In: Patten, B.C., Jørgensen, S.E. (Eds.), Complex Ecology: The Part-Whole Relation in Ecosystems. Prentice Hall PTR, pp. 549–560.

Zhang, J., Gurkan, Z., Jørgensen, S.E., 2010. Application of eco-energy for assessment of ecosystem health and development of structurally dynamic models. Ecological Modelling 221, 693–702.

CHAPTER

7

Artificial Neural Networks: Multilayer Perceptron for Ecological Modeling

Y.-S. Park*,[1], S. Lek[§]

*Kyung Hee University, Seoul, Republic of Korea and [§]UMR CNRS-Université Paul Sabatier, Université de Toulouse, Toulouse, France
[1]Corresponding author: E-mail: parkys@khu.ac.kr

OUTLINE

7.1 INTRODUCTION

Artificial neural networks (ANNs) are neural computation systems which were originally proposed by McCulloch and Pitts (1943) and Metropolis et al. (1953). ANNs were widely used in the 1980s thanks to significant developments in computational techniques based on self-organizing properties and parallel information systems. Rumelhart et al. (1986) proposed a new learning procedure, backpropagation (BP), for networks of neuronlike units. The procedure repeatedly adjusts the weights of the connections in the network so as to minimize a measure of the difference between the actual output vector of the network and the desired output vector. This study contributed to the widespread use of ANNs in various research fields consequent to the development of the error BP rule in parallel distributed information processing frameworks. ANNs were also used extensively in the late 1980s to interpret complex and nonlinear phenomena in machine intelligence (Lippmann, 1987; Wasserman, 1989; Zurada, 1992; Haykin, 1994).

The development of ANNs was inspired by the characteristic functioning of the human brain, but they are only remotely related to their biological counterparts. ANNs do not approach the complexity of the brain, but there are two key similarities between biological neural networks and ANNs. First, the building blocks of both networks are simple computational devices that are highly interconnected. Second, the connections between neurons determine the function of the network. A human brain consists of approximately 10^{10} neurons, computing elements, which communicate through a connection network (approximately 10^{4} connections per element). ANNs function as parallel distributed computing networks and are analogous to biological neural systems in some basic characteristics (Fig. 7.1) (Wasserman, 1989; Lek and Park, 2008). As shown in Fig. 7.1B, there are many input signals $[X = (x_1, x_2, \ldots, x_n)]$ to neurons. Each input is given a relative weight $[W = (w_1, w_2, \ldots, w_n)]$ which affects the impact of that input. This is analogous to the varying synaptic strengths of the biological neurons—some inputs are more important than others in the way they combine to produce an impulse. Weights are adaptive coefficients within the network that determine the intensity of the input signal. The output signal of a neuron is produced by the summation block, corresponding roughly to the biological cell body, which adds all of the weighted inputs algebraically (Wasserman, 1989; Zurada, 1992).

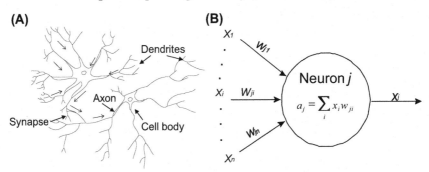

FIGURE 7.1   Schematic diagrams of (A) a biological neuron and (B) an artificial neuron as a basic processing element in a network. Each input value (x_i) is associated with a weight (w_{ji}), and the output value (X_j) can be transmitted to other units. *Arrows* indicate the direction of information flow.

ANNs are able to adjust their inner structures to provide optimal solutions, given enough data and proper initialization. If appropriate inputs are applied to an ANN, it can acquire knowledge from the environment, mimicking the functioning of a brain, and users can later recall this knowledge.

Several different types of ANNs have been developed, but two main categories can be easily recognized, according to their learning process (Lek and Park, 2008):

1. *Supervised* learning: This category utilizes a "teacher" who, in the learning phase, "tells" the ANN how well it performs or what the correct behavior should be.
2. *Unsupervised* learning: This category of ANNs autonomously analyzes the properties of the dataset and learns to reflect these properties in its output.

Both categories of ANNs have been used in ecology, with special attention to self-organizing map (SOM) for the unsupervised learning and multilayer perception (MLP) with a BP algorithm for the supervised learning.

7.2 MULTILAYER PERCEPTRON

7.2.1 Structure of MLPs

Multilayer perceptrons (MLPs) with BP learning algorithms, also called multilayer feed-forward neural networks, are very popular and are used more than other neural network types for a wide variety of problems. MLPs are based on a supervised procedure, i.e., the network builds a model based on examples in data with known outputs. An MLP has to extract this relation solely from the presented examples, which together are assumed to implicitly contain the necessary information for this relation.

An MLP comprises three layers (input, hidden, and output) with nonlinear computational elements (also called neurons and processing units). The information flows from input layer to output layer through the hidden layer (Fig. 7.2). All neurons from one layer are fully

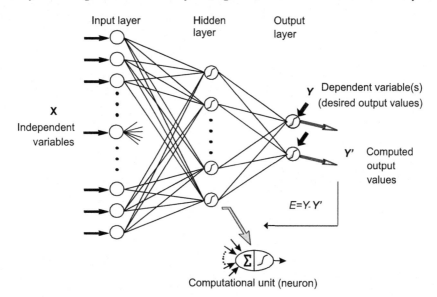

FIGURE 7.2 Three-layered feed-forward neural network with one input layer, one (or more) hidden layer(s), and one output layer: X, independent variables; Y, dependent variables; and Y', values computed from the model.

connected to neurons in the adjacent layers. These connections are represented as weights (connection intensity) in the computational process. The weights play an important role in propagation of the signal in the network. They contain the knowledge of the neural network about the problem–solution relation. The number of neurons in the input layer depends on the number of independent variables in the model, whereas the number of neurons in the output layer is equal to the number of dependent variables. The number of output neurons can be single or multiple. In ecological modeling, environmental variables are generally given to the input layer as independent variables to predict biological variables, which are given to the output layer as target values corresponding to the given input values. Further, both the numbers of hidden layers and their neurons are dependent on the complexity of the model and are important parameters in the development of the MLP model.

7.2.2 Learning Algorithm

An MLP is trained/learned to minimize errors between the desired target values and the values computed from the model. If the network gives the wrong answer, or if the errors are greater than a given threshold, the weights are updated to minimize them. Thus, errors are reduced and, as a result, future responses of the network are likely to be correct. In the learning procedure, datasets of input and desired target pattern pairs are presented sequentially to the network. The learning algorithm of an MLP involves a forward-propagation step followed by a backward-propagation step. MLP algorithms are discussed extensively in the literature, including by Rumelhart et al. (1986) and Lin and Lee (1996) for fundamental concepts, and Lek and Guégan (2000) and Park et al. (2003) for ecological application.

An MLP learns with a algorithm in the following two phases (Lek and Guégan, 2000; Lek and Park, 2008):

7.2.2.1 Forward Propagation

The forward-propagation phase begins with presentation of an input pattern to the input layer. As in biological neural systems, in which dendrites receive signals from neurons and send them to the cell body (Fig. 7.1), an MLP receives information through input and output neurons and summarizes the information. MLP training is based on an iterative gradient algorithm to minimize error between the desired target and the computed model output.

The net input to neuron j of the hidden layer for pattern p ($NET_{p,j}$) is calculated as the summation of each output of the input layer ($x_{p,i}$; input value) multiplied by weight ($v_{p,ji}$). An activation function is applied to calculate the output of neuron j of the hidden layer ($z_{p,j}$) and the output of neuron k of the output layer ($o_{p,k}$) as follows (Chon et al., 2000):

$$f(NET) = \frac{1}{1 + \exp(-\lambda NET)} \tag{7.1}$$

where λ is an activation function coefficient, and NET is expressed either in $z_{p,j}$ or $o_{p,k}$ as follows:

$$z_{p,j} = f\left(\sum_i x_{p,i} v_{p,ji}\right) \tag{7.2}$$

$$o_{p,k} = f\left(\sum_j z_{p,j} w_{p,kj} \right)$$ (7.3)

where $v_{p,ji}$ and $w_{p,kj}$ are the weight of the connections between neuron i of the input layer and neuron j of the hidden layer, and that between neuron j of the hidden layer and neuron k of the output layer for pattern p, respectively. Weights are initialized as small random numbers.

Various transfer functions, such as linear functions, threshold functions, and sigmoid functions, can be used (Fig. 7.3). A sigmoid function is often used, because of its nonlinearity.

The learning algorithm modifies the weights ($v_{p,ji}$ and $w_{p,kj}$) to minimize the error. The sum of the errors (E_p) in each neuron in pattern p is calculated as follows:

$$E_p = \frac{1}{2} \sum_k \left(d_{\rho,k} - o_{\rho,k} \right)^2$$ (7.4)

$$TE = \sum_\rho E_\rho$$ (7.5)

where $d_{p,k}$ is the target value corresponding to pattern p at neuron k, and TE is the total error during one iteration.

The forward-propagation phase continues as activation level calculations propagate forward to the output layer through the hidden layer(s). In each successive layer, every neuron sums its inputs and then applies a transfer function to compute its output. The output layer of the network then produces the final response, i.e., the estimated target value.

7.2.2.2 Backward Propagation

The error value associated with each neuron reflects the amount of error associated with that neuron. Consequently, the neuron is backpropagated for the appropriate weight adjustment. The weights at output neurons are updated as follows (Chon et al., 2000):

$$\delta_{p,k(o)} = o_{p,k}\left(1 - o_{p,k}\right)\left(d_{p,k} - o_{p,k}\right)$$ (7.6)

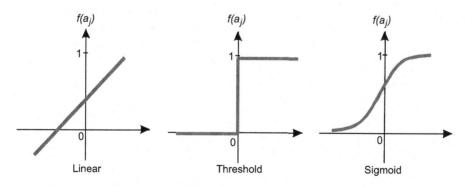

FIGURE 7.3 Three types of transfer functions commonly used in ANN models.

$$\Delta w_{p,kj}(t+1) = \eta \delta_{p,k(o)} o_{p,k} + \alpha \Delta w_{p,kj}(t) \tag{7.7}$$

$$w_{p,kj}(t+1) = w_{p,kj}(t) + \Delta w_{p,kj}(t+1) \tag{7.8}$$

where $\delta_{p,k(o)}$ is the error signal at neuron k of the output layer for pattern p, η is the learning rate coefficient, and α is the momentum coefficient.

Proper learning rate and momentum coefficient are chosen from experience. Both values have a range between zero and one. A large value for η can lead to instability in the network and unsatisfactory learning, whereas an overly small value can lead to excessively slow learning. Typically, the learning rate is varied to produce efficient learning in the network. For example, to obtain a better learning performance, the value of η may be high at the beginning and then decreases during the learning session. The momentum coefficient is defined to avoid oscillating weight changes. It contributes the impact of previous weights to calculate the current weight.

Adjustment of weights at the hidden layer is conducted in a similar manner to that at the output layer except for the error signal (δ), which is produced by summing all such products and then multiplying the derivative of the squashing function as follows (Wasserman, 1989):

$$\delta_{p,j(h)} = z_{p,j}\left(1 - z_{p,j}\right) \sum_{k} \delta_{p,k(o)} w_{p,kj} \tag{7.9}$$

where $\delta_{p,j(h)}$ is the error signal at neuron j of the hidden layer for pattern p.

The learning algorithm can be summarized as shown in Box 7.1 (Lin and Lee, 1996; Chon et al., 2000); the details of the algorithm can be found in Rumelhart et al. (1986), Lippmann (1987), Wasserman (1989), Zurada (1992), Lin and Lee (1996), Lek and Guégan (2000), and Chon et al. (2000).

7.2.3 Validation of Models

In the development of data-driven models, the dataset has to be split into two subsets which are respectively used for training and validation/testing. The proportion may be 1: 1, 2:1, 3:1, etc. for these two sets (Fig. 7.4). However, the training set still has to be large enough to be representative of the problem and the test set has to be large enough to facilitate correct validation of the network. This procedure of partitioning the data is called *k-fold cross-validation*, sometimes also called the *hold-out procedure* (Utans and Moody, 1991; Efron and Tibshirani, 1995; Kohavi and Wolpert, 1996; Friedman, 1997; Lek and Guégan, 2000).

Both datasets contain input/output pattern pairs taken from real data. The training set is used to train the network, whereas the test set is used to assess the performance of the network. The training phase can be time-consuming depending on such factors as the network structure (numbers of input and output neurons, number of hidden layers, and number of neurons in each of the hidden layers), the number of samples in the training set, and the number of iterations.

The learning process of the model is followed by a test phase to ascertain the performance of the model and the model properties. In the test phase, the input data are fed into the network and the desired target values are compared with values computed by the network. The agreement or disagreement of these two sets is used to evaluate the performance of the model.

BOX 7.1
_____

LEARNING ALGORITHM IN AN MLP

Step 0 (Input):

- Input a set of training pairs to the network

Step 1 (Initialization):

- Initialize the weights to small random values.
- Set parameters such as learning rate, momentum coefficient, and TE_{max} (maximum tolerable error).

Step 2 (Training loop):

- Apply the network input pattern to the input layer.

Step 3 (Forward propagation):

- Propagate the signal forward through the network.
- Calculate the network output vector.

Step 4 (Measurement of output error):

- Calculate the errors for each of the outputs, the difference between the desired target, and the network output.

Step 5 (Error backpropagation):

- Propagate the errors backward to adjust the weights in a manner that minimizes the error.

Step 6 (One iteration loop):

- Check whether the entire set of training data has been cycled once.
- Repeat Steps 2 through 5 for the entire training dataset.

Step 7 (Check total error):

- Check whether the current total error (TE) is acceptable.
- If $TE < TE_{max}$, then terminate the training process and output the final weights; otherwise, repeat Steps 2 through 6 until the total error for the entire system is acceptably low, or the predefined number of iterations is reached.

The test phase is conducted in two steps (validation and test) when an appropriate model has to be selected from rival approaches. In this case, the dataset is subdivided into three different parts: training, validation, and testing. The models are evaluated using the validation dataset, and the best performing model is selected. Then, the accuracy of the selected model is estimated.

If the number of samples is not sufficient to permit splitting of the dataset into representative training and test sets, other strategies, such as cross-validation, may be used. In such a case, the dataset is divided into two parts with $n-1$ samples in one part and *one* sample in the other. The MLP is then trained with $n-1$ samples and tested with the last sample. The same network structure is repeated to use every sample in the n procedures. The results of these tests together facilitate determination of the performance of the model. This is called the *leave-one-out* or *Jackknife* procedure (Efron, 1979; Kohavi, 1995). This strategy is often used in ecology when only a small database is available (Lek and Guégan, 2000).

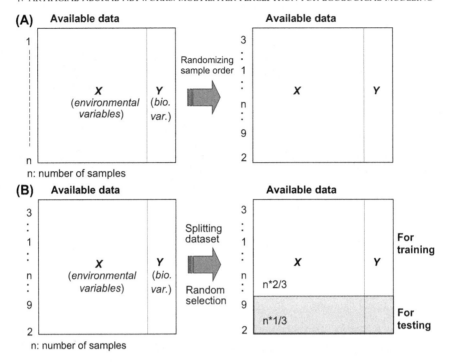

FIGURE 7.4   (A) Randomization of sample order in the dataset preparation procedure. (B) Splitting of the dataset into representative training and test sets. When the number of samples is sufficient, the dataset is divided into three parts if the validation procedure is involved in the analysis. X and Y represent independent variables and dependent variables, respectively.

7.2.4 Data Preprocessing

Ecological data are bulky, nonlinear, and complex—with noise, redundancy, internal relations, and outliers. Therefore, data-driven models are influenced by the characteristics of their datasets, although most models cover nonlinearity and complexity. To avoid these difficulties, datasets for learning the network should be carefully prepared by dealing with missing values, outliers and extremes, transformation, etc. Missing values are often replaced with mean values of the variables or estimated values from prediction models. Outliers and extremes are evaluated with box plot approaches.

Data transformation can be conducted in various ways according to the nature of the data variables. This is particularly important when variables have high variation and different units. In ecological modeling, (1) *logarithmic transformation* is most commonly used. This is especially useful when there is a high degree of variation within variables or a high degree of variation among attributes within a sample. To avoid the problem of log (0) being undefined, the number one (1) is added to all data points before applying the transformations. Meanwhile, (2) *standardization* (also called variance normalization) is very useful for studying environmental variables in ecological studies. It is a linear transformation carried out by scaling the values with mean = 0 and variance = 1 and making the data dimensionless without units. Another popular transformation is (3) *range normalization* (also called *min−max normalization*)—a linear transformation which scales the values between zero and one in the range of the data.

Note that the same transformation should be applied to both training dataset and the test dataset for meaningful results.

Details on the diverse transformation methods are available in McCune and Grace (2002) and Legendre and Legendre (2012).

7.2.5 Overfitting

It is important to avoid overfitting (or overtraining) problems in the development of the model because in such cases the model loses generalization and is no longer applicable to other datasets which were not used to train the network. When a network is overfitted, it has a good memory for specific data. In such a case, the network cannot learn the general features inherently present in the training, but it can learn progressively more of the specific details of the training dataset perfectly. Several rules have been developed by various researchers regarding approximate determination of the required network parameters to avoid overfitting. Two parameters are responsible for this phenomenon: number of epochs and number of hidden layers (along with corresponding number of neurons for each hidden layer) (Lek and Guégan, 2000). Determination of an appropriate number for each of these parameters is the most crucial matter in MLP modeling. The optimum size of epochs, hidden layers, or hidden nodes is determined by trial-and-error using the training and test datasets. If the error in the training set decreases constantly according to the learning process, the error in the test set can increase after minima values. The training procedure must be terminated when the error in the test dataset is lowest. Otherwise, generalization of the model would no longer be possible. Geman et al. (1992) provide an excellent summary of the issues affecting generalization in neural networks.

7.2.6 Sensitivity Analysis

Sensitivity analysis is used to evaluate the behavior of a model and to ascertain the dependence of its outputs on its input parameters (Koda, 1993; Saltelli and Scott, 1997). Sensitivity analysis can be used to determine the following: (1) resemblance of a model to the system or processes being studied, (2) the main factors contributing to output variability, (3) the importance of the model parameters, (4) whether there are regions in the space with factors for which the model variation is maximum, (5) the optimal regions within the space of the factors for use in a subsequent calibration study (global optimization), and (6) if and which factors interact with each other (Saltelli et al., 2000).

MLPs are highly recognized for their high prediction power and applicability to extraction of information from complex, nonlinear relationships (Lek et al., 1996). However, the mechanisms that occur within an MLP network are often ignored. Therefore, an MLP is often considered a blackbox. Various methods have been explored to illustrate the role of variables in MLP models (Gevrey et al., 2003). They include (1) the *Partial derivatives method*, in which the partial derivatives of the output are calculated according to the input variables (Dimopoulos et al., 1995, 1999); (2) the *Weights method*, in which computation is carried out using the connection weights (Garson, 1991; Goh, 1995); (3) the *Perturbation method*, which corresponds to a perturbation of the input variables (Scardi and Harding, 1999); (4) the *Profile method*, in which one input variable is successively varied while the others are kept constant at a fixed value (Lek et al., 1996); (5) the *Classical stepwise method*, in which the change in the error value is observed when an addition (forward) or an elimination (backward) step of the input variables is performed (Balls et al., 1996; Maier and Dandy, 1996); (6) the *Improved stepwise a* method, which uses the same principle as the *Classical stepwise method*, but the elimination of the input occurs when the network is trained and the connection weights corresponding to the input variable studied are also eliminated; and (7) the *Improved stepwise b* method, which also involves the network being trained and fixed step by step with one input variable at its mean value to note the consequences on the error. In their comparison of the methods, Gevrey et al. (2003) reported that the *Partial derivatives method* was the most useful as it gives the most complete results, followed by the *Profile* method, which gives the contribution profile of the input variables. Conversely, Olden et al. (2004) compared different methods by using Monte Carlo simulations with data exhibiting defined numeric relationships, and showed that (8) *a Connection weight approach* that uses raw input-hidden and hidden-output connection weights in the neural network provides the best methodology for accurately quantifying variable importance. However, Fischer (2015) reported that Garson's weight method is preferable to the connection weight method (Olden et al., 2004). Recently, Giam and Olden (2015) proposed a new (9) *Permutational R^2-based variable importance metric* that estimates the proportion of the total variance in the response variable which is uniquely associated with each predictor variable in both linear and nonlinear data contexts.

To conduct sensitivity analysis in an MLP, the network first has to be trained.

7.3 MLPs IN ECOLOGICAL MODELING

Following the pioneering works on ANNs, learning algorithms were applied in diverse ways in ecological modeling (Colasanti, 1991). Ever since MLPs were used to estimate biodiversity and the relationships between ecological community and their environment by Komatsu et al. (1994) and Lek et al. (1995) in ecological modeling, MLPs have been extensively implemented in various ecological studies, including nutrient flows (Nour et al., 2006; Schmid and Koskiaho, 2006), population dynamics (Elizondo et al., 1994; Baran et al., 1996; Aoki and Komatsu, 1997; Recknagel et al., 1997; Stankovski et al., 1998; Aussem and Hill, 1999), community changes (Tan and Smeins, 1996; Park et al., 2001b), determination of community types (Gevrey et al., 2004; Tison et al., 2005), remote sensing and GIS data analysis (Kimes et al., 1996; Keiner and Yan, 1998; Gross et al., 1999), effects of climate change (Lusk et al., 2001; Pearson et al., 2002), habitat and suitability and evaluation of environmental variables (Lek et al., 1996; Wagner et al., 2000; Park and Chung, 2006), and prediction of time series data (Chon et al., 2001).

Comprehensive overviews of ANN applications in ecological modeling have been compiled by Lek and Guégan (2000), Lek et al. (2005), and Recknagel (2006). There are also many valuable papers in the Special Issues of Ecological Modeling: Volume 120 (2–3) in 1999, Volume 146 (1–3) in 2001, Volume 160 (3) in 2003, Volume 195 (1–2) in 2006, Volume 203 (1–2) in 2007, and also in the Special Issues of Ecological Informatics: Volume 1 (3) in 2006, Volume 2 (3) in 2007, and Volume 29 in 2015.

7.4 ADVANTAGES AND DISADVANTAGES OF MLPs

7.4.1 Advantages of MLPs

1. MLPs only have a few parameters; thus, they can be used without prior knowledge, and the algorithms can be implemented easily.
2. MLPs create the required decision function directly through the learning process with a given dataset.
3. The learning process employed is adaptive and so MLPs can learn how to find the solution directly from the data being modeled.
4. MLPs can be applied in a wide range of fields to find solutions.
5. MLPs are used for discrimination, pattern recognition, empirical modeling, and many other tasks.
6. MLPs often provide more efficient results than conventional statistical methods when applied to the same tasks.
7. Whereas traditional linear models are inadequate for modeling data containing nonlinear characteristics, MLPs can represent both linear and nonlinear relationships.
8. Hybrid models that utilize both unsupervised and supervised learning algorithms also exist.
9. MLPs are effective for feature extraction of structure or patterns from static data as well as dynamic data (in both space and time), which are common in ecology.
10. These properties of MLPs can assist in the development of strategic tools for the management of the ecosystem.

7.4.2 Disadvantages of MLPs

1. An MLP's network requires a large number of patterns and a large number of iterations for effective learning.
2. The convergence of its learning process is dependent on the characteristics of the dataset.
3. Defining the number of neurons and layers in the hidden layer of an MLP is difficult. Consequently, many trials are required with different conditions to find the best combinations.
4. New training overwrites the properties of the existing network if existing data are not included in the new training process.
5. The network has to be retrained with both old and new data to reflect the properties of the new data to the trained network.
6. The problem of objectivity exists with MLPs, because the network is based on random effects and iterative calculations.
7. Different configurations of the network may have different convergences, depending on the initial training.
8. Because an MLP is a blackbox, it does not provide causality for events occurring in the system. However, some sensitivity analysis can help to evaluate the contribution of input variables on the model output.

7.4.3 Recommendations for Ecological Modeling

1. An MLP is a highly flexible function approximator for nonlinear and complex data which are commonly observed in ecological studies.
2. Consequently, it is a powerful tool for ecological modeling, particularly when the model is developed with limited information on the relationships between variables.
3. MLPs are applicable to both temporal and spatial data.
4. They can be useful for the prediction and discrimination of biological variables with their environmental variables.
5. They are useful for the assessment of species habitat suitability in conservation as well as restoration ecology.
6. They can be applied when large datasets are available.

7.5 EXAMPLE OF MLP USAGE IN ECOLOGICAL MODELING

7.5.1 Problem

Prediction of fish species richness in streams according to their environment gradients and evaluation for the contribution of their environmental variables.

7.5.2 Ecological Data

Fish community data and their environmental variables were used to illustrate application of MLP. The data were collected from 191 sample sites in the Adour—Garonne river network in France (Park et al., 2006). In the dataset, 34 species were recorded, and species richness was counted at each site. Among the environmental factors determining the spatial distribution of stream fish species, the upstream—downstream gradient is one of the key factors influencing stream fish assemblages, reflecting an increase in fish species richness with increasing stream size (Matthews, 1998). Water quality factors were not considered because the sampling sites were not heavily polluted. Therefore, two environmental factors, reflecting the position of each sampling site along the upstream—downstream gradient, were chosen: altitude and distance from source. In addition, the proportion (%) of various land use types in the basin surface area of each sampling site was used: urbanized artificial surface area and forest area. The proportion of agricultural area has a highly negative correlation with that of the forest area. Therefore, it was excluded from the analysis. The characteristics of the variables used in the model are given in Table 7.1.

7.5.3 Data Preparation

Variables have various units and variations. Therefore, independent variables (environmental variables) were transformed by variance normalization (standardization), resulting in unitless dimensions, and dependent variables (species richness) were transformed by using min—max normalization in the range zero to one. A dataset consisting of 191 samples was split into two subdatasets comprising a training dataset with 111 samples and a testing dataset with 80 samples.

TABLE 7.1 Characteristics of the Variables Used in the Model

Variable		N	Mean	SD	Range (Min−Max)	
Independent	Altitude (m)	191	255.4	233.3	1.0	1190.0
	Distance from source (km)	191	91.9	94.4	2.0	438.0
	Urban area (%)	191	1.0	1.2	0.0	10.6
	Forest area (%)	191	49.4	28.7	0.1	100.0
Dependent	Species richness	191	8.7	4.9	1.0	22.0

7.5.4 Model Training

An MLP model was trained with four inputs (independent) variables and one output (dependent) variable. One hidden layer with three neurons was used. Thus, the model had a 4-3-1 structure. Sum of square errors (SSEs)—differences between the desired target value and the computed model output values—abruptly decreased after 100 iterations, followed by oscillations (Fig. 7.5). Subsequently, the model stabilized at 490 iterations of the learning process. Consequently, the training was terminated at 490 iterations.

7.5.5 Results from the Example

Fish species richness was appropriately predicted with their four environmental factors through the learning process of MLP. The regression determination coefficients (R^2) were 0.70 and 0.62 for the training dataset and the testing dataset, respectively (Fig. 7.6). The

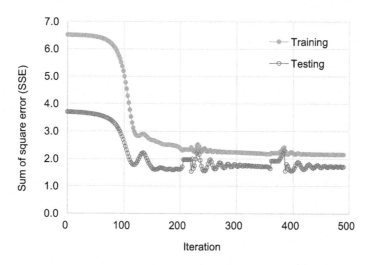

FIGURE 7.5 Changes in sum of square errors during the learning process of the model.

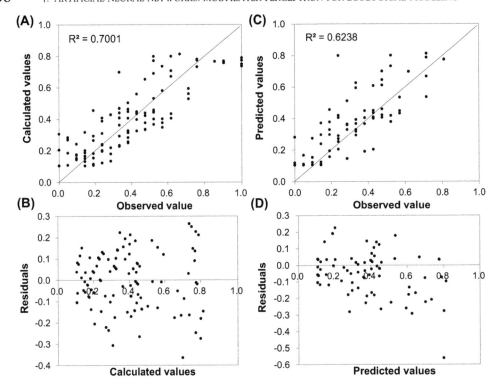

FIGURE 7.6   Relationships between (A,C) observed (desired) outputs and (B,D) values calculated by the model and their residuals. (A,B) Training dataset and (C,D) testing dataset.

residuals—the differences between desired output and calculated values—in the model were distributed evenly around zero.

7.5.6 Contribution of Variables

Following prediction of species richness with environmental factors, the problem of evaluation of the contribution of each environmental factors now had to be solved. Although an MLP is considered as a blackbox in the learning system, the contribution of its input variables can be evaluated with several proposed methods. In this paper, we use a partial derivatives method as an example.

The partial derivative values of model response mainly responded negatively to altitude, indicating that species richness is negatively influenced by altitude (Fig. 7.7A). In contrast, the partial derivatives were positive as a function of distance from source (Fig. 7.7B), reflecting an increase in fish species richness with increasing stream size. In general, fish species richness is higher downstream than upstream. This is well recognized in fish ecology.

The partial derivatives are positive at less than 2% of the urban area. The derivatives are also positive at up to 40% of forest area but are negative at a higher proportion of the forest area (Figs. 7.7C and D). These results are congruent with those for altitude and distance from

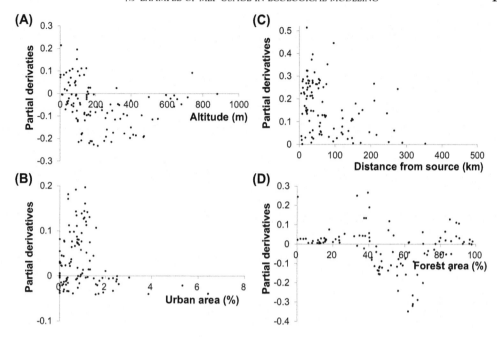

FIGURE 7.7   Partial derivatives of the MLP model response with respect to each independent variable.

source. Thus, they reflect the fact that fish species richness is lower at a high proportion of the forest area with high altitude. Altitude has a negative correlation with distance from source, but a positive correlation with proportion of forest area.

The relative contribution of each input variable was evaluated on the basis of the partial derivatives of the model responses. Among four input variables, distance from source is

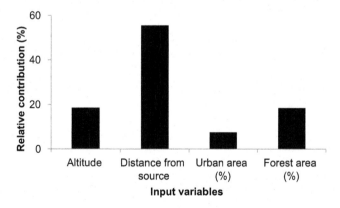

FIGURE 7.8   Relative contribution of input variables based on the partial derivatives of the model responses.

the most important in the determination of fish species richness in the streams studied, presenting a contribution of more than 55%. Altitude and forest areas show similar contribution, with each at 18%, and urban areas have the lowest contribution with less than 8% contribution (Fig. 7.8).

References

Aoki, I., Komatsu, T., 1997. Analysis and prediction of the fluctuation of the sardine abundance using a neural network. Oceanologica Acta 20, 81–88.

Aussem, A., Hill, D., 1999. Wedding connectionist and algorithmic modelling towards forecasting *Caulerpa taxifolia* development in the north-western Mediterranean sea. Ecological Modelling 120, 225–236.

Balls, G.-R., Palmer Brown, D., Sanders, G.-E., 1996. Investigating microclimatic influences on ozone injury in clover (*Trifolium subterraneum*) using artificial neural networks. New Phytologist 132, 271–280.

Baran, P., Lek, S., Delacoste, M., Belaud, A., 1996. Stochastic models that predict trout population density or biomass on a mesohabitat scale. Hydrobiologia 337, 1–9.

Chon, T.-S., Kwak, I.-S., Park, Y.-S., Kim, T.-H., Kim, Y., 2001. Patterning and short-term predictions of benthic macroinvertebrate community dynamics by using a recurrent artificial neural network. Ecological Modelling 146 (1–3), 181–193.

Chon, T.-S., Park, Y.-S., Kim, J.-M., Lee, B.-Y., Chung, Y.-J., Kim, Y., 2000. Use of an artificial neural network to predict population dynamics of the forest-pest pine needle gall midge (Diptera: Cecidomyiida). Environmental Entomology 29 (6), 1208–1215.

Colasanti, R.L., 1991. Discussions of the possible use of neural network algorithms in ecological modelling. Binary 3, 13–15.

Dimopoulos, I., Chronopoulos, J., Chronopoulou Sereli, A., Lek, S., 1999. Neural network models to study relationships between lead concentration in grasses and permanent urban descriptors in Athens city (Greece). Ecological Modelling 120, 157–165.

Dimopoulos, Y., Bourret, P., Lek, S., 1995. Use of some sensitivity criteria for choosing networks with good generalization ability. Neural Processing Letters 2, 1–4.

Efron, B., Tibshirani, R.J., 1995. Cross-Validation and the Bootstrap: Estimating the Error Rate of a Prediction Rule. Department of Statistics, Standford University, Standford. Technical Report 176.

Efron, B., 1979. Bootstrap methods: another look at the jackknife. Annals of Statistics 7 (1), 1–26.

Elizondo, D.A., McClendon, R.W., Hoongenboom, G., 1994. Neural network models for predicting flowering and physiological maturity of soybean. Transactions of the ASAE 37 (3), 981–988.

Fischer, A., 2015. How to determine the unique contributions of input-variables to the nonlinear regression function of a multilayer perceptron. Ecological Modelling 309–310, 60–63.

Friedman, J.H., 1997. On bias, variance, 0/1-loss and the curse-of-dimensionality. Data Mining and Knowledge Discovery 1, 55–77.

Garson, G.D., 1991. Interpreting neural network connection weights. Artificial Intelligence Expert 6, 47–51.

Geman, S., Bienenstock, E., Doursat, R., 1992. Neural networks and the bias/valance dilemma. Neural Computation 4, 1–58.

Gevrey, M., Dimopoulos, I., Lek, S., 2003. Review and comparison of methods to study the contribution of variables in Artificial Neural Network models. Ecological Modelling 160, 249–264.

Gevrey, M., Rimet, F., Park, Y.S., Giraudel, J.L., Ector, L., Lek, S., 2004. Water quality assessment using diatom assemblages and advanced modelling techniques. Freshwater Biology 49, 208–220.

Giam, X., Olden, J.D., 2015. A new R^2-based metric to shed greater insight on variable importance in artificial neural networks. Ecological Modelling 313, 307–313.

Goh, A.T.C., 1995. Back-propagation neural networks for modelling complex systems. Artificial Intelligence in Engineering 9, 143–151.

Gross, L., Thiria, S., Frouin, R., 1999. Applying artificial neural network methodology to ocean color remote sensing. Ecological Modelling 120, 237–246.

Haykin, S., 1994. Neural Networks: A Comprehensive Foundation. Prentice-Hall, Upper Saddle River, NJ.

Keiner, L.E., Yan, X.-H., 1998. A neural network model for estimating sea surface chlorophyll and sediments from Thematic Mapper imagery. Remote Sensing of Environment 66, 153–165.

Kimes, D.S., Holben, B.N., Nickeson, J.E., McKee, W.A., 1996. Extracting forest age in a Pacific Northwest forest from thematic mapped and topographic data. Remote Sensing of Environment 56, 133–140.

Koda, M., 1993. Sensitivity analysis of stochastic dynamical systems. International Journal of Systems Science 23 (12), 2187–2195.

Kohavi, R., Wolpert, D.H., 1996. Bias plus variance decomposition for zero-one loss functions. In: Saitta, L. (Ed.), Machine Learning: Proceedings of the Thirteenth International Conference. Morgan Kaufmann, Bari, Italy, pp. 275–283.

Kohavi, R., 1995. A study of the cross-validation and bootstrap for accuracy estimation and model selection. In: Proceeding of the 14th International Joint Conference on Artificial Intelligence (IJCAI), pp. 1137–1143. Montréal, Canada.

Komatsu, T., Aoki, I., Mitani, I., Ishii, T., 1994. Prediction of the catch of Japanese sardine larvae in Sagami Bay using a neural network. Fisheries Science 60 (4), 385–391.

Legendre, P., Legendre, L., 2012. Numerical Ecology, 3rd English edition. Developments in Environmental Modelling, Vol. 24. Elsevier Science BV, Amsterdam.

Lek, S., Guégan, J.F. (Eds.), 2000. Artificial Neuronal Networks: Application to Ecology and Evolution. Springer, Berlin.

Lek, S., Belaud, A., Baran, P., Dimopoulos, I., Delacoste, M., 1996. Role of some environmental variables in trout abundance models using neural networks. Aquatic Living Resources 9, 23–29.

Lek, S., Belaud, A., Dimopoulos, I., Lauga, J., Moreau, J., 1995. Improved estimation, using neural networks, of the food consumption of fish populations. Marine and Freshwater Research 46, 1229–1236.

Lek, S., Scardi, M., Verdonschot, P., Descy, J., Park, Y.-S. (Eds.), 2005. Modelling Community Structure in Freshwater Ecosystems. Springer, Berlin.

Lek, S., Park, Y.-S., 2008. Artificial Neural Networks. In: Jorgensen, S.E., Fath, B. (Eds.), Encyclopedia of Ecology. Elsevier, Amsterdam, pp. 237–245.

Lin, C.-T., Lee, C.S.G., 1996. Neural Fuzzy Systems: A Neuro-Fuzzy Synergism to Intelligent Systems. Prentice-Hall, NJ, Upper Saddle River.

Lippmann, R.P., April 1987. An Introduction to Computing With Neural Nets. IEEE ASSP, pp. 4–22.

Lusk, J.J., Guthery, F.S., DeMaso, S.J., 2001. Northern bobwhite (Colinus virginianus) abundance in relation to yearly weather and long-term climate patterns. Ecological Modelling 146, 3–15.

Maier, H.R., Dandy, G.C., 1996. The use of artificial neural networks for the prediction of water quality parameters. Water Resources Research 32, 1013–1022.

Matthews, W.J., 1998. Patterns in Freshwater Fish Ecology. Chapman and Hall, New York, p. 756.

McCulloch, W.S., Pitts, W., 1943. A logical calculus of the ideas imminent in nervous activity. Bulletin of Mathematical Biology 5, 115–133.

McCune, B., Grace, J.B., 2002. Analysis of Ecological Communities. MjM Software Design, Gleneden Beach, Oregon.

Metropolis, N., Rosenbluth, A.W., Rosenbluth, M.N., Teller, A.H., Teller, E., 1953. Equation of state calculation by fast computing machines. The Journal of Chemical Physics 21, 1087–1091.

Nour, M.H., Smith, D.W., El-Din, M.G., Prepas, E.E., 2006. Neural networks modelling of streamflow, phosphorus, and suspended solids: application to the Canadian Boreal forest. Water Science and Technology 53 (10), 91–99.

Olden, J.D., Joy, M.K., Death, R.G., 2004. An accurate comparison of methods for quantifying variable importance in artificial neural networks using simulated data. Ecological Modelling 178, 389–397.

Park, Y.-S., Céréghino, R., Compin, A., Lek, S., 2003. Applications of artificial neural networks for patterning and predicting aquatic insect species richness in running waters. Ecological Modelling 160, 265–280.

Park, Y.-S., Chon, T.-S., Kwak, I.S., Kim, J.-K., Jorgensen, S.E., 2001b. Implementation of artificial neural networks in patterning and prediction of exergy in response to temporal dynamics of benthic macroinvertebrate communities in streams. Ecological Modelling 146, 143–157.

Park, Y.-S., Chung, Y.-J., 2006. Hazard rating of pine trees from a forest insect pest using artificial neural networks. Forest Ecology and Management 222 (1–3), 222–233.

Park, Y.-S., Grenouillet, G., Esperance, B., Lek, S., 2006. Stream fish assemblages and basin land cover in a river network. Science of the Total Environment 365, 140–153.

Pearson, R.G., Dawson, T.P., Berry, P.M., Harrison, P.A., 2002. SPECIES: a spatial evaluation of climate impact on the envelope of species. Ecological Modelling 154, 289–300.

Recknagel, F., 2006. Ecological Informatics: Understanding Ecology by Biologically-Inspired Computation. Springer, Berlin.

Recknagel, F., French, M., Harkonen, P., Yabunaka, K.-I., 1997. Artificial neural network approach for modelling and prediction of algal blooms. Ecological Modelling 96, 11–28.

Rumelhart, D.E., Hinton, G.E., Williams, R.J., 1986. Learning representations by backpropagation error. Nature 323, 533–536.

Saltelli, A., Chan, K., Scott, M. (Eds.), 2000. Sensitivity Analysis. John Wiley & Sons, New York.

Saltelli, A., Scott, M., 1997. The role of sensitivity analysis in the corroboration of models and its links to model structural and parametric uncertainty. Special issue Reliability Engineering and System Safety 57 (1).

Scardi, M., Harding, L.W., 1999. Developing an empirical model of phytoplankton primary production: a neural network case study. Ecological Modelling 120, 213–223.

Schmid, B.H., Koskiaho, J., 2006. Artificial neural network modeling of dissolved oxygen in a wetland pond: the case of Hovi, Finland. Journal of Hydrologic Engineering 11, 188–192.

Stankovski, V., Debeljak, M., Bratko, I., Adamic, M., 1998. Modelling the population dynamics of red deer (*Cervus elaphus* L.) with regard to forest development. Ecological Modelling 108, 143–153.

Tan, S.S., Smeins, F.E., 1996. Predicting grassland community changes with an artificial neural network model. Ecological Modelling 84, 91–97.

Tison, J., Park, Y.-S., Coste, M., Wasson, J.G., Ector, L., Rimet, F., Delmas, F., 2005. Typology of diatom communities and the influence of hydro-ecoregions: a study at the French hydrosystem scale. Water Research 39, 3177–3188.

Utans, J., Moody, J.E., 1991. Selecting neural network architectures via the prediction risk: application to corporate bond rating prediction. In: Proceedings of the First International Conference on Artificial Intelligence Applications on Wall Street. IEEE Computer Society Press, Los Alamitos, CA.

Wagner, R., Dapper, T., Schmidt, H.H., 2000. The influence of environmental variables on the abundance of aquatic insects: a comparison of ordination and artificial neural networks. Hydrobiologia 422–423, 143–152.

Wasserman, P.D., 1989. Neural Computing: Theory and Practice. Van Nostrand Reinhold, New York.

Zurada, J.M., 1992. Introduction to Artificial Neural Systems. West Publishing Company, New York.

CHAPTER

8

Ecotoxicological Models

S.E. Jørgensen

Emeritus Professor, Copenhagen University, Denmark
E-mail: soerennorsnielsen@gmail.com

OUTLINE

8.1 APPLICATION OF ECOTOXICOLOGICAL MODELS

Ecotoxicological models are increasingly applied to assess the environmental risk of chemical emissions to the environment. We distinguish between fate models and effect models. Fate models provide the concentration of a chemical in one or more environmental compartments, for instance the concentration of a chemical compound in a fish or in a lake. Effects models translate a concentration or body burden in a biological compartment to an effect.

The effect models presume that we know the concentration of a chemical in a focal compartment, either by a model or by analytical determinations. The effect models translate the found estimated concentrations into an effect on either the growth of an organism, the development of a population or the community, the changes of an ecosystem or a landscape, or on the entire ecosphere.

The results of a fate model can be used to perform an ERA (environmental risk assessment): find the ratio, RQ, between the computed concentration, PEC = predicted environmental concentration, and the nonobserved-effect concentration, NOEC, which is determined through literature values or laboratory experiments. Further detail about the procedure for environmental risk assessment, ERA, and how to account for the uncertainty of the assessment (details see Jørgensen and Fath (2011)). It is also possible to merge fate models with effect models and thereby combine the two approaches. We could call such models FTE models, meaning fate-transport-effect models.

Many fate models, fewer effect models, and only a few FTE models have been applied to solve ecotoxicological problems and perform ERAs. The development is, however, toward a wider application of effect and FTE models.

A. Fate models may be divided into three classes:
    I. Models that map the fate and transport of a chemical in a region or a country. These models are sometimes called MacKay-type models after Donald Mackay, who first developed these models. A detailed discussion of the application of these models can be found in MacKay (1991) and SETAC (1995). This type of fate models is rarely calibrated and validated, although indicating the standard deviation of the results has been attempted; see SETAC (1995). This type will not be presented in this chapter as it is covered in Chapter 12 on fugacity models.
    II. Models that consider a specific case of toxic substance pollution, for instance a discharge of a chemical to a coastal zone from a chemical plant or a sewage treatment plant. This type of fate model must always be calibrated and validated. This type of models is a biogeochemical model that is applied on a toxic substance in the environment.
    III. Models that focus on a chemical that is used locally. It implies that an evaluation of the risk will require that we determine a typical concentration (which is much higher than the regional concentration that would be obtained from model type I) in a typical locality. A typical example is the application of pesticides, where the model will have to look into a typical application on an agriculture field close to a stream and with a ground water mirror close to the surface. This model type can be considered a hybrid of I and II. The conceptual diagram and the equations of the type III model are similar to model type II, but the interpretation of the model results are similar to model type I. This model type should always be calibrated and validated by data obtained for a typical case study, but the prognosis is most commonly applied for development of "a worst case situation" or "an average situation," which in general may be different from the case study applied for the calibration and validation. This model type is also biogeochemical models applied on toxic substances.

Examples of the two last model types II + III are presented in this chapter.

B. Effect models may be classified according to the hierarchical level of concern:

    I. Organism models, where the core of the model is the influence of a toxic substance on an organism, e.g., a relationship between the growth parameters and the concentration of a toxic substance.

    II. Population models, where the population models presented in Chapter 3, including individual-based models, may include relationships between toxic substance concentrations and the model parameters.

    III. An ecosystem model where the influences of a toxic substance on several parameters are included. The result of these impacts of a chemical is an ecosystem with a different structure and composition.

    IV. As ecosystems are open systems, the effects of chemicals may change several interrelated ecosystems. Landscape models can be used in these cases.

    V. Global models where the impacts of chemicals are the core of the model. A typical global model is a model of the ozone layer and its decomposition due to the discharge of chemicals (e.g., freon). Effect models types I + II are population dynamic models and they are not included in this chapter but are briefly touched on in Chapter 3.

FTE models can be any combination of fate and effect models, although the combinations of AII and AIII fate models with BII and BIII effect models will be most applied for ecotoxicological management. Types III, IV, and V are biogeochemical models with the modification that the parameters are changed according to a known effect of toxic substances, for instance decreased growth rate by increasing concentration of a toxic substance.

The effect models applied up to now are mainly of type I and II, although the effects on ecosystem levels may be of particular importance due to their frequent irreversibility. Ecosystems may, in some cases, change their composition and structure significantly due to discharge of toxic substances. In such cases, it is recommended to apply structural dynamic models, also called variable parameter models; see Chapter 7.

Ecotoxicological models are applied either for registration of chemicals, to solve site-specific pollution problems, or to follow ecosystem recovery after pollution abatement or remediation has taken place.

Type AI and III models are widely used for registration of chemicals. About 100,000 chemicals are registered, but only about 20,000 chemicals are used at a scale which may threaten the environment with high probability. It was the long-term goal to perform an ERA for all these 20,000 chemicals which were in use before 1984, where an ecotoxicological evaluation of all new chemicals became compulsory in the EU. Among the 20,000 chemicals, 2500 have been selected as high-volume chemicals which obviously are of most concern. Among the 2500 chemicals, 140 have been selected in the EU to be examined in detail, including performance of ERA which will require the application of models. They are named HERO chemicals (highly expected regulatory output chemicals). A proper ecotoxicological evaluation of the chemicals in use before 1984 is important; it will take 100 years before we have a proper ecotoxicological evaluation of the 2500 high-volume chemicals and 800 years before we have evaluated all chemicals in use!!!—by which time there will be many new chemicals.

In the EU law, a premarket testing of chemicals was introduced in 1980. This requirement of premarket testing was very reasonable, but a problem was that the chemicals already on market as mentioned above have not been tested at least not properly and that very often their properties were not uncovered sufficiently to be able to develop an applicable ERA. What about these chemicals? The obligation of testing existing substances was, however, not imposed on the industry but on public authorities. The major bulk of chemicals were introduced before 1980 and remain therefore untested, because it would require very long time and be very costly to perform the testing of this major bulk of chemicals. In addition, the chemicals introduced after 1980 had to be tested, which of course was a disadvantage for the new chemicals and posed a hindrance for innovation of more environmentally friendly and better chemicals. The REACH (Registration, Evaluation, Authorisation of CHemicals) reform was adapted in 2006 to try to solve these problems. The details of REACH reform can be found in Jørgensen (2016).

About 300–400 new chemicals are registered per year. These chemicals have to be evaluated properly, although it may be possible in some cases for the chemical manufacturers to postpone the evaluation and the final decision for a few years. AII fate models and BII, BIII, and in a few cases BIV effect models are applied, sometimes in combination as an FTE model to solve site-specific pollution problems caused by toxic substances or to make predictions on the recovery of ecosystems after the impacts have been removed. These applications are mainly carried out by environmental protection agencies and rarely by chemical manufacturers. It can be concluded from this short overview of model types and classes and their application in practical environmental management that there is an urgent need for good ecotoxicological models and for a wide experience in the applicability of these models. The application of ecotoxicological models up to now has been minor compared to the environmental management possibilities that these models offer. ERA uses extensively ecotoxicological models, but ERA is not included in this chapter. The performance of ERA can be found in Jørgensen and Fath (2011).

Section 8.2 presents the characteristics and structure of ecotoxicological models. Section 8.3 gives an overview of some of the most illustrative, ecotoxicological models published during the last 20 years. The description of the chemical, physical, and biological processes will, in general, be according to the equations presented in Chapter 2. Section 8.4 is devoted to parameter estimations methods, which are of particular importance in ecotoxicological models. The following sections are used to present ecotoxicological models of case studies. The case study in Section 8.5 covers an ecotoxicological model for relating contamination of agricultural products by cadmium with the heavy metal pollution of soil due to the content of cadmium in fertilizers, dry deposition, and sludge. Sections 8.2–8.5 build on the rational of Jørgensen and Fath (2011). Section 8.6 gives two recently published examples of type A II + III models where the models have been used as experimental tools. As pointed out in Chapter 1, it is a very important application of models in general.

8.2 CHARACTERISTICS OF ECOTOXICOLOGICAL MODELS

Toxic substance models are most often biogeochemical models because they attempt to describe the mass flows of the considered toxic substances, although there are effect models of the population dynamics, which include the influence of toxic substances on the birth rate and/or the mortality, and therefore should be considered toxic substance models.

Toxic substance models differ from other ecological models included biogeochemical models in that:

1. The need for parameters to cover all possible toxic substance models is great, and general estimation methods are therefore widely used. Section 8.4 is devoted to this question, which has also been touched on in Chapter 2.
2. The safety margin, assessment factors, should be high, for instance, expressed as the ratio between the predicted concentration and the concentration that gives undesired effects.
3. They require sometimes possible inclusion of an effect component, which relates the output concentration to its effect. It is easy to include an effect component in the model; it is, however, often a problem to find a well-examined relationship to base it on.
4. They need simple models due to points 1 and 2, and our limited knowledge of process details, parameters, sublethal effects, antagonistic effects, and synergistic effects is limited.

It may be an advantage to outline the approach before developing a toxic substance model according to the procedure presented in Section 2.3:

1. Obtain the best possible knowledge about the possible processes of the toxic substances in the ecosystem.
2. Attempt to get parameters from the literature and/or from own experiments (in situ or in the laboratory). There is a rather rich literature with information about the properties of toxic substances.
3. Estimate all parameters by all available methods to make the final selection of parameter values as certain as possible. The next section will present methods that are particularly applicable for toxic substance.
4. Compare the results from (2) and (3) and attempt to explain discrepancies.
5. Estimate which processes and state variables it is feasible and relevant to include in the model. When in doubt at this stage, it is better to include too many processes and state variables rather than too few.
6. Use a sensitivity analysis to evaluate the significance of the individual processes and state variables. This often may lead to further simplification. The use of sensitivity analysis for parameters (often the properties of toxic substances) is of course also important to assess which parameters would be beneficial to determine for instance by laboratory examinations.

To summarize, ecotoxicological models differ from biogeochemical ecological models by:

1. often being more simple conceptually—fewer state variables,
2. requiring more parameters,
3. a wider use of parameter estimation methods,
4. a possible inclusion of an effect component.

Ecotoxicological models may be divided into five classes according to their structure. The five classes illustrate also the possibilities of simplification which is urgently needed as already discussed a few times.

8.2.1 Food Chain or Food Web Dynamic Models

This class of models considers the flow of toxic substances through the food chain or food web. It can also be described as an ecosystem model focusing on the transfer of a toxic substance to ecological and nonecological components. Such models will be relatively complex and contain many state variables. The models will contain many parameters, which often have to be estimated by one of the methods presented in Section 8.4. This model type will typically be used when many organisms are affected by the toxic substance or the entire structure of the ecosystem is threatened by the presence of the toxic substance. Because of the complexity of these models, they have not been used widely. They are similar to the more complex biogeochemical eutrophication models that consider the nutrient flow through the food chain or even through the food web. Sometimes they are even constructed as submodels of a eutrophication model, see for instance Thomann et al. (1974). Fig. 8.1 shows a conceptual diagram of an ecotoxicological food chain model for lead. There is a flow of lead from atmospheric fallout and wastewater to an aquatic ecosystem, where it is concentrated through the food chain—by "bioaccumulation." A simplification is hardly possible

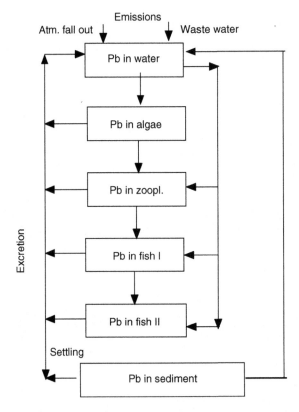

FIGURE 8.1   Conceptual diagram of the bioaccumulation of lead through a food chain in an aquatic ecosystem.

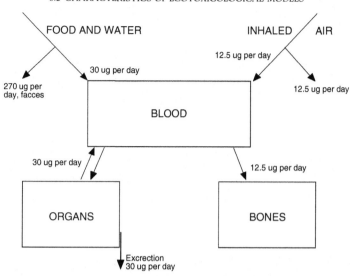

FIGURE 8.2   A static model of the lead uptake by an average Dane in 1980 before lead in the gasoline was banned.

for this model type because it is the aim of the model to describe and quantify the bioaccumulation through the food chain.

8.2.2 Static Models of Toxic Substance Mass Flows

If the seasonal changes are minor, or of minor importance, then a static model of the mass flows will often be sufficient to describe the situation and even to show the expected changes if the input of toxic substances is reduced or increased. This model type is based upon a mass balance as clearly seen from the example in Fig. 8.2. It will often, but not necessarily, contain more trophic levels, but the modeler is frequently concerned with the flow of the toxic substance through the food chain. The example in Fig. 8.2 considers only one trophic level. If there are some seasonal changes, then this type, which usually is simpler than type one, can still be an advantageous, for instance, if the modeler is concerned with the worst case or the average case and not with the seasonal changes.

8.2.3 A Dynamic Model of a Toxic Substance in One Trophic Level

It is often only the toxic substance concentration in one trophic level that is of concern. This includes the abiotic environment (sometimes called the zeroeth trophic level)—soil, water, or air. Fig. 8.3 gives an example. Here the main concern is the DDT concentration in fish, where there may be such high concentration of DDT that, according to the WHO standards, they are not recommended for human consumption. The model can be simplified by not including the entire food chain but only the fish. Some physical–chemical reactions in the water phase are still important and they are included as shown on the conceptual diagram (Fig. 8.3). As seen from these examples, simplifications are often feasible when the problem is well defined,

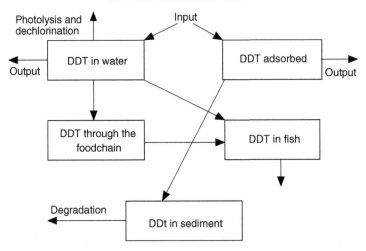

FIGURE 8.3 Conceptual diagram of a simple DDT model.

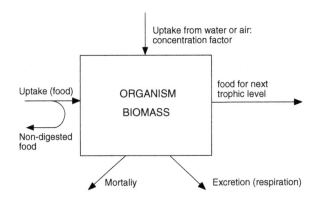

FIGURE 8.4 Processes of interest for modeling the concentration of a toxic substance at one trophic level.

including which component are the most sensitive to toxic matter, and which processes are most important for concentration changes. Fig. 8.4 shows the processes of interest for modeling the concentration of a toxic component at one trophic level. The inputs are uptake from the medium (water or air) and from digested food = total food − nondigested food. The outputs are mortality (transfer to detritus), excretion, and predation from the next level in the food chain.

8.2.4 Ecotoxicological Models in Population Dynamics

Although most ecotoxicological models are biogeochemical models, population dynamic models are also applied to solve the problems of toxic substances in the environment. Population models are biodemographic models and have the number of individuals or species as state variables. Simple population models consider only one population. Population growth is a result of the difference between natality and mortality, which may be influenced by the presence of toxic substances. This influence is taken into account in ecotoxicological population dynamic models. The usually applied population dynamic (simple) model is used to explain:

$$dN/dt = B \times N - M \times N = r \times N, \tag{8.1}$$

where N is the number of individuals, B is the natality, i.e., the number of new individuals per unit of time and per unit of population, M is the mortality, i.e., the number of organisms that die per unit of time and per unit of population; and r is the increase in the number of organisms per unit of time and per unit of population and is equal to B − M. B, N, and r are not necessarily constants as in the exponential growth equation but are dependent on N, the carrying capacity, and other factors. The concentration of a toxic substance in the environment or in the organisms may influence the natality and the mortality, and if the relation between a toxic substance concentration and these population dynamic parameters is included in the model, it becomes an ecotoxicological model of population dynamics.

Population dynamic models may include two or more trophic levels and ecotoxicological models will include the influence of the toxic substance concentration on natality, mortality, and interactions between these populations. In other words, an ecotoxicological model of population dynamics is a general model of population dynamics with the inclusion of relations between toxic substance concentrations and some important model parameters.

8.2.5 Ecotoxicological Models With Effect Components

Although class 4 models already may include relations between concentrations of toxic substances and their effects, these are limited to, for instance, population dynamic parameters not to a final assessment of the overall effect. In comparison, class 5 models include more comprehensive relations between toxic substance concentrations and effects. These models may include not only lethal and/or sublethal effects but also effects on biochemical reactions or on the enzyme system. The effects may be considered on various levels of the biological hierarchy from the cells to the ecosystems.

In many problems it may be necessary to go into more detail on the effect to answer the following relevant questions:

1. Does the toxic substance accumulate in the organism?
2. What will be the long-term concentration in the organism when uptake rate, excretion rate, and biochemical decomposition rate are considered?
3. What is the chronic effect of this concentration?
4. Does the toxic substance accumulate in one or more organs?
5. What is the transfer between various parts of the organism?
6. Will decomposition products eventually cause additional effects?

A detailed answer to these questions may require a model of the processes that take place in the organism and a translation of concentrations in various parts of the organism into effects. This implies that the intake = (uptake by the organism) × (efficiency of uptake) is known. Intake may either be from water or air, which also may be expressed (at steady state) by concentration factors, which are the ratios between the concentration in the organism and in the air or water.

But, if all the above-mentioned processes were taken into consideration for just a few organisms, the model would easily become too complex, contain too many parameters to calibrate, and require more detailed knowledge than it is possible to provide. Often we even do not have all the relations needed for a detailed model, as toxicology and ecotoxicology are not completely well understood. Therefore, most models in this class will not consider too many details of the partition of the toxic substances in organisms and their corresponding effects, but rather be limited to the simple accumulation in the organisms and their effects. Usually, accumulation is rather easy to model and the following simple equation is often sufficiently accurate:

$$dC/dt = (ef \times Cf \times F + em \times Cm \times V)/W - Ex \times C = (INT)/W - Ex \times C \qquad (8.2)$$

where C is the concentration of the toxic substance in the organism; ef and em are the efficiencies for the uptake from the food and medium, respectively (water or air); Cf and Cm are the concentration of the toxic substance in the food and medium, respectively; F is the amount of food uptake per day; V is the volume of water or air taken up per day; W is the body weight either as dry or wet matter; and Ex is the excretion coefficient (1/day). As can be seen from the equation, INT covers the total intake of toxic substance per day. The equation is based on the processes shown in Fig. 8.4.

This equation has a numerical solution and the corresponding plot is shown in Fig. 8.5:

$$C/C(max) = (INT \times (1 - exp(Ex \times t)))/(W \times Ex) \qquad (8.3)$$

where C(max) is the steady state value of C:

$$C(max) = INT/(W \times Ex) \qquad (8.4)$$

Synergistic and antagonistic effects have not been touched on so far. They are rarely considered in this type of model for the simple reason that we do not have much knowledge about these effects. If we have to model combined effects of two or more toxic substances,

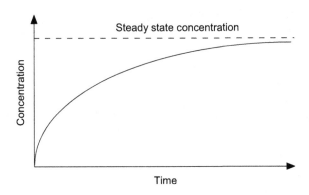

FIGURE 8.5   Concentration of a toxic substance in an organism versus time.

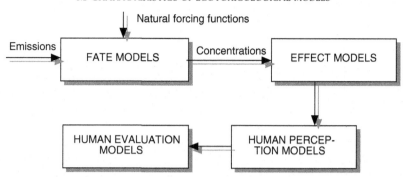

FIGURE 8.6   The four submodels of a total ecotoxicological model are shown.

then we can only assume additive effects, unless we can provide empirical relationships for the combined effect.

A complete solution of an ecotoxicological problem requires in principle four submodels, of which the fate model may be considered the first model in the chain; see Fig. 8.6. As seen in the figure the four components are (see Morgan, 1984):

1. A fate or exposure model which should be as simple as possible and as complex as needed, as already stressed.
2. An effect model, translating the concentration into an effect; see type 5 above and the different levels of effects.
3. A model for human perception processes. This model takes the sometimes irrational perception of an environmental problem or threat.
4. A model for human evaluation processes.

The first two submodels are in principle "objective," predictive models, corresponding to the structural model types 1–5 described above, or the classes described from an application point of view, described in Section 8.1. They are based upon physical, chemical, and biological processes. They are very similar to other environmental models and founded upon mass transfer, mass balances, physical, chemical, and biological processes. The submodels (3) and (4) are different from the generally applied environmental management models and are only touched on briefly below. A risk assessment component, associated with the fate model, comprises human perception and evaluation processes; see Fig. 8.6. These submodels are explicitly value-laden but must build on objective information concerning concentrations and effects. They are often considered in the ERA procedure by deciding on the assessment factor.

Factors that may be important to consider in this context are:

1. Magnitude and time constant of exposure.
2. Spatial and temporal distribution of concentration.
3. Environmental conditions determining the process rates and effects.
4. Translation of concentrations into magnitude and duration of effects.
5. Spatial and temporal distribution of effects.
6. Reversibility of effects.

The uncertainties relating to the information on which the model is based and the uncertainties related to the development of the models are crucial in the application of ecotoxicological models in environmental management including their application in risk assessment. The uncertainty assessment should consider the following five points:

1. Good direct knowledge and statistical evidence on the important components (state variables, processes, and interrelations of the variables) of the model is available.
2. Good knowledge and statistical evidence on the important submodels are available but the aggregation of the submodels is less certain.
3. No good knowledge of the model components for the considered system is available, but good data are available for the same processes from a similar system and it is estimated that these data may be applied directly or with minor modifications to the model development.
4. Some, but insufficient, knowledge is available from other systems. Attempts are made to use these data without the necessary transferability. Attempts are made to eliminate gaps in knowledge by using additional experimental data as far as it is possible within the limited resources available for the project.
5. The model is to a large extent or at least partly based on the subjective judgment of experts.

The acknowledgment of the uncertainty is of great importance and may be taken into consideration either qualitatively or quantitatively. Another problem is of course: Where to take the uncertainty into account? Should the economy or the environment benefit from the uncertainty? The ERA procedure has definitely facilitated the possibilities to consider the environment more than the economy.

Until about 15 years ago, researchers had developed very little understanding of the processes by which people actually perceive the exposures and effects of toxic chemicals, but these processes are just as important for the risk assessment as the exposures and effects processes themselves. The characteristics of risks and effects are important for the perceptions of people. These characteristics may be summarized in the following list:

Characteristics of risk:

Voluntary or involuntary?
Are the levels known to the exposed people or to science?
Is it novel or familiar?
Is it common or dreaded (for instance does it involve cancer)?
Does it involve death?
Are mishaps controllable?
Are future generations threatened?
What scale: global, regional, or local?
Function of time? How (whether for instance increasing or decreasing)?
Can it easily be reduced?

Characteristics of effects:

Immediate or delayed?
On many or a few people?
Global, regional, or local?

Involve death?
Are effects of mishaps controllable?
Observable immediately?
How are they a function of time?

A factor analysis was performed by Slovic et al. (1982) and it shows, among other results, a not surprising correlation between people's perception of dreaded and unknown risks. Broadly speaking there are two methods of selecting the risks we will deal with.

The first may be described as the "rational actor model," involving people that look systematically at all risks they face and make choices about which they will live with and at what levels. For decision making, this approach would use some single, consistent, objective functions and a set of decision rules.

The second method may be named the "political/cultural model." It involves interactions between culture, social institutions, and political processes for the identification of risks and determination of those which people will live with and at what level.

Both methods are unrealistic, as they are both completely impractical in their pure form. Therefore, we must select a strategy for risk abatement founded on a workable alternative based on the philosophy behind both methods.

Several risk management systems are available, but no attempt will be made here to evaluate them. However, some recommendations should be given for the development of risk management systems:

1. Consider as many characteristics listed above as possible and include the human perceptions of these characteristics in the model.
2. Do not focus too narrowly on certain types of risks. This may lead to suboptimal solutions. Attempt to approach the problem as broadly as possible.
3. Choose strategies that are pluralistic and adaptive.
4. Benefit-cost analysis is an important element of the risk management model, but it is far from being the only important element and the uncertainty in evaluation of benefit and cost should not be forgotten. The variant of this analysis applicable to environmental risk management may be formulated as follows:

net social benefit $=$ social benefits of the project $-$ environmental costs of the project (8.5)

5. Use multiattribute utility functions, but remember that people in general have trouble in thinking about more than 2–3, at the most 4, attributes in each outcome.

The application of the estimation methods, presented in the next section, renders it feasible to construct ecotoxicological models, even when our knowledge of the parameters is limited. The estimation methods have a high uncertainty, but a great safety factor (assessment factor) helps in accepting this uncertainty. On the other hand, our knowledge about the effects of toxic substances is very limited—particularly at the ecosystem, the organism, and organ level. It must not be expected, therefore, that models with effect components give more than a first rough picture of what is known today in this area.

8.3 AN OVERVIEW: THE APPLICATION OF MODELS IN ECOTOXICOLOGY

A number of toxic substance models have been published the last about 40 years and several models are available in ecotoxicology today. During the last 10 years many of the models developed from 1975 to 2000 have still been applied in environmental management, while a more limited number of new models have been developed, probably because the spectrum of available toxic substance models was sufficient to cover almost all the relevant ecotoxicological problems. Many of the ecotoxicological models that were developed in the years 1980–2000 can still be applied because the same model structure is valid for the new problems, but the parameters (it means the properties of the toxic substances) have of course to be changed according to which toxic substances the models were applied for. Most models reflect the proposition that good knowledge of the problem and ecosystem can be used to make reasonable and still workable simplifications. Ecotoxicological modeling has been approached from two sides: population dynamics and biogeochemical flow analysis. As the second approach has been most applied in environmental management, it has been natural also to approach the toxic substance problems from this angle and use mostly biogeochemical models. The most difficult part of modeling the effect and distribution of toxic substances is to obtain the relevant knowledge about the behavior of the toxic substances in the environment and to use this knowledge to make the feasible simplifications. It gives the modeler of ecotoxicological problems a particular challenge by selection of the right and balanced complexity, and there are many examples of rather simple ecotoxicological models, which can solve the focal problem. Table 8.1 gives a comprehensive overview of the available toxic substance models. The table is not a result of a complete literature review but gives an idea of the wide spectrum of different toxic substance models that have been developed. It shows, furthermore, that all five classes of models are represented. The references are included in the table to facilitate the search for a relevant model to be used directly or indirectly to solve an ecotoxicological, environmental problem.

It can be seen from the overview in Table 8.1 that many ecotoxicological models have been developed during the recent decades. Before around 1975, toxic substances were hardly associated with environmental modeling, as the problems seemed straightforward. The many pollution problems associated with toxic substances could easily be solved simply by eliminating the source of the toxic substance. During the 1970s, it was acknowledged that the environmental problems of toxic substances are very complex problems due to the interaction of many sources and many simultaneously, interacting processes and components. Several accidental releases of toxic substances into the environmental have reinforced the need for models. The result has been that several ecotoxicological models have been developed in the period from the since 1970s.

8.4 ESTIMATION OF ECOTOXICOLOGICAL PARAMETERS

Slightly more than 100,000 chemicals are produced in such an amount that they threaten or may threaten the environment. They cover a wide range of applications: household chemicals, detergents, cosmetics, medicines, dye stuffs, pesticides, intermediate chemicals,

TABLE 8.1 Examples of Toxic Substance Models

Toxic Substance Model Class	Model Characteristics	References
Cadmium (1)	Food chain similar to a eutrophication model	Thomann et al. (1974)
Mercury (1)	Six state variables: Water, sediment, suspended matter, invertebrates, plant, and fish	Miller (1979)
Vinyl chloride (3)	Chemical processes in water	Gillett et al. (1974)
Methyl parathion (1)	Chemical processes in water and benzothiophenemicrobial degradation, adsorption, 2–4 trophic levels	Lassiter (1978)
Methyl mercury (4)	A single trophic level: food intake, excretion metabolism growth	Fagerstrøm and Aasell (1973)
Heavy metals (3)	Concentration factor, excretion, bioaccumulation	Aoyama et al. (1978)
Pesticides in fish DDT and methoxychlor (5)	Ingestion, concentration factor, adsorption on body, defecation, excretion, chemical decomposition, natural mortality	Leung (1978)
Zinc in algae (3)	Concentration factor, secretion hydrodynamical distribution	Seip (1978)
Copper in sea (5)	Complex formation, adsorption sublethal effect of ionic copper	Orlob et al. (1980)
Radionuclides in sediment (3)	Photolysis, hydrolysis, oxidation, biolysis, volatilization, and resuspension	Onishi and Wise (1982)
Metals (2)	A thermodynamic equilibrium model	Felmy et al. (1984)
Sulfur deposition (3)	Box model to calculate deposition of sulfur	McMahon et al., (1976)
Radionuclides (3)	Distribution of radionuclides from a nuclear accident release	ApSimon et al. (1980)
Sulfur transport (3)	Long-range transmission of sulfur pollutants	Prahm and Christensen (1976)
Lead (5)	Hydrodynamics, precipitation, toxic effects of free ionic lead on algae, invertebrates, and fish	Lam and Simons (1976)
Radionuclides (3)	Hydrodynamics, decay, uptake, and release by various aquatic surfaces	Gromiec and Gloyna (1973)
Radionuclides (2)	Radionuclides in grass, grains, vegetables, milks, eggs, beef, and poultry are state variables	Kirschner and Whicker (1984)
SO_2, NO_x, and heavy metals (5)	Threshold model for accumulation effect of on spruce–fir pollutants. Air and soil in forests	Kohlmaier et al. (1984)
Toxic environmental chemicals (5)	Hazard ranking and assessment from physicochemical data and a limited number of laboratory tests	Bro-Rasmussen and Christiansen (1984)

(Continued)

TABLE 8.1   Examples of Toxic Substance Models—cont'd

Toxic Substance Model Class	Model Characteristics	References
Heavy metals (3)	Adsorption, chemical reactions, ion exchange	Several authors
Polycyclic aromatic hydrocarbons (3)	Transport, degradation, bioaccumulation	Bartell et al. (1984)
Persistent toxic organic substances (3)	Groundwater movement, transport, and accumulation of pollutants in groundwater	Uchrin (1984)
Cadmium, PCB (2)	Hydraulic overflow rate (settling), sediment interactions, steady state food chain submodel	Thomann (1984)
Mirex (3)	Water–sediment exchange processes, adsorption, volatilization, bioaccumulation	Halfon (1983, 1984)
Toxins (aromatic hydrocarbons, Cd) (3)	Hydrodynamics, deposition, resuspension, volatilization, photooxidation, decomposition, adsorption, complex formation (humic acid)	Harris et al. (1984)
Heavy metals (2)	Hydraulic submodel, adsorption	Nyholm et al. (1984)
Oil slicks (3)	Transport and spreading, influence of surface tension, gravity, and weathering processes	Nihoul (1984)
Acid rain (soil) (3)	Aerodynamic, deposition	Kauppi et al. (1986)
Persistent organic chemicals (5)	Fate, exposure, and human uptake	Mackay (1991)
Chemicals, general (5)	Fate, exposure, ecotoxicity for surface water and soil	Matthies et al. (1987)
Toxicants, general (4)	Effect on populations of toxicants	de Luna and Hallam (1987)
Chemical hazard (5)	Basinwide ecological fate	Morioka and Chikami (1986)
Pesticides (4)	Effects on insect populations	Schaalje et al. (1989)
Insecticides (2)	Resistance	Longstaff (1988)
Mirex and Lindane (1)	Fate in Lake Ontario	Halfon (1986)
Acid rain (5)	Effects on forest soils	Kauppi et al. (1986)
Acid rain (5)	Cation depletion of soil	Jørgensen et al. (1995)
pH, calcium and aluminum (4)	Survival of fish populations	Breck et al. (1988)
Photochemical smog (5)	Fate and risk	Wratt et al. (1992)
Nitrate (3)	Leaching to groundwater	Wuttke et al. (1991)
Oil spill (5)	Fate	Jørgensen et al. (1995)

(Continued)

TABLE 8.1   Examples of Toxic Substance Models—cont'd

Toxic Substance Model Class	Model Characteristics	References
Toxicants (4)	Effects on populations	Gard (1990)
Pesticides (3)	Loss rates	Jørgensen et al. (1995)
TCDD (3)	Photodegradation	Jørgensen et al. (1995)
Toxicants (4)	Effects general on populations	Gard (1990)
Pesticides and surfactants (3)	Fate in rice fields	Jørgensen et al. (1997)
Toxicants (3)	Migration of dissolved toxicants	Monte (1998)
Growth promoters (3)	Fate, agriculture	Jørgensen et al. (1998)
Toxicity (3)	Effect on eutrophication	Legovic (1997)
Pesticides (3)	Mineralization	Fomsgaard (1997)
Mecoprop (3)	Mineralization in soil	Fomsgaard and Kristensen (1999)
Pesticides (1)	Ecological assessment, pollution management	Jisng and Wsn (2009)
Cadmium (3)	Crop contamination	Chen et al. (2009)
Insect pheromone (3)	Dispersion within forest canopies	Strand et al. (2009)
PCB (1)	Risk assessment in Baiyangdian Lake, China	Zhang et al. (2013)
Pesticides (1)	Linking exposure and spatial dynamics	Liu et al. (2013)
PCB (1)	PCB in Baltimore Harbor	Shen et al. (2012)
Heavy metals (3)	Exposure by springtails	Meli et al. (2013)
Toxins (4)	Long-term effect on population size	Gledhill and Kirk (2011)
Copper (3)	Copper speciation	Richard et al. (2011)

auxiliary chemicals in other industries, additives to a wide range of products, chemicals for water treatment, and so on. They are viewed as mostly indispensable in modern society, resulting in increased production of chemicals about 40-fold during the last four decades. A minor or even major proportion of these chemicals reaches the environment through their production, transport, application, or disposal. In addition, the production or use of chemicals may cause more or less unforeseen waste or by-products, for instance chloro-compounds from the use of chlorine for disinfection. As we would like to have the benefits of using the chemicals but cannot accept the harm they may cause, this conflict raises several urgent questions which we already have discussed in this chapter. These questions cannot be answered without models, and we cannot develop models without knowing the most important parameters, at least within some ranges. OECD has made a review of the properties that we should know for all chemicals. We need to know the boiling point and melting point to know the chemical form (as solid, liquid, or gas) found in the environment. We must know the distribution of the chemicals in the five spheres: hydrosphere, atmosphere, lithosphere, biosphere, and technosphere. This will require knowing the solubility in water; the partition coefficient water/lipids; Henry's constant; the vapor pressure; the rate of degradation by hydrolysis, photolysis, chemical oxidation, and microbiological processes; and the adsorption equilibrium between water and soil—all as a function of the temperature. We need to discover the interactions between living organisms and the chemicals, which implies that we should know the biological concentration factor (BCF), the magnification through the food chain, the uptake rate, and the excretion rate by the organisms, and where in the organisms, the chemicals will be concentrated, not only for one organism but for a wide range of organisms. We must also know the effects on a wide range of different organisms. It means that we should be able to find the LC_{50} and LD_{50} values and the MAC (maximum allowable concentration) and NEC (noneffect concentrations) values as well as the relationship between the various possible sublethal effects and concentrations, the influence of the chemical on fecundity, and the carcinogenic and teratogenic properties. We should also know the effect on the ecosystem level. How do the chemicals affect populations and their development and interactions, i.e., the entire network of the ecosystem?

Table 8.2 gives an overview of the most relevant physical–chemical properties of organic compounds and their interpretation with respect to the behavior in the environment, which should be reflected in the model.

The mostly applied toxicological parameters are summarized with the definitions in Table 8.3.

TABLE 8.2 Overview of the Most Relevant Environmental Properties of Organic Compounds and Their Interpretation

Property	Interpretation
Water solubility	High water solubility corresponds to high mobility.
K_{ow}	High K_{ow} means that the compound is lipophilic. It implies that it has a high tendency to bioaccumulate and be sorbed to soil sludge and sediment. BCF and K_{oc} are correlated with K_{ow}.
Biodegradability	This is a measure of how fast the compound is decomposed to simpler molecules. A high biodegradation rate implies that the compound will not accumulate in the environment, while a low biodegradation rate may create environmental problems related to the increasing concentration in the environment and the possibilities of a synergistic effect with other compounds.
Volatilization, vapor pressure	High rate of volatilization (high vapor pressure) implies that the compound will cause an air pollution problem.
Henry's constant, K_H	K_H determines the distribution between the atmosphere and the hydrosphere.
pK	If the compound is an acid or a base, pH determines whether the acid or the corresponding base is present. As the two forms have different properties, pH becomes important for the properties of the compounds.

TABLE 8.3 Some Ecotoxicological Properties and Their Definitions

Parameter (Property)	Definition
LC_{50}	Lethal concentration. 50 indicates the % mortality, other mortality may be applied. Often also the duration of the experiment indicated 48 or 96 h are usually applied
LD_{50}	Lethal doses. 50 indicates the % mortality, other mortality may be applied. Often is also the duration of the experiment indicated 48 or 96 h are usually applied
MAC	Maximum allowable concentration
EC	Effect concentration. The effect is indicated, for instance no growth and also the % of organisms affected
NC	Narcotic concentration effect. The % of test organisms affected is indicated
HC	Hazardous concentration with indication of the % of test organisms affected
NEC	Noneffect concentration

Development of ecotoxicological models requires a wide knowledge of the properties of the focal chemical compounds (see Table 8.2, where the most important ones are listed), which again implies in the first hand an extensive literature search and/or selection of the best feasible estimation procedure. In addition to "Beilstein" it can be recommended to have at hand the following very useful handbooks of environmental properties of chemicals and methods for estimation of these properties in case literature values are not available:

Jørgensen et al. (1991). Handbook of Ecological Parameters and Ecotoxicology, Elsevier, 1991. Year 2000 published as a CD called Ecotox. It contains three times the amount of parameter in the 1991 book edition. See also Chapter 2 for further details about Ecotox.
P.H. Howard et al. (1991). Handbook of Environmental Degradation Rates. Lewis Publishers.
K. Verschueren, Several editions have been published, the latest in 2007. Handbook of Environmental Data on Organic Chemicals. Van Nostrand Reinhold.
D. Mackay, W.Y. Shiu and K.C. Ma. Illustrated Handbook of Physical-Chemical Properties and Environmental Fate for Organic Chemicals. Lewis Publishers.
Volume I. Mono-aromatic Hydrocarbons. Chloro-benzenes and PCBs. 1991.
Volume II. Polynuclear Aromatic Hydrocarbons, Polychlorinated Dioxins, and Dibenzofurans. 1992.
Volume III. Volatile Organic Chemicals. 1992.
Jørgensen et al. (1997a). Handbook of Estimation Methods in Environmental Chemistry and Ecotoxicology. Lewis Publishers.

These handbooks are still very useful, but today with the possibilities of using the internet, there is access to several important databases. Beilstein and Gmelin's Handbook are both available on internet and also PubChem should be mentioned. In addition, there are several encyclopedias that are able to provide information about ecotoxicological parameters. It is hardly possible to give an overview of all the possibilities on internet and today with the first search on internet, it is only recommendable to go to search and certainly you will find.

ERAs also require information about the chemical properties regarding their interactions with living organisms. It might not be necessary to know the properties with the very high accuracy that can be provided by literature as presented above or by measurements in a laboratory, but unfortunately only a small percentage of the parameters (properties) needed to develop models for about the 100,000 chemicals that we are using in modern society can be found in the literature. It would therefore be beneficial to try to estimate the properties that we cannot find in the literature with sufficient accuracy to make it possible to utilize the many applicable models for management and for risk assessments. Consequently, estimation methods have been developed as an urgently needed alternative to measurements. These are, to a great extent, based on the structure of the chemical compounds, the so-called QSAR and SAR methods, but it may also be possible to use allometric principles to transfer rates of interaction processes and concentration factors between a chemical and one or a few organisms to other organisms. This section focuses on these methods and attempts to give a brief overview on how these methods can be applied and what approximate accuracy they can offer. A more detailed overview of the methods can be found in Jørgensen et al. (1997a).

It may be interesting here to discuss the obvious question: why is it sufficient to estimate a property of a chemical in an ecotoxicological context with for instance 20%, or sometimes with 50% or even higher uncertainty? Ecotoxicological assessment usually gives an uncertainty of the same order of magnitude, which means that the indicated uncertainty may be sufficient from the modeling view point, but can results with such an uncertainty be used at all? The answer is often "yes" because in most cases we want to assure that we are (very) far from a harmful or very harmful level. We use often a safety factor of 10−1000 (most often 50−100). When we are concerned with very harmful effects, such as the complete collapse of an ecosystem or a health risk for a large human population, we will inevitably select a safety factor which is very high. In addition, our lack of knowledge about synergistic effects and the presence of many compounds in the environment at the same time forces us to apply a very high safety factor. In such a context, we will usually go for a concentration in the environment which is magnitudes lower than corresponding to a slightly harmful effect or considerably lower than the NEC. It is analogous to civil engineers constructing bridges. They make very sophisticated calculations (develop models) that account for wind, snow, temperature changes, and so on, and afterward they multiply the results by a safety factor of 2−3 to ensure that the bridge will not collapse. They use safety factors because the consequences of a bridge collapse are unacceptable.

The collapse of an ecosystem or a health risk to a large human population is also completely unacceptable. So, we should use safety factors in ecotoxicological modeling to account for the uncertainty. Due to the complexity of the system, the simultaneous presence of many compounds, and our present knowledge or rather lack of knowledge, we should as indicated above use 10−100 or even sometimes 1000 as safety factor. If we use safety factors that are too high, the risk is only that the environment will be less contaminated at maybe a higher cost. Besides, there are no alternatives to the use of safety factors. We can step-by-step increase our ecotoxicological knowledge, but it will take decades before it may be reflected in considerably lower safety factors. A measuring program of all processes and components is impossible due to the high complexity of the ecosystems. This does not imply that we should not use the information of measured properties available today. Measured data will almost always be more accurate than the estimated data. Furthermore, the use of measured data within the network of estimation methods will improve the accuracy of estimation methods. Several handbooks and internet databases on ecotoxicological parameters are fortunately available. References to the most important have already been given above. Estimation methods for the physical—chemical properties of chemical compounds were already applied 40—60 years ago, as they were urgently needed in chemical engineering. They are to a great extent based on contributions to a focal property by molecular groups and the molecular weight: the boiling point, the melting point, and the vapor pressure as function of the temperature are examples of properties that frequently were estimated in chemical engineering by these methods. In addition, a number of auxiliary properties results from these estimation methods, such as the critical data and the molecular volume. These properties may not have a direct application as ecotoxicological parameters in environmental risk assessment but are used as intermediate parameters which may be used as a basis for estimation of other parameters.

The water solubility, the partition coefficient octanol—water, K_{ow}, and Henry's constant are crucial parameters in our network of estimation methods because many other parameters (properties) are well correlated with these two parameters. The three properties can be found for a number of compounds or be estimated with reasonably high accuracy using knowledge of the chemical structure, i.e., the number of various elements, the number of rings, and the number of functional groups. In addition, there is a good relationship between water solubility and K_{ow}; see Fig. 8.7. Recently, many good estimation methods for these three core properties have been developed.

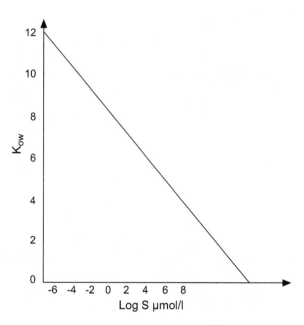

FIGURE 8.7　Relationship between water solubility (unit: μmol/L) and octanol—water distribution coefficient.

TABLE 8.4   Regression Equations for Estimation of the Biological Concentration Factor, BCF

Indicator	Relationship	Correlation Coefficient	Range (Indicator)
K_{ow}	$\log BCF = -0.973 + 0.767 \log K_{ow}$	0.76	$2.0 \times 10^{-2} - 2.0 \times 10^6$
K_{ow}	$\log BCF = 0.7504 + 1.1587 \log K_{ow}$	0.98	$7.0 - 1.6 \times 10^4$
K_{ow}	$\log BCF = 0.7285 + 0.6335 \log K_{ow}$	0.79	$1.6 - 1.4 \times 10^4$
K_{ow}	$\log BCF = 0.124 + 0.542 \log K_{ow}$	0.95	$4.4 - 4.2 \times 10^7$
K_{ow}	$\log BCF = -1.495 + 0.935 \log K_{ow}$	0.87	$1.6 - 3.7 \times 10^6$
K_{ow}	$\log BCF = -0.70 + 0.85 \log K_{ow}$	0.95	$1.0 - 1.0 \times 10^7$
K_{ow}	$\log BCF = 0.124 + 0.542 \log K_{ow}$	0.90	$1.0 - 5.0 \times 10^7$
S ($\mu g/l$)	$\log BCF = 3.9950 - 0.3891 \log S$	0.92	$1.2 - 3.7 \times 10^7$
S ($\mu g/l$)	$\log BCF = 4.4806 - 0.4732 \log S$	0.97	$1.3 - 4.0 \times 10^7$
S ($\mu mol/l$)	$\log BCF = 3.41 - 0.508 \log S$	0.96	$2.0 \times 10^{-2} - 5.0 \times 10^3$

During the last couple of decades, several correlation equations have been developed based upon a relationship between the water solubility, K_{ow} or Henry's constant on the one hand and physical, chemical, biological, and ecotoxicological parameters for chemical compounds on the other. The most important of these parameters are the adsorption isotherms soil-water; the rate of the chemical degradation processes: hydrolysis, photolysis, and chemical oxidation; the biological concentration factor (BCF); the ecological magnification factor (EMF, the magnification through the food chain); the uptake rate, excretion rate; and a number of other ecotoxicological parameters. Both the ratio of concentrations in the sorbed phase and in water at equilibrium, K_a, and BCF, defined as the ratio of the concentration in an organism and in the medium (water for aquatic organisms) at steady state presuming that both the medium and the food are contaminated, may often be estimated with a relatively good accuracy from expressions like K_a, K_{oc}, or $BCF = a \log K_{ow} + b$. K_{oc} is the ratio between the concentration in soil consisting of 100% organic carbon and in water at equilibrium between the two phases. Numerous expressions with different a and b values have been published (see Jørgensen et al., 1991, 1997, 2000; Jørgensen, 2000). Some of these relationships are shown in Table 8.4 and Fig. 8.8.

Biodegradation in waste treatment plants is often of particular interest, in which case the %BOD may be used. It is defined as the 5-day BOD as percentage of the theoretical BOD. It may also be indicated as the BOD_5 fraction. For instance, a BOD_5 fraction of 0.7 will mean that BOD_5 corresponds to 70% of the theoretical BOD. It is also possible to find an indication of BOD_5 percentage removal in an activated sludge plant.

Biodegradation is, in some cases, very dependent on the concentration of microorganisms. Therefore, it may be beneficial to indicate it as rate coefficient relative to the biomass of the active microorganisms in the units mg/(g dry wt 24 h).

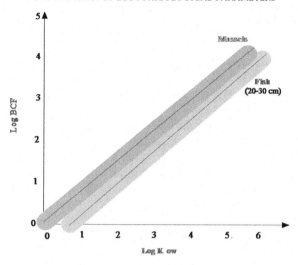

FIGURE 8.8 Two applicable relationships for octanol—water distribution coefficient and the biological concentration factor for fish and mussels.

In the microbiological decomposition of xenobiotic compounds, an acclimatization period from a few days to 1—2 months should be foreseen before the optimum biodegradation rate can be achieved. We distinguish between primary and ultimate biodegradation. Primary biodegradation is any biologically induced transformation that changes the molecular integrity. Ultimate biodegradation is the biologically mediated conversion of an organic compound to inorganic compound and products associated with complete and normal metabolic decomposition.

The biodegradation rate is expressed by a wide range of units:

1. As a first-order rate constant (1/24 h)
2. As half-life time (days or hours)
3. mg per g sludge per 24 h (mg/(g 24 h))
4. mg per g bacteria per 24 h (mg/(g 24 h))
5. mL of substrate per bacterial cell per 24 h (mL/(24 h cells))
6. mg COD per g biomass per 24 h (mg/(g 24 h))
7. mL of substrate per gram of volatile solids inclusive microorganisms (mL/(g 24 h))
8. BOD_x/BOD_8, i.e., the biological oxygen demand in x days compared with complete degradation (−), named the BOD_x coefficient.
9. BOD_x/COD, i.e., the biological oxygen demand in x days compared with complete degradation, expressed by means of COD (−)

The biodegradation rate in water or soil is difficult to estimate because the number of microorganisms varies several orders of magnitude from one type of aquatic ecosystem to the next and from one type of soil to the next.

Models enlisting artificial intelligence have been used as a promising tool to estimate this important parameter. However, a (very) rough, first estimation can be made on the basis of the molecular structure and the biodegradability. The following rules can be used to set up these estimations:

1. Polymer compounds are generally less biodegradable than monomer compounds. 1 point for a molecular weight >500 and =1000, 2 points for a molecular weight >1000.
2. Aliphatic compounds are more biodegradable than aromatic compounds. 1 point for each aromatic ring.
3. Substitutions, especially with halogens and nitro groups, will decrease the biodegradability. 0.5 points for each substitution, although 1 point if it is a halogen or a nitro group.
4. Introduction of double or triple bond will generally mean an increase in the biodegradability (double bonds in aromatic rings are of course not included in this rule). 1 point for each double or triple bond.
5. Oxygen and nitrogen bridges ($-O-$ and $-N-$ (or $=$)) in a molecule will decrease the biodegradability. 1 point for each oxygen or nitrogen bridge.
6. Branches (secondary or tertiary compounds) are generally less biodegradable than the corresponding primary compounds. 0.5 point for each branch.

Sum the total number of points and use the following classification:

=1.5 points: the compound is readily biodegraded. More than 90% will be biodegraded in a biological treatment plant.
2.0—3.0 points: the compound is biodegradable. Probably about 10—90% will be removed in a biological treatment plant. BOD_5 is 0.1—0.9 of the theoretical oxygen demand.
3.5—4.5 points: the compound is slowly biodegradable. Less than 10% will be removed in a biological treatment plant. $BOD_{10} = 0.1$ of the theoretical oxygen demand.
5.0—5.5 points: the compound is very slowly biodegradable. It will hardly be removed in a biological treatment plant and a 90% biodegradation in water or soil will take = 6 months.
=6.0 points: the compound is refractory. The half-life time in soil or water is counted in years.

Several useful methods for estimating biological properties are based upon the similarity of chemical structures. The idea is that if we know the properties of one compound, it may be used to find the properties of similar compounds. If, for instance, we know the properties of phenol, which is named the parent compound, then it may be used to give more accurate estimation of the properties of monochloro-phenol, dichloro-phenol, trichloro-phenol, and so on and for the corresponding cresol compounds. Estimation approaches based on chemical similarity give generally more accurate estimation but, of course, are also more cumbersome to apply, as they cannot be used generally in the sense that each estimation has a different starting point, namely the parent compound, with known properties.

Allometric estimation methods may also be used for development of ecotoxicological models. They presume (Peters, 1983) that there is a relationship between the value of a biological parameter and the size of the affected organism. The toxicological parameters LC_{50}, LD_{50}, MAC, EC, and NEC can be estimated from a wide spectrum of physical and chemical parameters, although these estimation equations generally are more inaccurate than the estimation methods for physical, chemical, and biological parameters. Both molecular connectivity and chemical similarity usually offer better accuracy for estimation of toxicological parameters.

The various estimation methods may be classified into two groups:

A. General estimation methods based on an equation of general validity for all types of compounds, although some of the constants may be dependent on the type of chemical compound or they may be calculated by adding contributions (increments) based on chemical groups and bonds.

B. Estimation methods valid for a specific class of chemical compounds for instance aromatic amines, phenols, aliphatic hydrocarbons, and so on. The property of at least one key compound is known. Based upon the structural differences between the key compounds and all other compounds of the considered type—for instance two chlorine atoms have substituted hydrogen in phenol to get 2,3-dichlorophenol—and the correlation between the structural differences and the differences in the considered property, the properties for all compounds of the considered class can be found. These methods are therefore based on chemical similarity.

Methods of Class B are generally more accurate than methods of Class A, but they are more cumbersome to use as it is necessary to find the right correlation for each chemical type. Furthermore, the requested properties should be known for at least one key component which sometimes may be difficult when a series of properties are needed. If estimation of the properties for a series of compounds belonging to the same chemical class is required, then it is tempting to use a suitable collection of class B methods.

Methods of Class A form a network which facilitates possibilities of linking the estimation methods together in a computer software system, like for instance estimation of ecotoxicological parameters (EEP) which contains many estimation methods. The relationship between the two properties is based on the average result obtained from a number of different equations found in the literature. There is, however, a price for using such "easy to go" software. The accuracy of the estimations is not as good as with the more sophisticated methods based upon similarity in chemical structure, but in many, particularly modeling, contexts the results found by EEP can offer sufficient accuracy. In addition, it is always useful to come up with a first intermediate guess. EEP is downloadable with the book by Jørgensen (2016).

The software also makes it possible to start the estimations from the properties of the chemical compound already known. The accuracy of the estimation from use of the software can be improved considerably by having knowledge about a few key parameters such as the boiling point and Henry's constant. As it is possible to get software which is able to estimate Henry's constant and K_{ow} with generally higher accuracy than EEP, a combination of separate estimations of these two parameters before using EEP can be recommended. Another possibility would be to estimate a couple of key properties using chemical similarity methods and then use these estimations as known values in EEP. These methods for improving the accuracy will be discussed in the next section. The network of EEP as an example of these estimation networks is illustrated in Fig. 8.9. As it is a network of Class A methods, it should not be expected that the accuracy of the estimations is as high as it is possible to obtain by the more specific Class B methods. By EEP it is, however, possible to estimate the most pertinent properties directly and relatively from the structural formula. The last version of EEP contains an estimation of the biodegradation based on a further development of the system presented above.

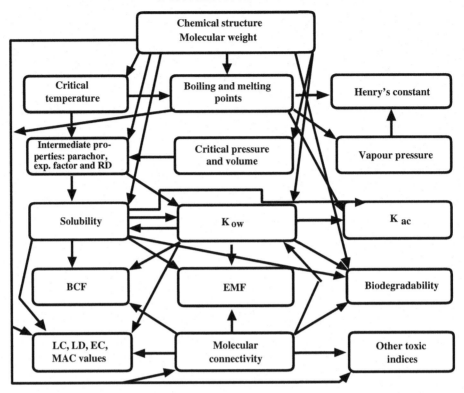

FIGURE 8.9 The network of estimation methods in EEP is shown. An arrow represents a relationship between two or more properties.

EEP is based on average values of results obtained by simultaneous use of several estimation methods for most of the parameters. It implies increased accuracy of the estimation, mainly because it gives a reasonable accuracy for a wider range of compounds. If several methods are used in parallel, then a simple average of the parallel results have been used in some cases, while a weighted average is used in other cases where it has been found beneficial for the overall accuracy of the program. While parallel estimation methods give the highest accuracy for different classes of compounds, use of weighting factors seems to offer a clear advantage. It is generally recommended to apply as many estimation methods as possible for a given case study to increase the overall accuracy of the average value. If the estimation by EEP can be supported by other recommended estimation methods, then it is strongly recommended to do so.

8.5 ECOTOXICOLOGICAL CASE STUDY I: CONTAMINATION OF AGRICULTURAL PRODUCTS BY CADMIUM AND LEAD

Agricultural products are contaminated by lead and cadmium originating from air pollution, the application of sludge from municipal wastewater plant as a soil conditioner, and from the use of fertilizers.

The uptake of heavy metals from municipal sludge by plants has previously been modeled (see Jørgensen, 1976). This model can briefly be described as follows: Depending on the soil composition it is possible to find for various heavy metal ions a distribution coefficient, i.e., the fraction of the heavy metal that is dissolved in the soil—water relative to the total amount. The distribution coefficient was found by examining the dissolved heavy metals relative to the total amount for several different types of soil. Correlation between pH, the concentration of humic substances, clay, and sand in the soil on the one hand, and the distribution coefficient on the other, was also determined. The uptake of heavy metals was considered a first-order reaction of the dissolved heavy metal.

This model does, however, not consider:

1. the direct uptake from atmospheric fallout onto the plants.
2. the other contamination sources such as fertilizers and the long-term release of heavy metal bound to the soil and the not harvested parts of the plants.

The objective of the model is to include these sources in a model for lead and cadmium contamination of plants. It is a fate model type A3 (see Section 8.1). Published data on lead and cadmium contamination in agriculture are used to calibrate and validate the model which is intended to be used for a more generally applicable risk assessment for the use of fertilizers and sludge that contains cadmium and lead as contaminants. The structure of the model is according to type 3; see Section 8.2.

The basis for the model is the lead and cadmium balance for average Danish agricultural land. Figs. 8.10 and 8.11 give the balances, modified from Andreasen (1985) and Knudsen and Kristensen (1987), to account for the changes of the mass balances year 2005. The atmospheric fallout of lead has gradually been reduced during the last 30 years due to reduction of the

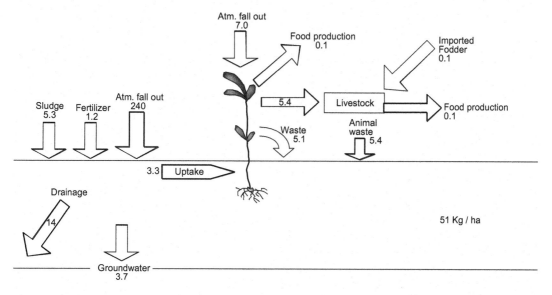

FIGURE 8.10   Lead balance of average Danish agriculture land. All rates are g Pb/ha year.

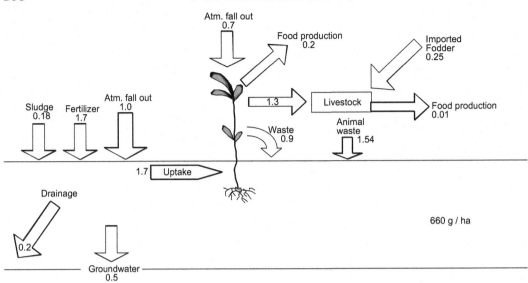

FIGURE 8.11 Cadmium balance of average Danish agriculture land. All rates are g Cd/ha y.

lead concentration in gasoline, while the most important source of cadmium contamination is fertilizer. The latter can only be reduced by using less contaminated sludge and phosphorus ore for the production of phosphorus fertilizer. The amounts of lead and cadmium coming from domestic animals and plant residues after harvest are not insignificant contributions.

8.5.1 The Model

Fig. 8.12 shows a conceptual diagram of the Cd model. STELLA software was used to construct a model with four state variables: Cd-bound, Cd-soil, Cd-detritus, and Cd-plant. An attempt was made to use one or two state variables for cadmium in the soil, but to get acceptable agreement between data and model output three state variables were needed. This can be explained by the presence of several soil components that bind the heavy metal differently; see Christensen (1981) and (1984), EPA, Denmark (1979), Hansen and Tjell (1981), Jensen and Tjell (1981), and Chubin and Street (1981). Cd-bound covers the cadmium bound to minerals and to more or less refractory material, Cd-soil covers the cadmium bound by adsorption and ion exchange, and Cd-detritus is the cadmium bound to organic material with a wide range of biodegradability.

The forcing functions are: airpoll, Cd-air, Cd-input, yield, and loss.

The atmospheric fallout is known, and the allocation of this source to the soil (airpoll) and to the plants (Cd-air) follows Hansen and Tjell (1981) and Jensen and Tjell (1981). Cd-input covers the heavy metal in the fertilizer, which comes as a pulse on day 1 and afterward with a frequency of every 180 days (Table 8.4). The yield corresponds to the part of the plants that is harvested, which is also expressed as a pulse function at day 180, and afterward with an occurrence every 360 days. Here, it is 40% of the plant biomass (Table 8.4).

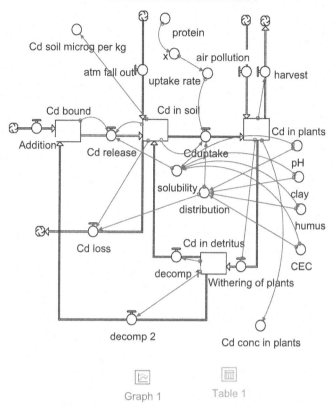

FIGURE 8.12   Conceptual diagram of the model as developed in STELLA software.

The loss covers transfer to the soil and groundwater below the root zone. It is expressed as a first-order reaction with a rate coefficient dependent on the distribution coefficient that is found from the soil composition and pH, according to the correlation found by Jørgensen (1976). Furthermore the rate constant is dependent on the hydraulic conductivity of the soil. Here in Table 8.5 the constant 0.01 reflects the dependence of the hydraulic conductivity.

The transfer from Cd-bound to Cd-soil indicates the slow release of cadmium due to a slow decomposition of the more or less refractory material to which cadmium is bound. The cadmium uptake by plants is expressed as a first-order reaction, where the rate is dependent on the distribution coefficient, as only dissolved cadmium can be taken up. It is furthermore dependent on the plant species (particularly their protein content). As will be seen, the uptake is a step function that is x (dependent on the protein content of the plants during the growing season) and zero after the harvest until the next growing season starts. Cd-waste covers the transfer of plant residues to detritus after harvest. It is a pulse function, which here is 60% of the plant biomass, as the remaining 40% has been harvested Table 8.5.

TABLE 8.5   Model equations

Cd-detritus = Cd-detritus + dt × (Cd-waste − mineralization − minquick)

INIT(Cd-detritus) = 0.27

Cd-plant = Cd-plant + dt × (Cduptake − yield − Cd-waste + Cd-air)

INIT(Cd-plant) = 0.0002

Cd-soil = Cd-soil + dt × (Cduptake − loss + transfer + minquick + airpoll)

INIT(Cd-soil) = 0.08

Cdtotal = Cdtotal + dt × (Cd-input − transfer + mineralization)

INIT(Cdtotal) = 0.19

Airpoll = 0.0000014

Cd-air = 0.0000028 + STEP(−0.0000028,180) + STEP(+0.0000028,360) + STEP(−0.0000028,540) + STEP(+0.0000028,720) + STEP(−0.0000028,900)

Cd-input = PULSE(0.0014,1,180)

Cduptake = distributioncoeff × Cd-soil × uptake rate

Cd-waste = PULSE(0.6 × Cd-plant,180,360) + PULSE(0.6 × Cd-plant,181,360)

CEC = 33

Clay = 34.4

Distributioncoeff = 0.0001 × (80.01 − 6.135 × pH-0.2603 × clay-0.5189 × humus-0.93 × CEC)

Humus = 2.1

Loss = 0.01 × Cd-soil × distributioncoeff

Mineralization = 0.012 × Cd-detritus

Minquick = IF TIME_180 THEN 0.01 × Cd-detritus ELSE 0.0001 × Cd-detritus

pH = 7.5

Plantvalue = 3000 × Cd-plant/14

Protein = 47

Solubility = 10^(+6.273 − 1.505 × pH + 0.00212 × humus + 0.002414 × CEC) × 112.4 × 350

Transfer = IF Cd-soil < solubility THEN 0.00001 × Cdtotal ELSE 0.000001 × Cdtotal

Uptake rate = x + STEP(−x,180) + STEP(x,360) + STEP(−x,540) + STEP(x,720) + STEP(−x,900)

x = 0.002157 × (−0.3771 + 0.04544 × protein)

Yield = PULSE(0.4 × Cd-plant,180,360) + PULSE(0.4 × Cd-plant,181,360)

Cd-detritus covers a wide range of biodegradable matter and mineralization is accounted for in the model by two mineralization processes: one for Cd-soil and one for Cd-total.

8.5.2 Model Results

Data from Jensen and Tjell (1981) and Hansen and Tjell (1981) were used for model calibration and validation. This phase of the modeling procedure revealed that three state variables for heavy metal in soil were needed to get acceptable results. It was particularly difficult to obtain the right values for heavy metal concentrations the second and third year after municipal sludge had been used as a soil conditioner. This use of models may be called experimental mathematics or modeling, where simulations with different models are used to deduce which model structure should be preferred. The results of experimental mathematics must be explained by examining the processes involved and here can be referred to the references given above.

The results of the validation demonstrate an acceptable agreement between observations and model prediction (Fig. 8.13), especially considering the low model complexity. Wider use of the model would require more data from experiments with many plant species to test the model applicability. It may be concluded from these results that the model structure must account for at least three state variables for the heavy metal in soil to cover the ability of different soil components to bind the heavy metal differently.

The problem modeled is very complex and many processes are involved. On the other hand, an ecotoxicological management model should be somewhat simple and not involve too many parameters. The model can obviously be improved, but it gives at least a first rough

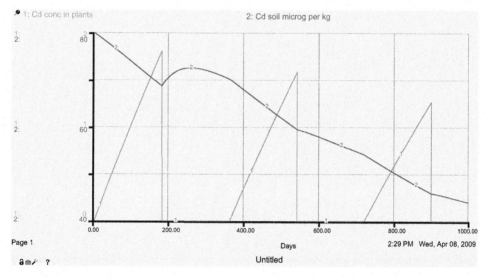

FIGURE 8.13   The graph shows the cadmium concentration in plants and soil in μg/kg dry matter. The harvest takes place at day 180, 540, and 900. The cadmium concentration according to observations was found at the three harvests to be, respectively, 1.7, 1.1, and 0.8 μg/kg dry matter. The cadmium in soil is reduced over the simulation period from about 80 μg/kg dry matter in soil to about 45 μg/kg dry matter in soil.

picture of the important factors in the contamination of agricultural crops. Mostly, it is not possible to get very accurate results with toxic substance models but, on the other hand, as we want to use somewhat large safety factors, the need for high accuracy is not pressing. In this light, the model results are acceptable.

8.6 ILLUSTRATIONS OF ECOTOXICOLOGICAL MODELS USED AS EXPERIMENTAL TOOLS

A model was developed with focus on the heavy metal concentration in fish living in an aquaculture fed with wastewater treated by waste stabilization ponds in Ghana (Azanu et al., 2016). Five heavy metals were considered: Cd, Cu, Pb, Cr, and Hg. The developed model using STELLA diagrams is shown in Fig. 8.14. The model is simple as it considers the growth of the fish corresponding to the feeding minus the loss by respiration. The feeding is expressed as a feeding rate times the weight of the fish in the exponent 2/3 and the respiration is proportional to the weight in the exponent 0.75. The model was calibrated varying the feeding rate and the proportion of feeding that are organisms in the water and organisms that are living in the sediment. The concentration of heavy metals in the fish is determined by uptake directly from the water through the gills and from the feeding.

The simple metal model was working acceptably well for Pb, Cu, and Cd but not working properly for chromium and mercury. Additional processes, including precipitation of chromium (transfer of chromium from the water to the sediment according to the solubility product of the only slightly soluble chromium(III) hydroxide) and biomagnification of methylmercury were introduced to explain concentration of chromium and mercury in fish. Methylmercury is known to be formed in the sediment and it is able to biomagnify.

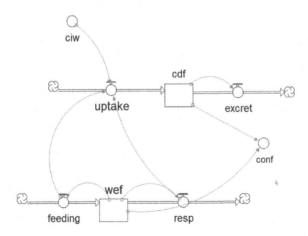

FIGURE 8.14 The simple and general heavy metal model shown by use of a STELLA conceptual diagram. The growth and fish and the concentration of heavy metals are followed two months. The feeding is expressed as a feeding rate times the weight in the exponent 2/3 and the growth is the feeding minus the respiration which is expressed as a respiration rate which is known for fish times the weight in the exponent 3/4. The uptake of heavy metals is determined by the direct uptake through the gills from the water and the uptake from the contaminated feed.

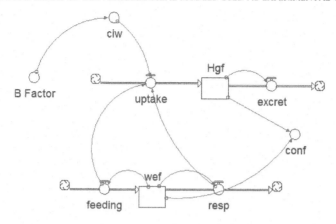

FIGURE 8.15   The model has as additional process that biomagnifications can take place for the methyl mercury dissolved in the water. The validation shows that the general model with the extra processes for Cr and Hg gives acceptable results.

The model was applied in this context as an experimental tool by examination of which causal process expressions should be added to obtain an acceptable calibration. The applied mercury model is shown Fig. 8.15. The general models for Pb, Cu, and Cd and for Cr with transfer according to the solubility product and for Hg with an extra process (biomagnifications) were validated and gave an R^2 value of 0.9 indicating a good agreement between the model predictions and the experimental measurements. The finding suggests that the simple metal model has an acceptable accuracy and is useful for predicting uptake and chemical processes of the examined heavy metal, provided it is modified as described for Cr and Hg. Notice that the modifications are in accordance with causal processes that we know take place and that the modifications were found by use of the model as an experimental tool.

Antibiotics in wastewater is a growing problem in urban areas in developing countries as a result of increased use and misuse of antibiotics. A simple dynamic model that describes the most important removal processes of antibiotic from a wastewater stabilization pond system (WSP) in Morogoro, Tanzania, was developed; see C. Christmas Møller et al. (2016). Concentrations of trimethoprim were measured in the dry season and the rainy season for development of the model. To determine the model's applicability to simulate the removal of trimethoprim a calibration was performed using concentrations from the dry season and a validation was performed using concentrations from the rainy season. Both calibration and validation gave acceptable results as the standard deviation between modeled and measured values were 18% and 1%, respectively.

The model was developed under the assumption that settling, biodegradation, hydrolysis, and photolysis were the only removal processes other than outflow; see the conceptual diagrams Figs. 8.16 and 8.17. Biodegradation, hydrolysis, and photolysis were described by a first-order reaction and the rate coefficients were respectively measured and found in the literature for all the examined antibiotics. The removal by settling was expressed by the following equation:

$$Settling_x = \left(Koc_x \times Csusp_x \times 10^{-3} \times Ksett_x\right) \times (Pond_x) \qquad (8.6)$$

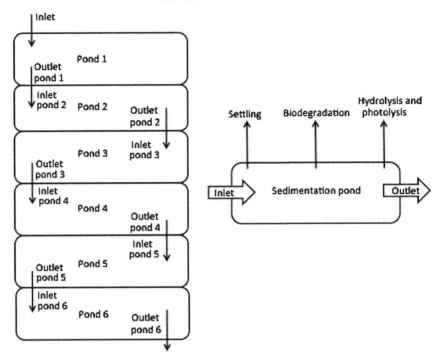

FIGURE 8.16   A conceptual diagram of the removal of antibiotic from the WSP. Left: A schematic overview of the water flow through the WSP. Right: A schematic overview of a single sedimentation pond showing the removal processes.

Where Koc_x is the soil adsorption coefficient, $Csusp_x$ is the concentration of suspended matter and $Ksett_x$ is the settling constant (1/24 h) and $(Pond_x)$ represents the concentration of antibiotics at time "t" in a specific tank (see Fig. 8.16). This equation expresses that the antibiotics that is adsorped by the suspended matter is removed by settling and the Koc_x is the ratio between the concentration on the suspended matter and in the water. This ratio is usually denoted Ka and can be found from what is denoted K_{oc} by the following equation $K_a = K_{oc} \times fc$, where K_{oc} is the ratio provided that the suspended matter has a concentration of organic C = 100% and fc is the fraction of organic carbon in the suspended matter. In this case, the concentration of suspended matter and the settling rate are calibrated and fc could of course be included in this calibration, which implies that we actually find which number we have to use to multiply K_{oc} to be able to find (predict) the best value of the concentration in the water as function of time and the ponds.

It would be beneficial when we are using the model for other antibiotics because we can in most cases find K_{oc} for other organic chemicals in the literature (or we could determine the value by measurements). Notice that the model is calibrated to cover a specific WSP with respect to retention time in the ponds, the concentration of suspended matter, and the settling rate of this suspended matter, while the model results in addition is dependent on the properties of the chemicals (antibiotics) that contaminates the water in the WSP. It is very important

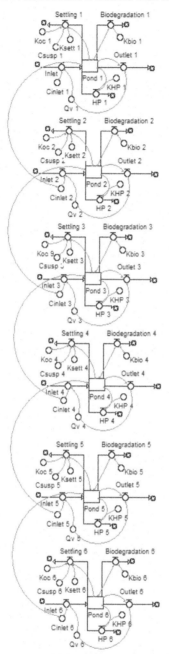

FIGURE 8.17   A STELLA diagram of the model. HP is the process of removal by hydrolysis and photolysis, K_{oc} is the soil adsorption coefficient, K_{HP} is the first-order rate constant of hydrolysis and photolysis measured at 30°C, K_{bio} is the biodegradation constant expressed as a first order rate constant, C_{susp} is the amount of suspended matter, K_{sett} is the settling constant, and C_{inlet} is the concentration of the compound at the inlet of the pond. Ponds 1–6 are state variables. Settling, biodegradation, HP, inlet, and outlet are processes. Qv is the ratio of flow rate and volume of water per day.

because it makes it possible to use the model once erected for a specific WSP for many relevant organic chemicals by use of the properties of a specific chemical. We will therefore use the model as an experimental tool to see if it is possible.

To test the model's capacity to simulate the removal of other antibiotics than trimethoprim, a second validation was performed for three other antibiotics: metronidazole, sulfamethoxazole, and ciprofloxacin. A two-tailed t-test with a confidence interval of 95% showed no significant difference ($P = 0.7819$) between the values given by the model (CSIM) and the values measured, and the standard deviation (SD) between two sets of values was 1%. The major removal processes for sulfamethoxazole were through settling and the outlet. Ciprofloxacin was mainly removed by settling in the first pond, while metronidazole was mainly removed through the outlet, but settling and hydrolysis/photolysis also played a role. Trimethoprim was removed through settling and the outlet. The model experiment was successful in the sense that it was easy when the model was calibrated and validated for one antibiotics to use for other organic chemicals and it was possible to assess the importance of the possible removal processes.

References

Andreasen, I., 1985. A General Ecotoxicological Model for the Transport of Lead Through the System: Air-Soil(Water)-Grass-Cow-Milk (Thesis at DIA-K). Technical University of Denmark, 57 pp.

Aoyama, I., Yos. Inoue, Yor. Inoue, 1978. Simulation analysis of the concentration process of trace heavy metals by aquatic organisms from the viewpoint of nutrition ecology. Water Research 12, 837–842.

ApSimon, H., Goddard, A.J.H., Wrigley, J., 1976. Estimating the possible transfrontier consequences of accidental releases: the MESOS model for long range atmospheric dispersal. In: Seminar on Radioactive Releases and Their Dispersion in the Atmosphere Following a Hypothetical Reactor Accident, Rise, Denmark, April 1980. CEC, Luxembourg, pp. 819–842.

Azanu, D., Jørgensen, S.E., Darko, G.M., Styrishave, B., 2016. Simple metal model for predicting uptake and chemical processes in sewage-fed aquaculture ecosystem. Ecological Modelling 319, 130–136.

Bartell, S.M., Gardner, R.H., O'Neill, R.V., 1984. The fates of aromatics model. Ecological Modelling 22, 109–123.

Breck, J.E., DeAngelis, D.L., Van Winkle, W., Christensen, S.W., 1988. Potential importance of spatial and temporal heterogeneity in pH, Al and Ca in allowing survival of a fish population: a model demonstration. Ecological Modelling 41, 1.

Bro-Rasmussen, F., Christensen, K., 1984. Hazard assessment Ð a summary of analysis and integrated evaluation of exposure and potential effects from toxic environmental chemicals. Ecological Modelling 22, 67–85.

Chen, W., et al., 2009. Assessing the effect of long-term crop cultivation on distribution of Cd in the root zone. Ecological Modelling 220, 1836–1843.

Christensen, T.H., 1981. The Application of Sludge as Soil Conditioner, vol. 3. Polyteknisk Forlag, Copenhagen, pp. 19–47.

Christensen, T.H., 1984. Cadmium soil sorption at low concentrations, 1) Effect of time, cadmium load, pH and calcium and 2) Reversibility, effect of changes in solute composition, and effect of soil ageing. Water, Air and Soil Pollution 21, 105–125.

Christmas-Møller, C., et al., 2016. Modelling antibiotics transport in a waste stabilization pond system in Tanzania. Ecological Modelling 319, 137–146.

Chubin, R.G., Street, J.J., 1981. Adsorption of cadmium on soil constituents in the presence of complexing agents. Journal of Environmental Quality 10, 225–228.

De Luna, J.T., Hallam, T.G., 1987. Effect of toxicants on populations: a qualitative approach IV. Resource-consumer-toxicant models. Ecological Modelling 35, 249.

EPA, Denmark, 1979. The Lead Contamination in Denmark. Copenhagen. 145 pp.

Fagerstrøm, T., Aasell, B., 1973. Methyl mercury accumulation in an aquatic food chain. A model and implications for research planning. Ambio 2, 164–171.

Felmy, A.R., Brown, S.M., Onoshi, Y., Yabusaki, S.B., Argo, R.S., Girvin, D.C., Jenne, E.A., 1984. Modelling the Transport, Speciation, and Fate of Heavy Metals. Aquatic Systems, Project Summary, EPA-600/S3-84-033, April 1984. US EPA, Environmental Research Laboratory, Athens, Georgia, 4 pp. (EPA Project Officer: R.B. Ambrose).

Fomsgaard, I., 1997. Modelling the mineralisation kinetics for low concentrations of pesticides in surface and sub-surface soil. Ecological Modelling 102, 175—208.

Fomsgaard, I., Kristensen, K., 1999. Influence of microbial activity, organic carbon content, soil texture and soil depth on mineralisation rates of low concentrations of 14-C mecoprop — development of a predictive model. Ecological Modelling 122, 45—68.

Gard, T.C., 1990. A stochastic model for the effect of toxicants on populations. Ecological Modelling 51, 273—280.

Gillett, J.W., et al., 1974. A Conceptual Model for the Movement of Pesticides Through the, Environment. National Environmental Research Center, U.S. Environmental Protection Agency, Corvallis, OR, p. 79. Report EPA 600/3-74-024.

Gledhill, M., Van Kirk, R.W., 2011. Modeling the effect of toxin exposure in fish on long-term population size. Ecological Modelling 222, 3587—3597.

Gromiec, M.J., Gloyna, E.F., 1973. Radioactivity Transport in Water. Final Report No. 22 to U.S. Atomic Energy Commission, Contract AT (11—1)-490.

Harris, J.R.W., Bale, A.J., Bayne, B.L., Mantoura, R.C.F., Morris, A.W., Nelson, L.A., et al., 1984. A preliminary model of the dispersal and biological effect of toxins in the Tamar estuary, England. Ecological Modelling 22, 253—285.

Halfon, E., 1983. Is there a best model structure? Comparing the model structures of different fate models. Ecological Modelling 20, 153—163.

Halfon, E., 1984. Error analysis and simulation of Mirex behavior in Lake Ontario. Ecological Modelling 22, 213—253.

Halfon, E., 1986. Modelling the fate of Mirex and Lindane in Lake Ontario, off the Niagara River Mouth. Ecological Modelling 33, 13.

Hansen, J.Aa, Tjell, J.C., 1981. The Application of Sludge as Soil Conditioner, vol. 2. Polyteknisk Forlag, Copenhagen, pp. 137—181.

Howard, P.H., et al., 1991. Handbook of Environmental Degradation Rates. Lewis Publishers, New York.

Jensen, K., Tjell, J.C., 1981. The Application of Sludge as Soil Conditioner, vol. 3. Polyteknisk Forlag, Copenhagen, pp. 121—147.

Jiang, J., Wan, N., 2009. A model for ecological assessment to pesticide pollution management. Ecological Modelling 220, 1844—1851.

Jørgensen, S.E., 1976. An ecological model for heavy metal contamination of crops and ground water. Ecological Modelling 2, 59—67.

Jørgensen, S.E., 2016. Application of Environmental Chemistry and Ecotoxicology in Environmental Management. Book in print. CRC, Boca Raton, 380 pp.

Jørgensen, S.E., Nors Nielsen, S., Jørgensen, L.A., 1991. Handbook of Ecological Parameters and Ecotoxicology. Elsevier, Amsterdam. Published as CD under the name ECOTOX, with L.A. Jørgensen as first editor in year 2000.

Jørgensen, S.E., Halling-Sørensen, B., Nielsen, S.N., 1995. Handbook of Environmental and Ecological Modeling. CRC Lewis Publishers, Boca Raton, New York, London, Tokyo, 672 pp.

Jørgensen, S.E., Halling-Sørensen, B., Mahler, H., 1997a. Handbook of Estimation Methods in Ecotoxicology and Environmental Chemistry. Lewis Publishers, Boca Raton, Boston, London, New York, Washington, D.C., 230 pp.

Jørgensen, S.E., Marques, J.C., Anastácio, P.M., 1997. Modelling the fate of surfactants and pesticides in a rice field. Ecological Modelling 104 (2—3), 205—213.

Jørgensen, S.E., Lutzhøft, Halling Sørensen, B., 1998. Development of a model for environmental risk assessment of growth promoters. Ecological Modelling 107, 63—72.

Jørgensen, S.E., 2000. Pollution Abatement. Elsevier, Amsterdam, p. 488.

Jørgensen, S.E., Patten, B.C., Straskraba, M., 2000. Ecosystem emerging: 4. Growth. Ecological Modelling 126, 249—284.

Jørgensen, S.E., Fath, B., 2011. Fundamentals of Ecological Modelling, fourth ed. Elsevier, Amsterdam. 400 pp.

Kauppi, P., KšmŠri, J., Posch, M., Kauppi, L., Matzner, E., 1986. Acidification of forest soils: model development and application for analysing impacts of acidic deposition in Europe. Ecological Modelling 33, 231—253.

Kirchner, T.B., Whicker, F.W., 1984. Validation of PATHWAY, a simulation model of the Transport of radionuclides through agroecosystems. Ecological Modelling 22, 21—45.

Knudsen, G., Kristensen, L., 1987. Development of a Model for Cadmium Uptake by Plants (Master thesis). University of Copenhagen.

Kohlmaier, G.H., Sire, E.O., Brohl, H., Kilian, W., Fishbach, U., Plochl, M., Muller, T., Ynsheng, J., 1984. Dramatic development in the dying of German spruce-fir forests: in search of possible cause effect relationships. Ecological Modelling 22, 45–65.

Lam, D.C.L., Simons, T.J., 1976. Computer model for toxicant spills in Lake Ontario. In: Nriago, J.O. (Ed.), Metals Transfer and Ecological Mass Balances, Environmental Biochemistry, vol. 2. Ann Arbor Science, pp. 537–549.

Lassiter, R.R., 1978. Principles and constraints for predicting exposure to environmental pollutants. U.S. Environmental Protection Agency, Corvallis. OR Report EPA 118–127519.

Legovic, T., 1997. Toxicity may affect predictability of eutrophication models in coastal sea. Ecological Modelling 99, 1–6.

Leung, D.K., 1978. Modelling the Bioaccumulation of Pesticides in Fish. Center for Ecological Modelling, Polytechnic Institute, Troy, NY. Report 5.

Liu, C., et al., 2013. Linking pesticide exposure and spatial dynamics. Ecological Modelling 248, 92–102.

Longstaff, B.C., 1988. Temperature manipulation and the management of insecticide resistance in stored grain pests. A simulation study for the rice weevil, Sitophilus oryzae. Ecological Modelling 43, 303–313.

Mackay, D., 1991. Multimedia Environmental Models. The Fugacity Approach, 257 pp.. Lewis Publishers, Boca Raton, Ann Arbor, London and Tokyo.

Matthies, M., Behrendt, H., Mynzer, B., 1987. EXSOL Modell fÿr den Transport und Verbleib von Stoffen im Boden. GSF-Bericht 23/87 Neuherberg.

McMahon, T.A., Denison, D.J., Fleming, R., 1976. A long distance transportation model incorporating washout and dry deposition components. Atmospheric Environment 10, 751–760.

Meli, M., et al., 2013. Population-levels consequences of spatial heterogeneous exposure to heavy metals in soil. Ecological Modelling 250, 338–351.

Miller, D.R., 1979. Models for total transpon. In: Butler, G.C. (Ed.), Principles of Ecotoxicology Scope, vol. 12. Willey, New York, pp. 71–90.

Monte, L., 1998. Prediction the migration of dissolved toxic substances from catchments by a collective model. Ecological Modelling 110, 269–280.

Morgan, M.G., 1984. Uncertainty and quantitative assessment in risk management. In: Rodricks, J.V., Tardiff, R.G. (Eds.), Assessment and Management of Chemical Risks. Chapter 8. ACS Symposium Series 239. American Chemical Society, Washington, D.C.

Morioka, T., Chikami, S., 1986. Basin-wide ecological fate model for management of chemical hazard. Ecological Modelling 31, 267.

Nihoul, J.C.J., 1984. A non-linear mathematical model for the transport and spreading of oil slicks. Ecological Modelling 22, 325–341.

Nyholm, N., Nielsen, T.K., Pedersen, K., 1984. Modelling heavy metals transpon in an arctic fjord system polluted from mine tailings. Ecological Modelling 22, 285–324.

Onishi, Y., Wise, S.E., 1982. Mathematical Model, SERA TRA, for Sediment-Contaminant Transpon in Rivers and Its Application to Pesticide Transpon in Four Mile and Wolf Creeks in Iowa. EPA-60013-82-045, Athens, Georgia, 56 pp.

Orlob, G.T., Hrovat, D., Harrison, F., 1980. Mathematical model for simulation of the fate of copper in a marine environment. American Chemical Society, Advances in Chemistry Series 189, 195–212.

Peters, R.H., 1983. The Ecological Implications of Body Size. Cambridge University Press, Cambridge, 329 pp.

Prahm, L.P., Christensen, O.J., 1976. Long-Range Transmission of Sulphur Pollutants Computed by the Pseudospectral Model. Danish Meteorological Institute, Air Pollution Section, Lyngbyvej, DK-2100 Copenhagen. Prepared for the ECE Task Force for the Preparation of a Co-operative Programme for the Monitoring and Evaluation of the Long-range Transmission of Air Pollutants in Europe, October 1976, Lillestrom, Norway.

Richards, R., et al., 2011. Modelling the effects of Coastal acidification on copper speciation. Ecological Modelling 222, 3559–3567.

Schaalje, G.B., Stinner, R.L., Johnson, D.L., 1989. Modelling insect populations affected by pesticides with application to pesticide efficacy trials. Ecological Modelling 47, 223.

Seip, K.L., 1978. Mathematical model for uptake of heavy metals in benthic algae. Ecological Modelling 6, 183–198.

SETAC, 1995. The Multi-Media Fate Model: A Vital Tool for Predicting the Fate of Chemicals.

Thomann, R.V., et al., 1974. A food chain model of cadmium in western Lake Erie. Water Research 8, 841–851.

Thomann, R.V., 1984. Physico-chemical and ecological modelling the fate of toxic substances in natural water systems. Ecological Modelling 22, 145–170.

Shen, J., et al., 2012. Modeling the PCBs in Baltimore Harbor. Ecological Modelling 242, 54–68.

Slovic, P., Fischhoff, B., Lichtenstein, S., 1982. Facts and fears: understanding perceived risk. In: Schwing, R.C., Albers Jr., W.A. (Eds.), Societal Risk Assessment: How Safe Is Safe Enough? Plenum Press, New York.

Strand, T., et al., 2009. A simple model for simulation of insect pheromone dispersion within forest canopies. Ecological Modelling 220, 640–656.

Uchrin, C.G., 1984. Modelling transport processes and differential accumulation of persistent toxic organic substances in groundwater systems. Ecological Modelling 22, 135–144.

Wratt, D.S., Hadfield, M.G., Jones, M.T., Johnson, G.M., McBurney, I., 1992. Power stations, oxides of nitrogen emissions, and photochemical smog: a modelling approach to guide decision makers. Ecological Modelling 64, 185–204.

Wuttke, G., Thober, B., Lieth, H., 1991. Simulation of nitrate transpon in groundwater with a three-dimensional groundwater model run as a subroutine in an agroecosystem model. Ecological Modelling 57, 263–276.

Zhang, L., et al., 2013. Applying AQUATOX in determining the ecological risk assessment of PCBs in Baiyangdian Lake, North China. Ecological Modelling 265, 239–249.

9

Fugacity Models

X.-Z. Kong, F.-L. Xu[1], W. He, W.-X. Liu, B. Yang

Peking University, Beijing, China

[1]Corresponding author: E-mail: xufl@urban.pku.edu.cn

9.1 FUGACITY MODELS: DEVELOPMENT AND APPLICATIONS

9.1.1 Introduction of Fugacity Models

Multimedia models were mathematical models for description of chemical behavior that was being rapidly developed since the 1980s (Cohen, 1984). These models were built based on the principles that the fate, transport, and transformation of the chemicals were determined by both environmental factors and physical−chemical properties. The most important processes for transport and transformation within and between compartments are incorporated. As a result, multimedia models for chemicals were applied abundantly in both ecological and environmental researches, such as biogeochemical studies of pollutants, ecological risk assessment, and environmental management. Multiple types of multimedia models were proposed. Representatives of these models are TOXIC model (Schnoor and Mcavoy, 1981), EXAMS model (Burns et al., 1982; Yoshida et al., 1987), and PRZM model (Carsel et al., 1984). Among all these multimedia models, fugacity models are one of the most successful types due to its simplicity in structure and accessible in parameterization (Mackay, 2001).

9.1.2 Development and Applications of Fugacity Models

Fugacity evaluates the equilibrium distribution and tendency of chemicals to escape between phases (Mackay and Paterson, 1981). Fugacity was firstly introduced into chemical modeling in the beginning of the 1980s (Mackay, 1979; Mackay and Paterson, 1981, 1982b). There are four levels in fugacity model in total (Levels I−IV) with increasing complexity (Mackay and Paterson, 1982b). Level I model is for equilibrium, stable, and nonflowing system; Level II is equilibrium, stable, and flowing system; Level III is nonequilibrium, stable, and flowing system; and Level IV is nonequilibrium, nonstable, and flowing system.

Much works have been done during the development of fugacity models, particularly by the group of Donald Mackay. A Level III model for aquatic ecosystem, including lakes and rivers, was developed and applied to predict the fate and residual levels of organic pollutants in air, water, and sediment (Mackay et al., 1983a,b). The model was subsequently utilized for both organic and inorganic compounds in lakes (Mackay and Diamond, 1989). Moreover, an explicitly model description of air−water interactions was developed (Mackay et al., 1986). As the processes in fugacity model became more precise, the model was validated by more case studies. For example, fate of six chemicals in four environmental compartments (air, water, soil, and sediment) was modeled simultaneously by a Level III fugacity model (Mackay and Paterson, 1991), and the model was subsequently applied in a much larger regional scale (Mackay et al., 1992).

In addition, fugacity models were widely used and continuously modified by researchers around the world, which also played an important role in the model development. Earlier works focused on incorporating major components into fugacity models, most of which were Level III steady state models. These components include biota in food chain and food web in aquatic ecosystem (Connolly and Pedersen, 1988). The bioconcentration, bioaccumulation, and biomagnification effects of organic pollutants were nicely modeled. Moreover, efforts were taken in extending the scales of fugacity models, such as application to a large group of chemicals (45 priority contaminants from USEPA) with different physical—chemical properties (Edwards et al., 1999), which led to a clearer linkage between chemical properties and their environmental behavior. Deep insights were obtained by considering the effects of spatial heterogeneity in Level III fugacity model ranged from regional (Tao et al., 2003) to global (Ballschmiter, 1992; Wania and Mackay, 1995) scales. It was suggested that the incorporation of spatial variations would decrease the model uncertainty (Tao et al., 2003). In general, applications of Level III models were abundant, focusing on either terrestrial (Parajulee and Wania, 2014; Wang et al., 2002; Wania et al., 2006) or aquatic (Baek and Park, 2000; Wang et al., 2012; Xu et al., 2013) study sites.

Following studies attempted to use Level IV model, the algorithm of which was primarily investigated as the basis for further research (Paraiba et al., 2002, 1999). Level IV fugacity models were applied in both short-term simulation with seasonal variations and long-term simulation with changes in emission intensity. For the short-term, seasonal variations of contaminants were successfully predicted by Level IV fugacity models with a relative low model uncertainty, particularly for polycyclic aromatic hydrocarbon (PAHs) in terrestrial systems (Lang et al., 2007; Wang et al., 2011) and hexachlorocyclohexanes (HCHs) in aquatic systems (Kong et al., 2014, 2012). For the long-term, impact of the changes in organochlorine pesticides (OCPs) application intensity on their environmental residual levels since the 1960s in China have been revealed by fugacity models in regional scales (Ao et al., 2009; Cao et al., 2007; Dong et al., 2009; Kong et al., 2014; Liu et al., 2007; Tao et al., 2006). These applications of fugacity models, which showed a significant model response to the changes in emission rate and a relative good fit of model predictions to measured data, fall within the objectives of Level IV fugacity model (Mackay and Paterson, 1982a). Serving as critical implications, fugacity models are able to predict the residual levels of certain chemicals in the coming decades, such that the concentrations of γ-HCH was predicted to decrease by 1.7—1.9 orders of magnitude in 2020 in Tianjin, China (Tao et al., 2006). Furthermore, it is interesting that spatial and temporal variations are addressed simultaneously in one fugacity model study (Wang et al., 2011). Factors influencing distributions of POPs in a global scale was also illustrated by fugacity models (Wania and Mackay, 1995). This study showed that the temperature was a significant factor, whereas parameters strongly affected by temperature were still poorly modeled. It is also interesting to note that fugacity models were integrated with atmospheric transportation model to account for global distribution and health risk from inhalation exposure of PAHs (Zhang et al., 2009).

Recent studies for fugacity models took great efforts on incorporating submodels for biota phases from different trophic levels, which would lead to a more simple structure of food web accumulation model. These studies were of great importance, because higher chemical concentrations in higher trophic levels may lead to catastrophic consequences (Sharpe and Mackay, 2000), such as the extinction of bird species due to polychlorinated biphenyls (PCBs) biomagnification (Gilbertson, 1996) and increasing health risks of human beings due to dioxin exposure from fish consumption (Thomas et al., 1998). The development of fugacity-based food web accumulation model was, on one hand, based on the quantitative evaluation of bioaccumulation model from earlier researches, which had demonstrated the potential bioaccumulation and biomagnification effects of certain chemicals. For example, relationship between bioaccumulation factor (K_B) and octanol/water partition coefficients (K_{ow}) was built ($K_B = 0.048K_{ow}$), which served as an essential basis for further model development (Mackay, 1982). On the other hand, it was unraveled that the fugacity ratio between biota and water could be higher than 1, indicating that the fugacity gradient along the trophic level is the key factor for potential bioaccumulation and biomagnification effects (Connolly and Pedersen, 1988). These findings are the theoretical foundation for food web fugacity model.

Fugacity model for food web started from fish (Mackay and Hughes, 1984), and a more detailed model for digestive system in fish was developed, by which digestion and food adsorption in gastrointestinal were both identified as the major mechanisms for biomagnification in food chain (Gobas et al., 1999; Gobas et al., 1993). Uncertainty was also addressed for fish fugacity model to increase accuracy of model prediction (Hauck et al., 2011). As a further step, fugacity models were extended to the whole food chain and food web, resulting in a more systematic description on transport and transformation of chemicals in food webs. The ultimate goal of the model is to predict chemical concentrations in biota from data in abiotic phases and to facilitate the ecological risk assessment of these chemicals. These approaches focused particularly on aquatic ecosystems, e.g., a food chain with six trophic levels, including zoobenthos, plankton, and fish (Diamond et al., 1996). A general framework for fugacity-based food web accumulation model was also developed, which could incorporate N groups in the model with a matrix of N dimensions (Campfens and Mackay, 1997). A case study on PCBs in the Great Lakes showed a good fit of model predictions to field data (deviations within three folds). The fugacity-based food web accumulation model was under continuous modifications and applications, with updated mechanisms embedded (Arnot and Gobas, 2004; Binelli and Provini, 2003; Gobas and Arnot, 2010). In addition, excretion of birds was found to play an important role in driving the fate of chemicals in aquatic ecosystems. Therefore, birds were included in the fugacity models and the framework of evaluation for chemicals in aquatic environment (Sharpe and Mackay, 2000). An interesting study in two high-latitude lakes in Norway showed that rather than food web structure and lake morphology, intensive bird's dropping in one lake was the major reason for the 10-fold higher PCBs concentrations in fish than those in the other lake. Overall, as a synthesis, a three-layer model framework was proposed for a rapid risk assessment of commercial chemicals, which considered the most important components and processes from earlier works via an analytic hierarchy process (Mackay and Fraser, 2000).

Terrestrial and aquatic vegetation were both involved into fugacity models much later than the other components, yet their important roles in driving the fate of chemicals were gradually realized. Vegetation is difficult to model because it has different parts with different functions. For terrestrial vegetation, fugacity-based models were development, as the vegetation was either considered as a whole (Calamari et al., 1987) or divided into leaf, stem, and root (Paterson et al., 1994). Among these modeling approaches, the one for agricultural herbaceous plant was crucial for health risk assessment as they are directly consumed by human beings (Hung and Mackay, 1997). An integrated study reviewed the model and experimental methods for estimating bioconcentration factors (BCFs) for vegetation and showed that uncertainties from the model were relatively larger (McKone and Maddalena, 2007). However, to date, studies focusing on aquatic vegetation are scarce, which remains as a challenge for both experiment and modeling researches (Diepens et al., 2014).

9.1.3 Future Perspectives

The previous section provides a brief review of the development and application of fugacity models. In summary, since the first report of the fugacity models, much effort has been made into the follow three aspects: (1) extending the temporary scale of the model from Levels I to III steady state model to Level IV dynamics model, including both short-term seasonal variations and long-term annual changes; (2) extending the spatial scale of the model from 0-dimension model to those with regional and global variations; (3) incorporating important compartments into the model. In addition to abiotic components such as air, soil, water, and sediment, biota in the system in different trophic levels (from primary producers such as vegetation to top predators such as birds) are incorporated into fugacity models. It has been a great success to combine the knowledge of food chain and food web theory with fugacity models to account for the effects of bioaccumulation and biomagnification of chemicals in the environmental system (Koelmans et al., 2001).

Nonetheless, to date, limitations still exist in the studies for fugacity models. Particularly, these studies were usually regional and were made for the chemical industry to assess whether new chemicals would cause environmental issues or not (Cowan, 1995), whereas models developed for a well-defined ecosystem, e.g., a lake, were rarely reported. However, these types of research are of great importance in that a model evaluation of chemicals behavior in a small scale would provide implications for local managers in environmental risk assessment. In addition, fugacity models were criticized for its poor prediction accuracy and large model uncertainty. Fate of chemicals is determined by many factors so that even one-order magnitude of deviations between model predictions and field data are still acceptable, while uncertainty is still difficult to determine (Cao et al., 2004).

Overall, future studies on fugacity models are suggested in the following three directions: (1) to develop models for a smaller scale. This would require concrete case studies for specific ecosystems (such as lakes), so that the model would be validated by field data and provide a detailed description of the chemical behavior in such systems; (2) to improve model prediction accuracy; ecological and human health risk assessment would benefit more from fugacity models with higher prediction accuracy; (3) to reduce model uncertainty. This would require efforts from two ways. One is incorporating more components into the fugacity models, such as vegetation (Diepens et al., 2014) and dissolved organic matter (Li et al., 2015), so that uncertainty from model structure would be reduced; the other is utilizing advanced statistical methods such as Bayesian-based Markov Chain Monte Carlo (Kong et al., 2014; Saloranta et al., 2008), so that the uncertainty from model parameters would not be overestimated.

9.2 FUGACITY MODEL FOR PAHs IN LAKE SMALL BAIYANGDIAN, NORTHERN CHINA: A CASE STUDY

In the following section, we provide a case study of Level III fugacity model for description of environmental behavior of 15 priority PAHs in a vegetation-dominant lake in China. Aquatic plants were included in the model. The reliability of the model estimates was evaluated by various means, including concentration validation, sensitivity, and uncertainty analysis. Therefore, this case study provides a nice example of fugacity model for a well-defined ecosystem within a significantly small scale, with many efforts in increasing model performance.

9.2.1 Polycyclic Aromatic Hydrocarbons

PAHs are globally concerned pollutants due to their widespread occurrence, strong persistence, long-range transportation potential, and carcinogenic toxicity (Xu et al., 2011). As one of the fastest growing countries in the world, China is suffering from severe contamination of PAHs from various sources (Zhang et al., 2007). The threat of PAH pollution to ecosystems and human health have become serious in China. It was reported that the atmospheric emissions of 16 priority PAHs in China in 2004 (114 Gg) accounted for about 22% of the total global emissions (520 Gg) of 16 priority PAHs set by USEPA (Zhang et al., 2009). It was estimated that 5.8% of China's land area, where 30% of the population lives, exceeded the national ambient benzo[a]pyrene (BaP) standard of 10 ng/m^3 and that the overall population attributable fraction for lung cancer caused by inhalation exposure to PAHs was 1.6% (Zhang et al., 2009). Therefore, it is meaningful to understand and predict the fate and transport of PAHs in various environmental media in China.

To date, two fugacity model studies included all 16 principal PAHs (Lang et al., 2007; Wang et al., 2011), while specific PAH components such as BaP and phenanthrene (Phe) were considered in other studies (Tao et al., 2003; Wang et al., 2002). However, the fate of PAH components in water and sediments were not well modeled in such studies at regional scales. To solve this limitation, a quantitative water–air sediment interaction (QWASI) fugacity model is strongly required (Mackay et al., 1983a,b). The QWASI fugacity model has been successively applied to predict the fate of heavy metals and organic chemicals such as PCBs and BaP in lake or river ecosystems (Diamond et al., 1996; Mackay et al., 1983a,b). However, priority PAH components other than BaP have not been included in the QWASI fugacity models. To understand properly and to compare the behaviors of individual PAH components in aquatic ecosystems, it is critical to include other priority PAH components excluding BaP, because the behaviors and physical–chemical features are varied for different PAH components (Zhang et al., 2005). Even for BaP, more case studies on using QWASI fugacity model are still needed to understand and predict its fate behaviors in lake ecosystems, especially in such a lake as Small Baiyangdian that has abundant aquatic plants. In the previous studies on the development and application of QWASI fugacity model, aquatic plants were not included.

9.2.2 Study Area and Measurements

Lake Baiyangdian, the largest freshwater lake in Northern China, is located at the central place of three big cities, Beijing, Tianjing, and Shijiazhuang (Fig. 9.1), one of the most seriously polluted areas in China for PAHs (Zhang et al., 2007). Lake Baiyangdian is one of the important locations of fish production in China. However, during the last decades, with the rapid economic development and population growth in the watershed and neighbor regions, the lake receives an increased loading of PAHs (Xu et al., 2011). The lake with total area of 366 km^2 is composed of 134 interconnected small lakes with different size areas. Lake

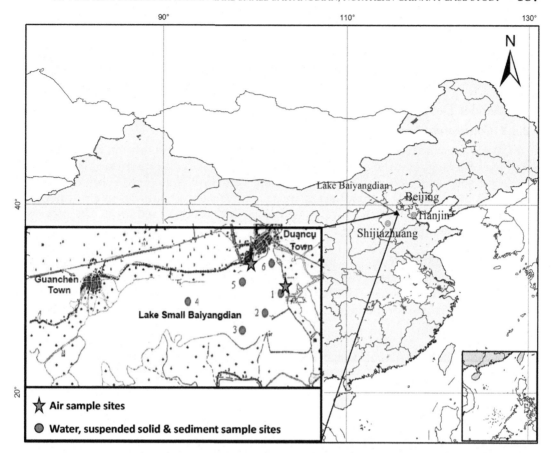

FIGURE 9.1   Location of Lake Small Baiyangdian and sampling sites.

Small Baiyangdian, with the area of 13.3 km^2, is the biggest one among 134 interconnected small lakes.

Sampling for water, suspended solids, sediment, and macrophytes at six sites (Fig. 9.1), and for fish at one site in Lake Small Baiyangdian was performed once on October 7, 2007. Air sampling including gaseous, particulate, and dust samples at two sites was carried out four times seasonally during autumn 2007 to summer 2008. Gaseous and particulate samples were collected using passive air sampler (Tao et al., 2009). Dust samples were collected by dust tank. Macrophytes samples included two species of floating plants, three species of submerged plants, and three species of emergent plant. Four species of commonly consumed freshwater fish including 15 individuals of crucian carp, and 10 individuals each of snakehead fish, grass carp, and silver fish were collected. Fifteen priority PAHs included acenaphthylene (Acy), acenaphthene (Ace), fluorene (Flo), Phe, anthracene (Ant), fluoranthene (Fla), pyrene (Pyr), chrysene (Chr), benzo[a]anthracene (BaA), benzo[b]fluoranthene (BbF), benzo [k]fluoranthene (BkF), BaP, indeno[1,2,3-cd]pyrene (IcdP), benzo[ghi]perylene (BghiP), and dibenz[a,h]anthracene (DahA) and were measured by GC–MS. Organic carbon contents in water, suspended solids, and sediments, and lipid contents in fish and macrophytes were also analyzed. The mean contents of PAHs, lipid, and organic carbon in the studied multimedia were calculated for the parameters and calibrated for the model.

9.2.3 Model Development

9.2.3.1 Conceptual Framework

A QWASI fugacity model was developed to characterize the multimedia fate of PAHs in Lake Small Baiyangdian. The conceptual diagram of the model is presented in Fig. 9.2. Air, water, and sediment were defined as three bulk compartments. Eight subcompartments included in the three bulk compartments are as follows: air and particles in air; water, suspended solids, plants, and fish in water; and water and solids in sediment. The processes taken into consideration are defined in Fig. 9.2 and additional details are given in Table 9.1.

9.2.3.2 Model Equations

The mass balance equations with fugacities as variables for air, water, and sediment bulk compartments are tabulated in Table 9.2. The equations for the transfer rate coefficients of the modeled processes and for the fugacity capacity of each bulk phase and subphase are listed in Tables 9.3 and 9.4.

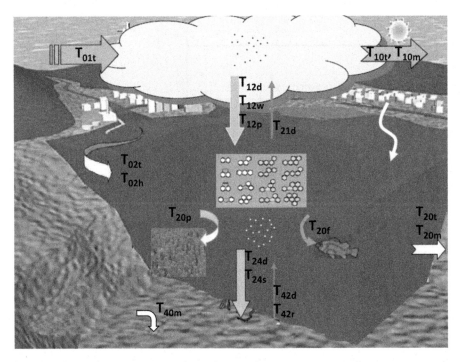

FIGURE 9.2 Conceptual diagram of the QWASI fugacity model for modeling multimedia fates of PAHs in Lake Small Baiyangdian.

TABLE 9.1 Transfer and Transformation Processes Defined in the QWASI Fugacity Model

Symbol	Transfer Process
T_{01t}, T_{10t}	Advective air flows into/out of the lake air bulk
T_{02t}, T_{20t}	Advective water flows into/out of the lake water bulk
T_{02h}	Local wastewater discharge to the lake water bulk
$T_{10m}, T_{20m}, T_{40m}$	Degradation in air, water, and sediment
T_{20p}, T_{20f}	Bioaccumulation in aquatic plants and fishes
$T_{12d}, T_{12p}, T_{12w}$	Diffusion, dry deposition, and wet precipitation from air to water
T_{21d}	Diffusion from water to air
T_{24d}, T_{42d}	Diffusion between water column and bottom sediment
T_{24s}	Sedimentation from column and bottom sediment

TABLE 9.2 Mass Balance Equations in the QWASI Fugacity Model

Phase	Mass Balance Equation[a]	Mass Balance Equation in Detail[a]
Air	$T_{01t} + T_{21d} = T_{10t} + T_{10m} + T_{12d} + T_{12p} + T_{12w}$	$Q_{01t}C_{01t} + D_{21d}f_2 = (D_{10t} + D_{10m} + D_{12d}$ $+ D_{12p} + D_{12w})f_1$
Water	$T_{02t} + T_{02h} + T_{12d} + T_{12p} + T_{12w} + T_{42d} = T_{20t}$ $+ T_{20m} + T_{20f} + T_{20r} + T_{21d} + T_{24d} + T_{24s}$	$Q_{02t}C_{02t} + Q_{02h}C_{02h} + (D_{12d} + D_{12p} + D_{12w})f_1$ $+ D_{42d}f_4 = (D_{20t} + D_{20m} + D_{20f} + D_{20r}$ $+ D_{21d} + D_{24d} + D_{24s})f_2$
Sediment	$T_{24d} + T_{24s} = T_{42d} + T_{40m}$	$(D_{24d} + D_{24s})f_2 = (D_{42d} + D_{40m})f_4$

[a]T_{ijk} are transfer processes, defined in Table 9.1. D_{ijk} are transfer rate coefficients for major transfer processes from the ith bulk phase to the jth bulk phase (see Table 9.3 for details). For system input, $T_{01t} = Q_{01t} \times C_{01t}$, $T_{02t} = Q_{02t} \times C_{02t}$, $T_{02h} = Q_{02h} \times C_{02h}$. For system output, $T_{10t} = D_{10t} \times f_1$, $T_{20t} = D_{20t} \times f_2$, $T_{10m} = D_{10m} \times f_1$, $T_{20m} = D_{20m} \times f_2$, $T_{40m} = D_{40m} \times f_4$, $T_{20p} = D_{20r} \times f_2$, $T_{20f} = D_{2f} \times f_2$. For air–water transfer, $T_{12d} = D_{12d} \times f_1$, $T_{21d} = D_{21d} \times f_2$, $T_{12p} = D_{12p} \times f_1$, $T_{12w} = D_{12w} \times f_1$. For water–sediment transfer, $T_{24d} = D_{24d} \times f_2$, $T_{42d} = D_{42d} \times f_4$, $T_{24s} = D_{24s} \times f_2$.

9.2.3.3 Model Parameters

The model parameters including environmental, physical, chemical, and process kinetic ones were determined in three different ways, i.e. literature review, laboratory experiments, and model calibration. The symbols, descriptions, values, and sources of all model parameters are presented in Tables 9.5 and 9.6. The mean values of collected data are used for the parameters after abnormal values are excluded. The Henry's law constant, saturation vapor pressure and fugacity ratio is the value under the temperature of 25°C. However, the average annual temperature in Lake Small Baiyangdian is 12.1°C. The necessary corrections for these parameters are performed by the Paasivirta's equation (Paasivirta et al., 1999).

TABLE 9.3 Equations for Calculating Transfer Rate Coefficients (D)

Process		Equations[a]	Remarks
Air(1)– water(2)	Diffusion	$D_{12d} = A_2/[1/(K_{12} \times Z_{11}) + 1/(K_{21} \times Z_{22})]$	$D_{21d} = D_{12d}$
	Dry deposition	$D_{12p} = A_2 \times K_p \times X_{13} \times Z_{13}$	
	Wet precipitation	$D_{12w} = A_2 \times K_w \times S_c \times X_{13} \times Z_{13}$	
Water(2) –sediment(4)	Diffusion	$D_{24d} = A_2/[1/(K_{24} \times Z_{22}) + L_4/(B_4 \times Z_{22})];$	$D_{42d} = D_{24d}$
	Deposition	$D_{24s} = A_2 \times K_s \times Z_{23}$	
Reaction	Degradation in air	$D_{10m} = K_{m1} \times A_1 \times h_1 \times Z_1$	–
	Degradation in water	$D_{20m} = K_{m2} \times A_2 \times h_2 \times (Z_{22} + Z_{23})$	
	Degradation in sediment	$D_{40m} = K_{m4} \times A_4 \times h_4 \times Z_4$	
Advection	Advective air flows	$D_{01t} = Q_{01t} \times Z_1$	–
	Advective water flows	$D_{02t} = Q_{02t} \times Z_2$	
Biota	Fish harvest	$D_{20f} = Y_f \times Z_{2f}/\rho_{2f}$	–
	Plants harvest	$D_{20p} = Y_p \times Z_{2p}/\rho_{2p}$	

[a]D_{ijk} are transfer rate coefficients for major transfer processes from the ith bulk phase to the jth bulk phase (1, 2, and 4 for air, water, and sediment, respectively, 0 for outside of the area). The subscript k indicates process category (t, d, p, w, f, r, s, and m for advective flow, diffusion, dry deposition, wet precipitation, fish bioaccumulation, pant bioaccumulation, sedimentation, and degradation, respectively). Z is fugacity capacity (see Table 9.4 for details). See Table 9.5 for the meanings of other parameters.

TABLE 9.4 Equations for Calculating Fugacity Capacities (Z)

Bulk Phase	Subphase	Equations for Calculating Z $(mol/m^3 \cdot Pa)$[a]	
Air	Air	$Z_{11} = 1/RT$	$Z_1 = X_{11}Z_{11} + X_{13}Z_{13}$
	Particle	$Z_{13} = 6E6/(P_SRT)/BP_S$	
Water	Water	$Z_{22} = 1/H$	$Z_2 = X_{22}Z_{22} + X_{23}Z_{23}$ $+ X_{2f}Z_{2f} + X_{2p}Z_{2p}$
	Suspended solids	$Z_{23} = O_{23}\rho_{23}K_{oc}/H$	
	Fish	$Z_{2f} = BCF_f/H$	
	Plant	$Z_{2p} = BCF_p/H$	
Sediment	Pore water	$Z_{42} = 1/H$	$Z_4 = X_{42}Z_{42} + X_{43}Z_{43}$
	Solids	$Z_{43} = O_{43}\rho_{43}K_{oc}/H$	

[a]Z_1, Z_2, and Z_4 are fugacity capacity for air, water, and sediment bulk compartments, respectively. Z_{11}, Z_{13}, Z_{22}, Z_{23}, Z_{2f}, Z_{2p}, Z_{42}, and Z_{43} are fugacity capacity for air and particle subphases in air, water, suspended solids, fish, plant subphases in water, pore water, and solids subphases in sediment, respectively. See Table 9.5 for the meanings of other parameters.

TABLE 9.5 Parameters for the QWASI Fugacity Model

Symbol	Unit	Parameters	Value	References
A_1, A_2, A_4	m^2	Interface areas of air–water and water–sediment	1.366×10^7	Zhao et al. (2007)
h_1	m	Thickness of air	7.00×10^2	HPEPB (2001) and Ma et al. (2007)
h_2	m	Thickness of water	1.87	HPEPB (2001) and Ma et al. (2007)
h_4	m	Thickness of sediment	1.00×10^{-1}	HPEPB (2001) and Ma et al. (2007)
X_{13}	v/v	Volume fractions of solids in air	9.84×10^{-11}	Mackay and Paterson (1991), note A
X_{23}	v/v	Volume fractions of solids in water	4.29×10^{-6}	Mackay and Paterson (1991), note A
X_{43}	v/v	Volume fractions of solids in sediment	3.00×10^{-1}	Mackay and Paterson (1991), note A
L4	m	Diffusion path lengths in sediment	5.00×10^{-3}	Mackay and Paterson (1991)
X_{2f}	v/v	Volume fractions of fish in water	4.08×10^{-5}	Zhao et al. (2005) and Zhao (1995)
X_{2p}	v/v	Volume fractions of plants in water	8.20×10^{-4}	Zhao et al. (2005) and Zhao (1995)
X_{42}	v/v	Volume fractions of water in sediment	7.00×10^{-1}	Mackay and Paterson (1991)
O_{23}	%	Contents of organic carbon in solids in water	4.41×10^{-1}	Note A
O_{43}	%	Contents of organic carbon in solids in sediment	2.89×10^{-2}	Note A
ρ_{23}	t/m^3	Densities of solids in water	1.89	Note A
ρ_{43}	t/m^3	Densities of solids in sediment	2.49	Note A
ρ_{2f}	t/m^3	Densities of fish in water	1.05	Davenport (1999), note A
ρ_{2p}	t/m^3	Densities of plants in water	8.83×10^{-1}	Davenport (1999), note A
Q_{01t}	m^3/h	Air advection flow into the lake area	1.13×10^{10}	HPEPB (2001), calculated
Q_{10t}	m^3/h	Air advection flow out of the lake area	1.13×10^{10}	HPEPB (2001), calculated
Q_{02t}	m^3/h	Water advection flow into the lake	3.00×10^4	Yin (2008)
Q_{20t}	m^3/h	Water advection flow out of the lake	2.50×10^4	Yin (2008)
Q_{02h}	m^3/h	Rate of local wastewater discharge	500	HPEPB (2001), calculated

(Continued)

TABLE 9.5 Parameters for the QWASI Fugacity Model—cont'd

Symbol	Unit	Parameters	Value	References
C_{02t}	mol/m^3	PAHs concentration in water advection flow	Note B	Bai (2008), note A
C_{02h}	mol/m^3	PAHs concentration in wastewater	Note B	Note A
Y_f	T/h	Harvest rate of fish	3.00	Zhao et al. (2005) and Zhao (1995)
Y_p	T/h	Harvest rate of plants	6.00×10^1	Zhao et al. (2005) and Zhao (1995)
T	K	Local average temperature	3.00×10^2	Zhao et al. (2007)
Ps_{25}	Pa	Local vapor pressure	Note B	Van Agreren et al. (1998), Mackay et al. (1997), Wang (1991), Wang (1993), and Jin (1990)
R	Pa·m^3/mol·K	The gas constant	8.314	Van Agreren et al. (1998), Mackay et al. (1997), and Karickhoff (1981)
F_{25}	—	Fugacity ratio at 25°C	Note B	Mackay et al. (1997)
H_{25}	Pa·m^3/mol	Henry's constant	Note B	Mackay et al. (1997), Wang (1991), Wang (1993), Jin (1990), and Ten Hulscher et al. (1992)
B_F	—	Fugacity ratio temperature correction factor	Note B	Paasivirta et al. (1999)
B_H	—	Henry's law constant temperature correction factor	Note B	Paasivirta et al. (1999)
BP_S	—	Saturation vapor pressure temperature correction factor	Note B	Paasivirta et al. (1999)
K_{oc}	m^3/t, 1/h	Adsorption coefficient	Note B	Mackay et al. (1997), Jin (1990), STF (1991), and US-EPA (1996)
K_{m1}	1/h	Degradation rate of PAHs in air	Note B	Mackay (2001) and Lang et al. (2008)
K_{m2}	1/h	Degradation rate of PAHs in water	Note B	Mackay (2001) and Lang et al. (2008)
K_{m4}	1/h	Degradation rate of PAHs in sediment	Note B	Mackay (2001), Lang et al. (2008)
BCF_f	m^3/t	Bioconcentration factor of fish	Note B	Lang et al. (2008), Duan (2005), and Tang et al. (2006), note A (calculated)
BCF_p	m^3/t	Bioconcentration factor of plants	Note B	Lang et al. (2008), Duan (2005), and Tang et al. (2006), note A (calculated)
B_1	m^2/h	Molecular diffusivities in air	Note B	Perry and Chilton (1973), US-EPA (1996), Shor et al. (2003), and Xu (1991)

(Continued)

TABLE 9.5   Parameters for the QWASI Fugacity Model—cont'd

Symbol	Unit	Parameters	Value	References
B_2	m^2/h	Molecular diffusivities in water	Note B	Perry and Chilton (1973); US-EPA (1996), Shor et al. (2003), and Xu (1991)
B_4	m^2/h	Molecular diffusivities in sediment	Note B	Perry and Chilton (1973), US-EPA (1996), Shor et al. (2003), and Xu (1991)
K_{12}	m/h	Air-side molecular transfer coefficient over water	3.00	Thibodeaux (1996) and Banks and Herrera (1997)
K_{21}	m/h	Water-side molecular transfer coefficient over air	3.00×10^{-2}	Thibodeaux (1996), Banks and Herrera (1997)
K_{24}	m/h	Water-side molecular transfer coefficient over sediment	1.00×10^{-2}	Thibodeaux (1996) and Banks and Herrera (1997)
K_P	m/h	Dry deposition velocity	5.69×10^{-1}	HPEPB (2001), Mackay and Paterson (1991), Beijing Statistics Bureau (2005), and Tainjin Environment Protection Bureau (1991, 1996, 2001)
K_w	m/h	Wet deposition velocity	6.51×10^{-5}	HPEPB (2001), Mackay and Paterson (1991), Beijing Statistics Bureau (2005), and Tainjin Environment Protection Bureau (1991, 1996, 2001)
K_S	m/h	Water sedimentation rates	4.60×10^{-6}	Chen et al. (2006) and Hu et al. (1998)
S_C	m/h	Rain scavenging rate	2.00×10^{5}	Mackay et al. (1986)

Note A: Data determined in our lab.
Note B: Presented in Table 9.6.

9.2.3.4 *Multimedia Modeling*

The concentrations of PAHs in the compartments and the transfer fluxes between adjacent compartments were modeled under a steady-state assumption. Measured concentrations in this study were used for model validation. Modeling was performed using Matlab v.6.5 (MathWorks, 2002). SPSS v.10.0 and MS Excel were employed for statistical analysis and data manipulation.

9.2.3.5 *Sensitivity Analysis*

An overview of the most sensitive components of the model can be determined through sensitivity analysis. This analysis provides a measure of the sensitivity of parameters, forcing functions or submodels to the state variables of greatest interest in the model. In practical modeling, the sensitivity analysis is carried out by changing the parameters, forcing functions and submodels, and the corresponding response of the selected state variables is observed (Jørgensen and Nielsen, 1994). In this study, the sensitivity analysis was performed

TABLE 9.6 Parameter Values for Different PAH Components in the QWASI Fugacity Model

Symbols	ACE	ACY	FIO	PHE	ANT	FLA	PYR	BaA	CHR	BbF	BkF	BaP	DahA	IcdP	BghiP
C_{01t}	5.27E-11	1.90E-11	3.31E-10	5.80E-10	2.61E-11	1.27E-10	6.48E-11	9.96E-12	1.18E-10	8.81E-11	5.83E-12	2.25E-11	1.93E-11	1.05E-11	3.64E-11
C_{02t}	1.71E-08	1.30E-08	4.21E-08	4.49E-08	6.17E-08	2.37E-08	1.14E-08	1.31E-09	4.38E-09	1.19E-09	7.930E-10	3.96E-10	1.09E-09	1.08E-09	0.00E+00
C_{02h}	8.35E-07	7.33E-07	1.88E-06	1.63E-06	3.69E-06	1.40E-06	5.93E-07	7.64E-08	2.51E-07	6.53E-08	4.71E-08	1.64E-08	6.38E-08	2.77E-08	0.00E+00
H_{25}	1.78E+01	1.16E+01	8.50E+00	4.00E+00	6.00E+00	1.49E+00	1.53E+00	8.61E-01	1.83E-01	5.40E-02	8.53E-02	9.00E-03	7.00E-03	7.00E-03	1.00E-03
P_{S25}	1.31E+00	2.12E+00	4.83E-01	5.98E-02	1.41E-02	2.34E-03	1.91E-03	5.74E-05	2.21E-05	4.45E-06	3.79E-06	3.22E-06	3.22E-06	3.22E-06	3.22E-06
K_{oc}	4.36E+03	6.76E+03	1.08E+04	1.81E+04	6.05E+04	2.41E+05	3.92E+04	4.04E+05	1.68E+06	1.68E+06	2.15E+06	5.25E+05	2.60E+06	4.89E+06	4.89E+06
BCF_f	4.81E+02	4.74E+02	9.65E+02	1.86E+03	1.93E+03	4.75E+03	3.43E+03	3.43E+03	2.02E+04	6.17E+04	7.78E+04	3.86E+04	3.86E+04	3.86E+04	3.86E+04
BCF_v	5.94E+01	8.91E+01	1.04E+02	3.32E+02	9.75E+02	1.00E+03	1.17E+03	7.12E+03	2.99E+03	4.26E+03	6.69E+03	7.33E+03	2.10E+04	3.72E+04	2.10E+04
K_{m1}	5.04E-03	5.04E-03	5.04E-03	5.04E-03	5.04E-03	4.08E-03	4.08E-03	4.08E-03	4.08E-03	4.08E-03	4.08E-03	4.08E-03	4.08E-03	4.08E-03	4.08E-03
K_{m2}	5.04E-04	5.04E-04	5.04E-04	5.04E-04	5.04E-04	4.08E-04	4.08E-04	4.08E-04	4.08E-04	4.08E-04	4.08E-04	4.08E-04	4.08E-04	4.08E-04	4.08E-04
K_{m4}	1.63E-05	1.63E-05	1.63E-05	1.63E-05	1.63E-05	1.26E-05	1.26E-05	1.26E-05	1.26E-05	1.26E-05	1.26E-05	1.26E-05	1.26E-05	1.26E-05	1.26E-05
B_1	1.66E-02	1.80E-02	1.55E-02	1.66E-02	1.47E-02	1.38E-02	1.34E-02	1.55E-02	1.25E-02	1.18E-02	1.18E-02	1.41E-02	1.11E-02	1.10E-02	1.33E-02
B_2	2.17E-06	1.88E-06	2.14E-06	1.85E-06	2.09E-06	1.85E-06	1.96E-06	2.10E-06	1.77E-06	1.64E-06	1.64E-06	2.05E-06	1.55E-06	1.60E-06	1.39E-06
B_4	4.54E-13	4.57E-13	4.37E-13	4.63E-12	7.55E-11	2.21E-11	2.88E-11	7.99E-12	6.87E-12	1.51E-12	1.89E-12	1.58E-12	3.15E-13	2.38E-13	6.26E-13
F_{25}	1.97E-01	3.38E-01	1.39E-01	2.08E-01	1.01E-02	7.62E-02	1.57E-01	5.07E-02	6.68E-02	6.68E-03	1.26E-02	3.11E-02	3.11E-03	3.11E-03	3.11E-03
B_F	1.13E+03	5.73E+02	1.02E+03	8.60E+02	1.51E+03	8.93E+02	9.87E+02	1.12E+03	1.37E+03	9.06E+02	1.45E+03	1.30E+03	9.09E+02	1.29E+03	1.12E+03
B_H	1.24E+03	2.18E+03	1.59E+03	2.12E+03	1.36E+03	2.48E+03	2.34E+03	2.64E+03	2.77E+03	3.56E+03	2.98E+03	3.28E+03	4.01E+03	3.21E+03	3.37E+03
BP_s	3.49E+03	3.32E+03	3.64E+03	3.84E+03	4.38E+03	4.26E+03	4.31E+03	4.88E+03	5.51E+03	5.37E+03	5.87E+03	5.82E+03	5.82E+03	5.82E+03	5.82E+03

Note: The meanings of the symbols are the same with them in Table 9.5.

only for the parameters. A change for the parameter at ±10% was chosen, and the sensitivity coefficient (S) was calculated by the following formula (Cao et al., 2004):

$$S = (Y_{1.1} - Y_{0.9})/(0.2 \cdot Y) \tag{12.1}$$

Y represents the model outputs of chemical concentrations. The terms, $Y_{1.1}$ and $Y_{0.9}$, represent the estimated concentrations when the tested parameter was changed at +10% and −10%, respectively. The greater the absolute value of sensitivity coefficient, the more sensitive the parameter.

9.2.3.6 Uncertainty Analysis

Both concentrations and fluxes estimated by the multimedia model are inherently variable (McKone, 1996). In addition to the inherent variability, there are also uncertainties in the parameters and estimates (Tao et al., 2003). For assessing the overall uncertainty and variability in predictions, Monte Carlo simulation was used to illustrate collective variance of the inputs through the model. Each input parameter was represented as a probability density function that defined both the range of values and the likelihood of the parameter having that value. All of the parameters were assumed to follow the log-normal distribution. The simulation was undertaken repeatedly 3000 times, with new values randomly selected for all parameters within the range of mean ± standard deviation. A built-in function of "randn" in Matlab was used to select the values randomly for each parameter (MathWorks, 2002). The model uncertainty was ascertained by statistical analysis on the output result. To quantify the differences, coefficients of variation (CVs) were calculated based on log-transformed data.

9.2.4 Results and Discussion

9.2.4.1 Modeled Concentration Distributions

The levels and distributions of calculated PAHs concentrations in the three bulk and seven subphases are presented in Table 9.7. The highest PAHs concentrations were found in the sediment phase, followed by the water and air phases (Fig. 9.3). The percentage ratios of individual PAH congeners ranged from 58.9% to 88.5%, 11.5 to 26.8%, and 0 to 14.3% for the sediment, water, and air phases, respectively (Fig. 9.3). This implies that the sediment would serve as the sink of PAHs. Among different PAHs congeners, low-molecular-weight PAHs (LMW-PAHs) predominated the distribution in three bulk phases. From LMW-PAHs to middle- and high-molecular-weight PAHs (MMW-PAHs and HMW-PAHs), the average percentage were increased from 64.8% to 66.1% and 87.0% in the sediment, and decreased from 26.3% to 22.9% and 13.0% in the water, and 11.0 to 8.9% and 0 in the air. This means that LMW-PAHs were in higher proportion in the water and air, while HMW-PAHs were in higher proportion in the sediment.

Fig. 9.4 illustrates that the different distribution patterns of LMW-PAHs, MMW-PAHs, and HMW-PAHs in seven subphases. In the air, LMW-PAHs predominated in the gaseous subphase, while HMW-PAHs were dominant in solid-particle subphase (Fig. 9.4A). In the water, the PAHs contents in subphases were in the declining order of suspended solids (C23) > fish (C2f) > aquatic plants (C2p) > dissolved subphase (C22); and PAHs contents in the dissolved

TABLE 9.7   Calculated PAHs Concentration in the Bulk and Subphases

	C1	C2	C4	C13	C22	C23	C2f	C2p	C42	C43
ACE	5×10^{-11}	5×10^{-9}	5×10^{-5}	1×10^{-14}	5×10^{-9}	2×10^{-5}	2×10^{-6}	3×10^{-7}	5×10^{-7}	2×10^{-4}
ACY	2×10^{-11}	1×10^{-19}	1×10^{-5}	5×10^{-15}	1×10^{-9}	5×10^{-6}	6×10^{-7}	1×10^{-7}	1×10^{-7}	5×10^{-5}
FlO	3×10^{-10}	2×10^{-8}	4×10^{-4}	1×10^{-13}	2×10^{-8}	2×10^{-4}	2×10^{-5}	2×10^{-6}	2×10^{-6}	1×10^{-3}
PHE	6×10^{-10}	3×10^{-8}	9×10^{-4}	3×10^{-12}	2×10^{-8}	3×10^{-4}	4×10^{-5}	7×10^{-6}	2×10^{-6}	3×10^{-3}
ANT	2×10^{-11}	6×10^{-10}	4×10^{-5}	3×10^{-14}	3×10^{-10}	1×10^{-5}	5×10^{-7}	3×10^{-7}	3×10^{-8}	1×10^{-4}
FLA	1×10^{-10}	1×10^{-9}	3×10^{-4}	7×10^{-12}	5×10^{-10}	9×10^{-5}	2×10^{-6}	5×10^{-7}	6×10^{-8}	1×10^{-3}
PYR	6×10^{-11}	4×10^{-9}	2×10^{-4}	8×10^{-12}	2×10^{-9}	6×10^{-5}	6×10^{-6}	2×10^{-6}	3×10^{-7}	7×10^{-4}
BaA	7×10^{-12}	4×10^{-10}	6×10^{-5}	5×10^{-12}	5×10^{-11}	2×10^{-5}	2×10^{-7}	3×10^{-7}	7×10^{-9}	2×10^{-4}
CHR	1×10^{-11}	2×10^{-10}	8×10^{-5}	6×10^{-12}	2×10^{-11}	2×10^{-5}	3×10^{-7}	5×10^{-8}	2×10^{-9}	3×10^{-4}
BbF	1×10^{-11}	3×10^{-10}	1×10^{-4}	1×10^{-11}	2×10^{-11}	3×10^{-5}	2×10^{-6}	1×10^{-7}	3×10^{-9}	4×10^{-4}
BkF	5×10^{-12}	1×10^{-10}	5×10^{-5}	5×10^{-12}	8×10^{-12}	1×10^{-5}	6×10^{-7}	5×10^{-8}	1×10^{-9}	2×10^{-4}
BaP	8×10^{-13}	5×10^{-11}	7×10^{-6}	6×10^{-13}	4×10^{-12}	2×10^{-6}	2×10^{-7}	3×10^{-8}	6×10^{-10}	2×10^{-5}
DahA	1×10^{-12}	3×10^{-11}	9×10^{-6}	8×10^{-13}	1×10^{-12}	2×10^{-6}	4×10^{-8}	2×10^{-8}	2×10^{-10}	3×10^{-5}
IcdP	8×10^{-13}	2×10^{-11}	7×10^{-6}	6×10^{-13}	5×10^{-13}	2×10^{-6}	2×10^{-8}	2×10^{-8}	7×10^{-11}	2×10^{-5}
BghiP	2×10^{-12}	3×10^{-11}	1×10^{-5}	1×10^{-12}	9×10^{-13}	4×10^{-6}	4×10^{-8}	2×10^{-8}	1×10^{-10}	5×10^{-5}

Note: C1, air phase; C2, water phase; C4, sediment phase; C13, solid particles in air phase; C22, dissolved in water phase; C23, solid particles in water phase; C2f, fish in water phase; C2p, aquatic plants in water phase; C42, pore water in sediments; C43, solid particles in sediment phase.

phase were much lower than these in the suspended solid, fish, and aquatic plant subphases. From LMW-PAHs to MMW-PAHs and HMW-PAHs, the contents were decreased in the dissolved subphase, and similar in the suspended solids, fish, and plant subphases (Fig. 9.4B). In the sediment, PAHs contents in the solid subphase (C43) were significantly higher than these in pore water subphase (C42) (Fig. 9.4C).

The different fate behaviors of LMW-PAHs, MMW-PAHs, and HMW-PAHs in the water, air, and sediment may be attributed to the difference in their physical and chemical properties. LMW-PAHs with higher vapor pressure and Henry's law constant are more volatile than HMW-PAHs with lower vapor pressure and Henry's law constant. On the other hand, LMW-PAHs are less lipophilic due to their lower K_{oc} values, so that their ability in binding to organic matter in suspended solids and sediments is obviously weaker than that of HMW-PAHs. The dissolved PAHs concentrations in the water and pore water were decreased with the increase of their molecular weight, probably due to the decreasing solubility.

9.2.4.2 Model Validation

The model was validated by the comparisons between calculated and measured PAHs concentrations in the subphases. As can be seen in Fig. 9.5, very similar distribution patterns

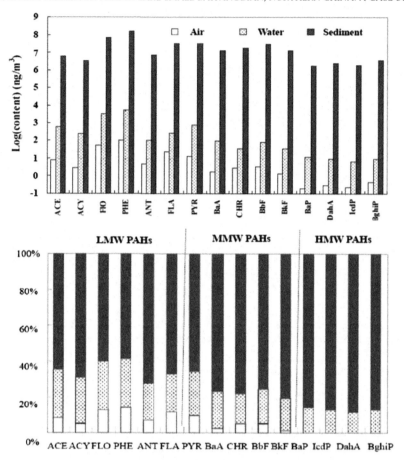

FIGURE 9.3   Distributions of calculated PAHs concentration and their percentage in three bulks.

in the subphases of the water and sediment would be found in both calculated and measured PAHs concentrations. The simulated values for the most of PAHs congeners were lower than their measured values, which may be attributed to the neglect of some input process such as soil erosion. The difference between the calculated and measured PAHs concentrations were different for both the subphases and PAH congeners. Among seven subphases, the best agreements with the difference less than one order of magnitude could be found for the PAHs concentrations in the solid subphase in the sediment (C43), while the worst agreements with the difference around two orders of magnitude were for the PAHs concentrations in the dissolved and solid subphase in water (C22 and C23). Among 15 PAH congeners, LMW-PAHs maintain better agreements in all subphases than HMW-PAHs. The best agreements with the difference less than one order of magnitude could be found for the LMW-PAHs except for Acy and ANT in the suspended solid subphases (C23) and ANT in

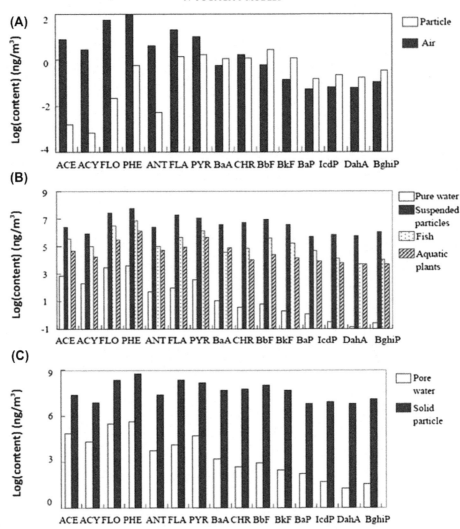

FIGURE 9.4   Calculated PAHs concentrations in the subphases in Lake Small Baiyangdian.

the fish subphases (C2f). The worst agreements with the difference around two orders of magnitude were for the HMW-PAHs in the suspended solid and fish subphases (C23 and C2f). The IcdP and DahA in the suspended solid and fish subphases as well as the BghiP in the suspended solid subphase were undetectable; however, their modeled values were relatively high.

The differences between the calculated and measured PAHs concentrations for seven subphases are attributable to the complexity of PAHs sources and the degree of influence by environmental changes; however, those for PAH congeners are attributed to their

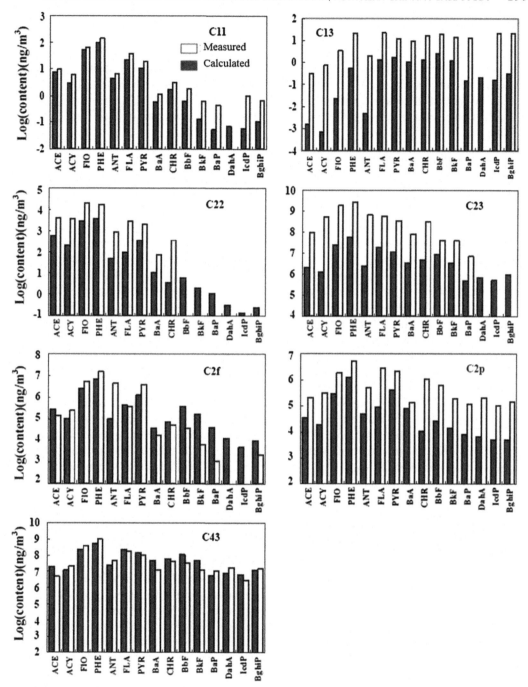

FIGURE 9.5   Comparisons between the calculated and measured concentration of PAHs in seven subphases.

different physical and chemical properties. The PAHs in the solid subphase in sediment were mainly from the sedimentation of suspended solids in water and were less influenced by ambient environmental changes. However, the PAHs in the dissolved and solid subphase in water were likely from the input of inflow rivers, surface runoff, dry and wet depositions, and they were easily influenced by ambient environmental changes. Compared with the concentrations of LMW-PAHs, the HMW-PAHs are much lower in concentration and are often undetectable or with large uncertainty (see Section 9.2.3.6 for details).

9.2.4.3 Transfer Fluxes of PAHs

The calculated transfer fluxes of PAHs are shown in Table 9.8. The fluxes into and out of the lake area as well as each compartment were well balanced. For instance, for the total flux into and out of the lake area, the average and maximum relative errors were 0.022% and 0.19% (Ant), respectively. The fluxes through advective air flows into and out of the lake (T_{01t}, T_{10t}) were predominant in the input and output fluxes of PAHs. The average bioaccumulation flux of 15 PAHs by plants (T_{2p}) was four times higher than that by fish (T_{2f}). The highest percentage of the average flux of 15 PAHs by degradation in the sediment (T_{40m}) was 61.81%, followed by the air (T_{10m}) (37.82%) and by the water (T_{20m}) (0.37%). From LMW-PAHs to MMW-PAHs and HMW-PAHs, the percentages of the average flux by degradation in the sediment (T_{40m}) were increased from 42.70% to 74.32% and 79.15%; however, those by degradation in the air (T_{10m}) were decreased from 56.60% to 25.56% and 20.78%. This meant that the degradation of LMW-PAHs mainly occurred in the air and that the degradation of MMW-PAHs and HMW-PAHs mainly happened in the sediment.

The calculated transfer fluxes of PAHs across the air–water and water–sediment interfaces are shown in Fig. 9.6. The transfer fluxes from air to water across the air–water interface and from water to sediment across the water–sediment interface were much higher than these from water to air and from sediment to water, respectively. This indicated that, in the air–water–sediment system, the transfer directions of PAHs were from air to water and to sediment. Air was the source of PAHs, while sediment could severe as the sink of PAHs. Among 15 PAH congeners, the highest transfer fluxes from air to water (T_{12}) and from water to sediment (T_{24}) could be found for LMW-PAHs (including 2- and 3-ring PAHs), followed by MMW-PAHs and HMW-PAHs.

Fig. 9.7 illustrates that contributions of the transfer fluxes of PAHs from air to water and from water to sediment though different processes were changed. From air to water (Fig. 9.7A), the transfer of LMW-PAHs by the diffuse process (T_{12d}) contributed the highest fluxes, followed by wet and dry precipitation processes (T_{12p} and T_{12w}). For MMW-PAHs and HMW-PAHs, the wet precipitation played the highest contribution of transfer fluxes (T_{12w}), followed by the diffuse and dry precipitation (T_{12d} and T_{12p}). From LMW-PAHs (including 2- and 3-ring PAHs) to MMW-PAHs and HMW-PAHs, the transfer fluxes through diffuse process (T_{12d}) were decreased. The transfer fluxes through dry and wet precipitations (T_{12p} and T_{12w}) were increased from 2-ring PAHs to 3- and 4-ring PAHs. From water to sediment (Fig. 9.7B), the transfer fluxes of PAHs were mainly depended on the sedimentation process (T_{24s}). The diffuse process (T_{24d}) could only have some contributions for 2-, 3-, and 4-ring PAHs.

TABLE 9.8 Calculated Transfer Fluxes In and Out of the Lake Area As Well As Each Compartment

	T_{01t}	T_{02t}	T_{02h}	T_{10t}	T_{20t}	T_{12d}	T_{12p}	T_{12w}	T_{21d}	T_{24d}	T_{24s}	T_{42d}	T_{10m}	T_{20m}	T_{40m}	T_{2p}	T_{2f}
ACE	3.74E+00	5.12E-04	4.17E-04	3.65E+00	9.51E-05	3.80E-02	2.16E-06	4.95E-05	1.36E-02	1.26E-10	2.32E-02	1.36E-08	6.64E-02	1.33E-03	2.32E-02	1.44E-05	5.77E-06
ACY	1.35E+00	3.89E-04	3.67E-04	1.31E+00	2.24E-05	1.66E-02	9.46E-07	2.17E-05	2.84E-03	4.68E-11	1.33E-02	5.07E-09	2.39E-02	4.94E-04	1.33E-02	7.97E-06	2.10E-06
FlO	2.35E+01	1.26E-03	9.38E-04	2.28E+01	3.25E-04	2.93E-01	2.83E-05	6.48E-04	3.15E-02	5.40E-10	2.56E-01	5.86E-08	4.15E-01	6.04E-03	2.56E-01	1.12E-04	5.16E-05
PHE	4.12E+01	1.35E-04	8.16E-04	3.99E+01	5.31E-04	5.65E-01	6.73E-04	1.54E-02	1.78E-02	7.41E-09	5.55E-01	8.05E-07	7.22E-01	8.02E-03	5.55E-01	4.64E-04	1.29E-04
ANT	1.86E+00	1.85E-03	1.84E-03	1.80E+00	6.79E-06	2.36E-02	5.98E-06	1.37E-04	4.10E-04	1.52E-09	2.34E-02	1.66E-07	3.22E-02	1.16E-04	2.34E-02	1.72E-05	1.69E-06
FLA	9.05E+00	7.12E-04	6.99E-04	8.77E+00	1.31E-05	1.20E-01	1.46E-03	3.33E-02	1.28E-04	7.38E-10	1.54E-01	1.04E-07	1.24E-01	2.37E-04	1.54E-01	2.92E-05	6.87E-06
PYR	4.60E+00	3.41E-04	2.96E-04	4.43E+00	4.48E-05	5.77E-02	1.74E-03	3.99E-02	5.39E-04	3.76E-09	9.81E-02	5.29E-07	6.52E-02	5.68E-04	9.81E-02	1.33E-04	1.94E-05
BaA	5.65E-01	3.94E-05	3.82E-05	5.30E-01	1.20E-06	2.73E-03	1.02E-03	2.33E-02	7.40E-06	2.79E-11	2.70E-02	3.93E-09	7.76E-03	3.26E-05	2.70E-02	2.17E-05	5.19E-07
CHR	8.36E+00	1.31E-04	1.26E-04	8.31E+00	5.86E-06	8.16E-03	1.18E-03	2.70E-02	4.91E-07	7.75E-12	3.63E-02	1.09E-09	1.37E-02	3.01E-05	3.63E-02	2.95E-06	9.87E-07
BbF	6.26E+00	3.57E-05	3.26E-05	6.18E+00	3.04E-06	2.84E-03	2.28E-03	5.23E-02	1.74E-07	2.69E-12	5.73E-02	3.79E-10	1.42E-02	4.76E-05	5.73E-02	6.63E-06	4.76E-06
BkF	4.14E-01	2.38E-05	2.36E-05	3.84E-01	2.06E-07	5.75E-04	1.01E-03	2.31E-02	1.13E-07	1.13E-12	2.46E-02	1.59E-10	5.61E-03	1.98E-05	2.46E-02	3.49E-06	2.01E-06
BaP	1.60E+00	1.19E-05	8.19E-06	1.59E+00	3.70E-06	2.36E-04	1.24E-04	2.85E-03	5.71E-09	5.03E-13	3.20E-03	7.09E-11	8.49E-04	3.49E-06	3.20E-03	2.04E-06	5.32E-07
DahA	1.37E+00	3.26E-05	3.19E-05	1.36E+00	6.47E-07	2.71E-04	1.64E-04	3.75E-03	9.06E-10	2.65E-14	4.19E-03	3.73E-12	1.08E-03	3.30E-06	4.19E-03	1.54E-06	1.41E-07
IcdP	7.45E-01	3.23E-05	1.38E-05	7.41E-01	1.85E-05	2.36E-04	1.25E-04	2.86E-03	4.91E-10	8.17E-15	3.22E-03	1.15E-12	8.52E-04	2.42E-06	3.22E-03	1.12E-06	5.74E-08
BghiP	2.58E+00	0	5.89E-07	2.57E+00	5.89E-07	4.35E-04	2.44E-04	5.59E-03	1.29E-10	4.18E-14	6.26E-03	5.89E-12	1.64E-03	4.71E-06	6.26E-03	1.23E-06	1.12E-07

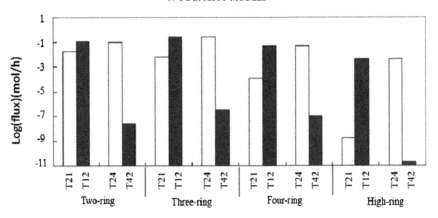

FIGURE 9.6 Transfer fluxes of PAHs across the air–water and water–sediment interfaces in Lake Small Baiyangdian.

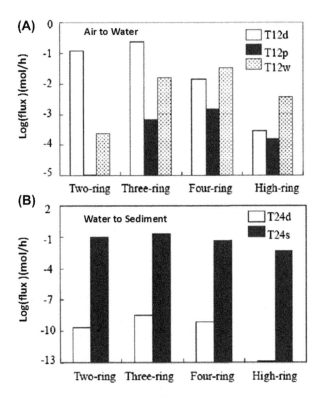

FIGURE 9.7 Transfer fluxes of PAHs through different processes from air to water (A) and from water to sediment (B) in Lake Small Baiyangdian.

9.2.4.4 *Sensitivities of Modeled Concentrations to Input Parameters*

The parameters with sensitivity coefficients higher than 0.5 are considered as more influential parameters in the model. Four compounds, Ace, Phe, Chr, and BaP, were chosen to represent 2-, 3-, 4-, and high-ring PAHs. The model outputs included the concentrations of four representative PAH compounds in the four subphases in the water including dissolved phase (C22), suspended solids (C23), fish (C2f), and aquatic plants (C2p), and in solids subphase in the sediment (C43). The results were summarized in Table 9.9. Among 54 parameters, there were only 17 parameters with sensitivity coefficients higher than 0.5. Temperature (T) was the most influential parameter in the model, and was more sensitive to the model concentrations for Chr and BaP than those for Ace and Phe in all the studied subphases (C22, C23, C2f, C2p, and C43). The number of sensitive parameters for BaP in the studied subphases was 9–13, however, that for Ace and Phe was only three to seven. There were only three parameters, T, K_s, and K_{12}, sensitive to Ace and Phe in C23. For the modeled concentration in a specific studied subphase, the parameters for BaP and Chr were more sensitive than those for Ace and Phe.

9.2.4.5 *Uncertainty of the Modeled Concentrations*

3000 Monte Carlo simulations were performed to simulate the concentrations of four representative PAHs (Ace, Phe, Chr, and BaP) in the seven subphases. Coefficients of variation (CVs) for Ace, Phe, Chr, and BaP in the seven subphases are shown in Fig. 9.8 and the heights of the bars indicate the perturbations of the calculated concentrations.

Fig. 9.8 illustrates that, among four representative PAH components, BaP had the highest CV values of 29% in the particulate subphase in air (C13) to 50% in the sediments (C43), followed by Chr with the CV ranging from about 4% in the particulate subphase in the air (C13) to 22% in the plant subphase in the water (C2p); however, all the CV values for Ace and Phe in seven subphases were less than 5%. This indicated that there were the highest uncertainty for the modeled BaP concentration, and very low uncertainty for the modeled Ace and Phe concentrations. In seven subphases studied, the variabilities for the modeled Ace and Phe concentrations were relatively similar to each other, and those for the modeled Chr and BaP concentrations were in a descending order of C2p > C22 ≈ C2f > C43 ≈ C11 > C23 ≈ C13, and of C43 > C2f > C23 ≈ C22 > C11 > C2p ≈ C13, respectively. The largest uncertainties of the calculated BaP concentrations are related to the most influential parameters identified in the sensitivity analysis (Table 9.9).

9.2.5 The Ecological Implications of the Proposed Model

Ecotoxicological models are increasingly applied to assess the fate and effect of chemical emissions to the environment, and they can be divided into three types, fate models, effect models, and fate-transport-effect models (FTE models; Jørgensen and Fath, 2011). Fate models provide the concentration of a chemical in one or more environmental compartments; effect models translate a concentration or body burden in a biological compartment to an effect either on an organism, a population, a community, an ecosystem, a landscape, or the entire ecosphere; and fate-transport-effect models are the merging of fate models with effect models (Jørgensen and Fath, 2011). So far, many fate models, fewer effect models, and only a

TABLE 9.9 Sensitivity Coefficients of the More Sensitive Parameters in the QWASI Fugacity Model (SCi > 0.5)

		X_{13}	X_{43}	r_{23}	K_{12}	h_4	O_{23}	T	K_s	Ps_{25}	K_{oc}	K_w	S_c	K_{m4}	F_{25}	BP_S	BCF_p	BCF_f
ACE	C22	—	—	-0.68	0.54	—	-0.68	-3.95	-0.67	—	-0.68	—	—	—	—	—	—	—
	C23	—	—	—	0.54	—	—	-3.95	-0.67	—	—	—	—	—	—	—	—	—
	C2f	—	—	-0.68	0.54	—	-0.68	-3.95	-0.67	—	-0.68	—	—	—	—	—	—	1
	C2p	—	—	-0.68	0.54	—	-0.68	-3.95	-0.67	—	-0.68	—	—	—	—	—	1	—
	C43	—	-1	—	0.54	-1.01	—	-3.95	—	—	—	—	—	-1.01	—	—	—	—
PHE	C22	—	—	-0.96	0.87	—	-0.96	-3.75	-0.96	—	-0.96	—	—	—	—	—	—	—
	C23	—	—	—	0.87	—	—	-3.75	-0.96	—	—	—	—	—	—	—	—	—
	C2f	—	—	-0.96	0.87	—	-0.96	-3.75	-0.96	—	-0.96	—	—	—	—	—	—	1
	C2p	—	—	-0.96	0.87	—	-0.96	-3.75	-0.96	—	-0.96	—	—	—	—	—	1	—
	C43	—	-1	—	0.87	-1.01	—	-3.75	—	—	—	—	—	-1.01	—	—	—	—
CHR	C22	0.78	—	-1	—	—	-1	-159.577	-1	-0.78	-1	0.74	0.74	—	0.78	1.51	—	—
	C23	0.78	—	—	—	—	—	-159.57	-1	-0.78	—	0.74	0.74	—	0.78	1.51	—	—
	C2f	0.78	—	-1	—	—	-1	-159.57	-1	-0.78	-1	0.74	0.74	—	0.78	1.51	—	1
	C2p	0.78	—	-1	—	—	-1	-159.57	-1	-0.78	-1	0.74	0.74	—	0.78	1.51	1	—
	C43	0.78	-1.01	—	—	-1.01	—	-159.57	—	-0.78	—	0.74	0.74	-1.01	0.78	1.51	—	—
BaP	C22	0.93	—	-1.01	0.73	—	-1.01	-267.82	-1.01	-0.94	-1.01	0.89	0.89	—	0.93	1.91	—	—
	C23	0.93	—	—	0.73	—	—	-267.82	-1.01	-0.94	—	0.89	0.89	—	0.93	1.91	—	—
	C2f	0.93	—	-1.01	0.73	—	-1.01	-267.82	-1.01	-0.94	-1.01	0.89	0.89	—	0.93	1.91	—	1
	C2p	0.93	—	-1.01	0.73	—	-1.01	-267.82	-1.01	-0.94	-1.01	0.89	0.89	—	0.93	1.91	1	—
	C43	0.93	-1.01	—	0.73	-1.01	—	-267.82	—	-0.94	—	0.89	0.89	-1.01	0.93	1.91	—	—

"—" means SCi < 0.5.

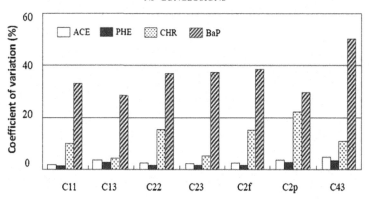

FIGURE 9.8   Comparisons of CV values for the four representative PAHs in the seven subphases.

few FTE models have been applied to solve ecotoxicological problems and perform ecological risk assessments; however, the development is toward a wider application of effect and FTE models (Jørgensen and Fath, 2011).

Through the QWASI fugacity model developed in the present study, the PAHs concentrations in main environmental compartments including the air (air, particulates), water (water, suspended solids, plants, and fishes), and sediment (water and solids) were derived. These results could be used for the ecological risk assessments of PAHs in the Lake Small Baiyangdian. The proposed multimedia fate model could serve as a fundamental for developing an effect model and FTE model to solve ecotoxicological problems and perform ecological risk assessments of PAHs in the Lake Small Baiyangdian. Further, the applications of the effect model and FTE model in the lake would promote the development of ecotoxicological model.

The modeling results in the present study point out that macrophytes play a very important role in maintaining a healthy lake ecosystem by taking up toxic substances and by creating a favorable environment for a variety of complex chemical, biological, and physical processes that contribute to the removal and degradation of toxic pollutants (Xu et al., 1999). Macrophytes growing in a lake are also crucial to regulate lake biological structure, because they limit algal growth by shading and competing for nutrients with algae and because they increase herbivorous fish biomass by providing food and a refuge (Xu et al., 1999).

9.3 CONCLUSIONS

A QWASI fugacity model was developed to characterize the fate and transfer of 15 priority PAHs in Lake Small Baiyangdian. The reliability of the model estimates was evaluated by various means including concentration validation, sensitivity, and uncertainty analysis. There was generally good agreement between the modeled and measured concentrations with the differences within an order of magnitude for the majority of PAH components. The fluxes into and out of the lake as well as each compartment were well balanced. The average bioaccumulation flux of PAHs by plants was four times higher than that by fishes. The transfer directions of PAHs were from air to water and to sediment. Temperature was the most influential parameter and was more sensitive to the modeled concentrations of middle- and high-molecular-weight PAHs that were considered as the source of the model uncertainty. Overall, the model developed in this study could well characterize the fate and transfer of PAHs in a well-defined ecosystem, i.e., a typical vegetation-dominant lake, which is rarely reported before. This approach provide an illustrative case for fugacity model study within a small scale and open for a wider application of this model approach for future studies.

Acknowledgments

The funding for this study was provided by National Project for Water Pollution Control (2012ZX07103-002), the National Science Foundation of China (NSFC) (41271462, 41030529). This work is also supported by a grant from the 111 Project (B14001).

References

Ao, J.T., Chen, J.W., Tian, F.L., Cai, X.Y., 2009. Application of a level IV fugacity model to simulate the long-term fate of hexachlorocyclohexane isomers in the lower reach of Yellow River basin, China. Chemosphere 74, 370–376.

Arnot, J.A., Gobas, F., 2004. A food web bioaccumulation model for organic chemicals in aquatic ecosystems. Environmental Toxicology and Chemistry 23, 2343–2355.

Baek, J.M., Park, S.J., 2000. Tracking the distribution of organic compounds using fugacity model. Korean Journal of Chemical Engineering 17, 12–16.

Bai, Y.J., 2008. Pollution characteristics of polycyclic aromatic hydrocarbons (PAHs) in surface water in western area around Bohai Bay. Ph.D. Thesis. Peking University, Beijing.

Ballschmiter, K., 1992. Transport and fate of organic-compounds in the global environment. Angewandte Chemie-International Edition in English 31, 487–515.

Banks, R.B., Herrera, F.F., 1997. Effect of Wind and Rain on Surface Reaeration. J. Environ. Engr. Div. ASCE 103 (EE3), 489–504.

Beijing Statistic Bureau, 2005. Beijing Statatistic Year Book. Chinese Statistic Press, Beijing.

Binelli, A., Provini, A., 2003. The PCB pollution of Lake Iseo (N. Italy) and the role of biomagnification in the pelagic food web. Chemosphere 53, 143–151.

Burns, L.A., Cline, D.M., Lassiter, R.R., 1982. Exposure Analysis Modeling System (EXAMS): User Manual and System Documentation. EPA-600/3-82-023. Environmental Research Laboratory, US EPA, Athens, CA.

Calamari, D., Vighi, M., Bacci, E., 1987. The use of terrestrial plant biomass as a parameter in the fugacity model. Chemosphere 16, 2359–2364.

Campfens, J., Mackay, D., 1997. Fugacity-based model of PCB bioaccumulation in complex aquatic food webs. Environmental Science & Technology 31, 577–583.

Cao, H.Y., Liang, T., Tao, S., Zhang, C.S., 2007. Simulating the temporal changes of OCP pollution in Hangzhou, China. Chemosphere 67, 1335–1345.

Cao, H.Y., Tao, S., Xu, F.L., Coveney, R.M., Cao, J., Li, B.G., Liu, W.X., Wang, X.J., Hu, J.Y., Shen, W.R., Qin, B.P., Sun, R., 2004. Multimedia fate model for hexachlorocyclohexane in Tianjin, China. Environmental Science & Technology 38, 2126–2132.

Carsel, R.F., Smith, C.N., Mulkey, L.A., 1984. User's Manual for the Pesticide Root Zone Model (PRZM). release 1, EPA-600/3-84-109. US EPA, Athens, GA.

Chen, J.G., Zhou, W.H., Deng, A.J., Sun, G.H., 2006. Formation and evolution of the longitudinal profile of the Lower Yellow River in modern times. Sediment Study 2 (1), 1–8.

Cohen, Y., 1984. Modeling of Pollutant Transport and Accumulation in a Multimedia Environment. In: The Conference on Geochemical and Hydrologic Processes and their Protection, Council on Environmental Quality, September 25, Washington, DC.

Connolly, J.P., Pedersen, C.J., 1988. A thermodynamic-based evaluation of organic-chemical accumulation in aquatic organisms. Environmental Science & Technology 22, 99–103.

Cowan, C.E., M, D., Feijtel, T.C.J., van de Meent, D., Di Guardo, A., Davies, J., Mackay, N., 1995. The Multi-Media Fate Model: A Vital Tool for Predicting the Fate of Chemicals. SETAC Press, Pensacola, FL.

Davenport, J., 1999. Swimbladder volume and body density in an armoured benthic fish, the streaked gurnard. Journal of Fish Biology 55, 527–534.

Diamond, M.L., MacKay, D., Poulton, D.J., Stride, F.A., 1996. Assessing chemical behavior and developing remedial actions using a mass balance model of chemical fate in the Bay of Quinte. Water Research 30, 405–421.

Diepens, N.J., Arts, G.H.P., Focks, A., Koelmans, A.A., 2014. Uptake, Trans location, and Elimination in Sediment-Rooted Macrophytes: A Model-Supported Analysis of Whole Sediment Test Data. Environmental Science & Technology 48, 12344–12353.

Dong, J.Y., Gao, H., Wang, S.G., Yao, H.J., Ma, M.Q., 2009. Simulation of the transfer and fate of HCHs since the 1950s in Lanzhou, China. Ecotoxicology and Environmental Safety 72, 1950–1956.

Duan, Y.H., 2005. Distribution and source–sink relationship of polycyclic aromatic hydrocarbons in topsoil from Tianjin. Ph.D. Thesis. Peking University, Beijing.

Edwards, F.G., Egemen, E., Brennan, R., Nirmalakhandan, N., 1999. Ranking of toxics release inventory chemicals using a Level III fugacity model and toxicity. Water Science and Technology 39, 83–90.

Gilbertson, M., 1996. Organochlorine contaminants in the Great Lakes. Ecological Applications 6, 966–971.

Gobas, F., Arnot, J.A., 2010. Food web bioaccumulation model for polychlorinated biphenyls in San Francisco bay, California, USA. Environmental Toxicology and Chemistry 29, 1385–1395.

Gobas, F., Wilcockson, J.B., Russell, R.W., Haffner, G.D., 1999. Mechanism of biomagnification in fish under laboratory and field conditions. Environmental Science & Technology 33, 133–141.

Gobas, F., Zhang, X., Wells, R., 1993. Gastrointestinal magnification - the mechanism of biomagnification and food-chain accumulation of organic-chemicals. Environmental Science & Technology 27, 2855–2863.

Hauck, M., Hendriks, H.W.M., Huijbregts, M.A.J., Ragas, A.M.J., van de Meent, D., Hendriks, A.J., 2011. Parameter uncertainty in modeling bioaccumulation factors of fish. Environmental Toxicology and Chemistry 30, 403–412.

HPEPB (Hebei Provincial Environmental Protection Bureau), 2001. Environmental Quality Statement 1996–2000. Hebei Provincial Environmental Protection Bureau, Hebei.

Hu, C.H., Ji, Z.W., Huang, Y.J., Chen, D., 1998. Analysis on dredging practice in rivers, lakes and reservoirs. Journal of Sediment Research 12 (4), 47–55.

Hung, H., Mackay, D., 1997. A novel and simple model of the uptake of organic chemicals by vegetation from air and soil. Chemosphere 35, 959–977.

Jørgensen, S.E., Fath, B.D., 2011. Fundamentals of Ecological Modelling, fourth ed. Elsevier, Amsterdam.

Jørgensen, S.E., Nielsen, S.N., 1994. Models of the structural dynamics in lakes and reservoirs. Ecological Modelling 74, 39–46.

Jin, X.C., 1990. Chemistry of Organic Compound Pollutant. Tsinghua University Press, Beijing.

Karickhoff, S.W., 1981. Semi-empirical estimation of sorption of hydrophobic pollutants on natural sediments and soils. Chemosphere 10, 833–846.

Koelmans, A.A., Van der Heijde, A., Knijff, L.M., Aalderink, R.H., 2001. Integrated modelling of eutrophication and organic contaminant fate & effects in aquatic ecosystems. A review. Water Research 35, 3517–3536.

Kong, X., He, W., Qin, N., He, Q., Yang, B., Ouyang, H., Wang, Q., Yang, C., Jiang, Y., Xu, F., 2014. Modeling the multimedia fate dynamics of γ-hexachlorocyclohexane in a large Chinese lake. Ecological Indicators 41, 65–74.

Kong, X.Z., He, W., Qin, N., He, Q.S., Yang, B., Ouyang, H.L., Wang, Q.M., Yang, C., Jiang, Y.J., Xu, F.L., 2012. Simulation of the Fate and Seasonal Variations of alpha-Hexachlorocyclohexane in Lake Chaohu Using a Dynamic Fugacity Model. Scientific World Journal. http://dx.doi.org/10.1100/2012/691539.

Lang, C., Tao, S., Wang, X.J., Zhang, G., Li, J., Fu, J.M., 2007. Seasonal variation of polycyclic aromatic hydrocarbons (PAHs) in Pearl River Delta region, China. Atmospheric Environment 41, 8370–8379.

Lang, C., Tao, S., Wang, X.J., Zhang, G., Fu, J.M., 2008. Modeling polycyclic aromatic hydrocarbon composition profiles of sources and receptors in the Pearl River Delta, China. Environmental Toxicology and Chemistry 27 (1), 4–9.

Li, Y.L., He, W., Liu, W.X., Kong, X.Z., Yang, B., Yang, C., Xu, F.L., 2015. Influences of binding to dissolved organic matter on hydrophobic organic compounds in a multi-contaminant system: coefficients, mechanisms and ecological risks. Environmental Pollution 206, 461–468.

Liu, Z.Y., Quan, X., Yang, F.L., 2007. Long-term fate of three hexachlorocyclohexanes in the lower reach of Liao River basin: Dynamic mass budgets and pathways. Chemosphere 69, 1159–1165.

Ma, Z.P., Zhao, J.H., Kang, X.J., Jing, A.Q., 2007. The wind-driven water circulation in Baiyangdian Lake, China and the implication to environmental remediation. Oceanologia et Limnologia Sinica 38 (5), 405–410.

Mackay, D., Shiu, W.Y., Ma, K.C., 1997. Illustrated Handbook of Physical–Chemical Properties and Environmental Fate for Organic Chemicals. In: Pesticide Chemicals, Volume V. Lewis Publishers, Boca Raton, FL.

Mackay, D., 1979. Finding fugacity feasible. Environmental Science & Technology 13, 1218–1223.

Mackay, D., 1982. Correlation of bioconcentration factors. Environmental Science & Technology 16, 274–278.

Mackay, D., 2001. Multimedia Environmental Models: The Fugacity Approach, second ed. Lewis Publishers, New York, USA.

Mackay, D., Diamond, M., 1989. Application of the QWASI (quantitative water air sediment interaction) fugacity model to the dynamics of organic and inorganic chemicals in lakes. Chemosphere 18, 1343–1365.

Mackay, D., Fraser, A., 2000. Bioaccumulation of persistent organic chemicals: mechanisms and models. Environmental Pollution 110, 375–391.

Mackay, D., Hughes, A.I., 1984. 3-parameter equation describing the uptake of organic-compounds by fish. Environmental Science & Technology 18, 439–444.

Mackay, D., Joy, M., Paterson, S., 1983a. A quantitative water, air, sediment interaction (QWASI) fugacity model for describing the fate of chemicals in lakes. Chemosphere 12, 981–997.

Mackay, D., Paterson, S., 1981. Calculating fugacity. Environmental Science & Technology 15, 1006–1014.

Mackay, D., Paterson, S., 1982a. Fugacity revisited. Environmental Science & Technology 16, 654A–660A.

Mackay, D., Paterson, S., 1982b. Fugacity revisited - the fugacity approach to environmental transport. Environmental Science & Technology 16, A654–A660.

Mackay, D., Paterson, S., 1991. Evaluating the multimedia fate of organic-chemicals - a Level-III fugacity model. Environmental Science & Technology 25, 427–436.

Mackay, D., Paterson, S., Joy, M., 1983b. A quantitative water, air, sediment interaction (QWASI) fugacity model for describing the fate of chemicals in rivers. Chemosphere 12, 1193–1208.

Mackay, D., Paterson, S., Schroeder, W.H., 1986. Model describing the rates of transfer processes of organic-chemicals between atmosphere and water. Environmental Science & Technology 20, 810–816.

Mackay, D., Paterson, S., Shiu, W.Y., 1992. Generic models for evaluating the regional fate of chemicals. Chemosphere 24, 695–717.

MathWorks, 2002. Using MATLAB Version 6. The MathWorks, Inc, Natick, MA, US.

McKone, T.E., 1996. Alternative modeling approaches for contaminant fate in soils: uncertainty, variability, and reliability. Reliability Engineering & System Safety 54, 165–181.

McKone, T.E., Maddalena, R.L., 2007. Plant uptake of organic pollutants from soil: bioconcentration estimates based on models and experiments. Environmental Toxicology and Chemistry 26, 2494–2504.

Paasivirta, J., Sinkkonen, S., Mikkelson, P., Rantio, T., Wania, F., 1999. Estimation of vapor pressures, solubilities and Henry's law constants of selected persistent organic pollutants as functions of temperature. Chemosphere 39, 811–832.

Paraiba, L.C., Bru, R., Carrasco, J.M., 2002. Level IV fugacity model depending on temperature by a periodic control system. Ecological Modelling 147, 221–232.

Paraiba, L.C., Carrasco, J.M., Bru, R., 1999. Level IV Fugacity Model by a continuous time control system. Chemosphere 38, 1763–1775.

Parajulee, A., Wania, F., 2014. Evaluating officially reported polycyclic aromatic hydrocarbon emissions in the Athabasca oil sands region with a multimedia fate model. Proceedings of the National Academy of Sciences of the United States of America 111, 3344–3349.

Paterson, S., Mackay, D., McFarlane, C., 1994. A model of organic-chemical uptake by plants from soil and the atmosphere. Environmental Science & Technology 28, 2259–2266.

Perry, R.H., Chilton, C.H., 1973. Chemical Engineers' Handbook, 5th edition. McGraw-Hill Book Co., New York, pp. 230–235.

Saloranta, T.M., Armitage, J.M., Haario, H., Naes, K., Cousins, I.T., Barton, D.N., 2008. Modeling the effects and uncertainties of contaminated sediment remediation scenarios in a Norwegian Fjord by Markov chain Monte Carlo simulation. Environmental Science & Technology 42, 200–206.

Schnoor, J.L., Mcavoy, D.C., 1981. Pesticide transport and bioconcentration model. J. of Environ. Engineering – American Society of Civil Engineering 107, 1229–1246.

Sharpe, S., Mackay, D., 2000. A framework for evaluating bioaccumulation in food webs. Environmental Science & Technology 34, 2373–2379.

Shor, L.M., Rockne, K.J., Taghon, G.L., Young, L.Y., Kosson, D.S., 2003. Desorption Kinetics for Field-Aged Polycyclic Aromatic Hydrocarbons from Sediments. Environmental Science & Technology 37 (8), 1535–1544.

STF (Soil Transport and Fate Database and Model Management System), 1991. Environmental Systems and Technologies. Blacksburg, USA.

Tao, S., Cao, H.Y., Liu, W.X., Li, B.G., Cao, J., Xu, F.L., Wang, X.J., Coveney, R.M., Shen, W.R., Qin, B.P., Sun, R., 2003. Fate modeling of phenanthrene with regional variation in Tianjin, China. Environmental Science & Technology 37, 2453–2459.

Tao, S., Cao, J., Wang, W., Zhao, J., Wang, W., Wang, Z., Cao, H., Xing, B., 2009. A Passive Sampler with Improved Performance for Collecting Gaseous and Particulate Phase Polycyclic Aromatic Hydrocarbons in Air. Environmental Science & Technology 43, 4124–4129.

Tao, S., Yang, Y., Cao, H.Y., Liu, W.X., Coveney, R.M., Xu, F.L., Cao, J., Li, B.G., Wang, X.J., Hua, J.Y., Fang, J.Y., 2006. Modeling the dynamic changes in concentrations of γ-hexachlorocyclohexane (γ-HCH) in Tianjin region from 1953 to 2020. Environmental Pollution 139, 183–193.

Tang, M.J., Xu, Z.X., Zuo, Q., Huang, M.H., Tao, S., 2006. Multimedia fate modeling of PAHs in Guangdong Hong Kong, and Macao. Ecology and Environment 15 (4), 670–673.

Ten Hulscher, T.E.M., Vandervelde, L.E., Bruggeman, W.A., 1992. Temperature dependence of Henry's law constants for selected chlorobenzenes, polychlorinated biphenyls and polycyclic aromatic hydrocarbons. Environmental Toxicology and Chemistry 11 (11), 1595–1603.

Tianjin Environmental Protection Bureau (TJEPB), 1991. Environmental Quality Statement. Tianjin Environmental Protection Bureau, Tianjin.

Tianjin Environmental Protection Bureau (TJEPB), 1996. Environmental Quality Statement. Tianjin Environmental Protection Bureau, Tianjin.

Tianjin Environmental Protection Bureau (TJEPB), 2001. Environmental Quality Statement. Tianjin Environmental Protection Bureau, Tianjin.

Thomas, G., Sweetman, A.J., Ockenden, W.A., Mackay, D., Jones, K.C., 1998. Air-pasture transfer of PCBs. Environmental Science & Technology 32, 936–942.

Thibodeaux, L.J., 1996. Environmental chemodynamics: movement of chemicals in air, water, and soil. John Wiley & Sons, INC.

US EPA, 1996. Soil Screening Guidance: Technical Background Document. US Environmental Protection Agency, Office of Emergency and Remedial Response, Washington, 9355.4–17A.

Van Agreren, M.H., Sytze, K., Dick, B.J., 1998. Handbook on Biodegradation and Biological Treatment of Hazardous Organic Compounds. Kluwer Academic Publishers, New York.

Wang, C., Feng, Y., Sun, Q., Zhao, S., Gao, P., Li, B.-L., 2012. A multimedia fate model to evaluate the fate of PAHs in Songhua River, China. Environmental Pollution 164, 81–88.

Wang, R., Cao, H.Y., Li, W., Wang, W., Wang, W.T., Zhang, L.W., Liu, J.M., Ouyang, H., Tao, S., 2011. Spatial and seasonal variations of polycyclic aromatic hydrocarbons in Haihe Plain, China. Environmental Pollution 159, 1413–1418.

Wang, L.S., 1991. Chemistry of Organic Pollutant. Science Press, Beijing.

Wang, L.S., 1993. Chemical Carcinogens. Chinese Environment Science Press, Beijing.

Wang, X.L., Tao, S., Xu, F.L., Dawson, R.W., Cao, J., Li, B.G., Fang, J.Y., 2002. Modeling the fate of benzo a pyrene in the wastewater-irrigated areas of Tianjin with a fugacity model. Journal of Environmental Quality 31, 896–903.

Wania, F., Breivik, K., Persson, N.J., McLachlan, M.S., 2006. CoZMo-POP 2-A fugacity-based dynamic multi-compartmental mass balance model of the fate of persistent organic pollutants. Environmental Modelling & Software 21, 868–884.

Wania, F., Mackay, D., 1995. A global distribution model for persistent organic-chemicals. Science of the Total Environment 160-61, 211–232.

Xu, F.L., Jørgensen, S.E., Tao, S., Li, B.G., 1999. Modeling the effects of ecological engineering on ecosystem health of a shallow eutrophic Chinese lake (Lake Chao). Ecological Modelling 117, 239–260.

Xu, F.L., Qin, N., Zhu, Y., He, W., Kong, X.Z., Barbour, M.T., He, Q.S., Wang, Y., Ou-Yang, H.L., Tao, S., 2013. Multimedia fate modeling of polycyclic aromatic hydrocarbons (PAHs) in Lake Small Baiyangdian, Northern China. Ecological Modelling 252, 246–257.

Xu, F.L., Wu, W.J., Wang, J.J., Qin, N., Wang, Y., He, Q.S., He, W., Tao, S., 2011. Residual levels and health risk of polycyclic aromatic hydrocarbons in freshwater fishes from Lake Small Bai-Yang-Dian, Northern China. Ecological Modelling 222, 275–286.

Xu, Z.H., Mao, Z.X., Wang, L.S., Pang, Y.L., 1991. Handbook of Chemistry Property Estimate Method (Environmental Property of Organic Compound). Chemical Industry Press, Beijing.

Yoshida, K., Shigeoka, T., Yamauchi, F., 1987. Multi-phase unsteady state equilibrium model for evaluation of environmental fate of organic chemicals. Toxicological and Environmental Chemistry 15, 159–183.

Yin, J.M., 2008. Chaos characterization of water level and adding solution of water in Lake Bai-Yang-Dian. Masters Thesis. Agricultural University of Hebei.

Zhang, X.L., Tao, S., Liu, W.X., Yang, Y., Zuo, Q., Liu, S.Z., 2005. Source diagnostics of polycyclic aromatic hydrocarbons based on species ratios: A multimedia approach. Environmental Science & Technology 39, 9109–9114.

Zhang, Y., Tao, S., Cao, J., Coveney Jr., R.M., 2007. Emission of polycyclic aromatic hydrocarbons in China by county. Environmental Science & Technology 41, 683–687.

Zhang, Y., Tao, S., Shen, H., Ma, J., 2009. Inhalation exposure to ambient polycyclic aromatic hydrocarbons and lung cancer risk of Chinese population. Proceedings of the National Academy of Sciences of the United States of America 106, 21063–21067.

Zhao, F., 1995. Baiyangdian large aquatic resource survey and the impact of eutrophication. Journal of Environmental Sciences 16, 21–24.

Zhao, X., Cui, B.S., Yang, Z.F., 2005. A study of the lowest ecological water level of Baiyangdian Lake. Acta Ecologica Sinica 25 (5), 1033–1040.

Zhao, Y.H., Lian, J.Y., Zhao, X.P., 2007. Protection for habitat security of biological resources in wetland of Baiyangdian Natural Reserve. Journal of Shijiazhuang Vocational Technology Institute 17 (2), 1–4.

Fuzzy Adaptive Management of Coupled Natural and Human Systems

T. Prato

Professor Emeritus, Department of Agricultural and Applied Economics,
University of Missouri, Columbia, MO, United States
E-mail: pratoa@missouri.edu

OUTLINE

10.1 INTRODUCTION

Coupled natural and human systems are systems for which natural and human elements interact (Liu et al., 2007). Three kinds of rules can be used to make management decisions for coupled systems: crisp, stochastic, and fuzzy (Prato, 2009). Crisp decision rules assume that a manager can make unambiguous assessments regarding the state of a coupled system based on measured or forecasted values for system attributes. An example of a crisp decision rule is that the system is strongly sustainable if $X_i \geq X_i^*$ for positive attributes and $X_j \leq X_j^*$ for negative attributes, where X_i and X_j are measured values and X_i^* and X_j^* are threshold values of attributes, respectively. Crisp decision rules do not account for sampling and measurement errors in attribute data and stochastic variability in attributes both of which can result in decision errors regarding the state of a coupled system. Stochastic decision rules account for stochastic variability in system attributes, and thereby reduce the likelihood of decision errors. An example of a stochastic decision rule is that a coupled system is strongly sustainable if $p\{X_i \geq X_i^*\} \geq \alpha_i$ for positive attributes and $p\{X_j \leq X_j^*\} \geq \beta_j$ for negative attributes of the system, where p stands for probability, $0 \leq \alpha_i \leq 1$, and $0 \leq \beta_j \leq 1$. Most stochastic decision rules require managers to specify probability distributions for system attributes, which is not possible when there is uncertainty about changes over time in those attributes. Fuzzy decision rules allow for sampling and measurement errors in the data used to estimate attributes and stochastic variability in system responses to drivers and do not require managers to specify probability distributions for system drivers or attributes.

Managing coupled systems for climate change and human impacts is challenging as evidenced by remarks by Ann Rodman of the Yellowstone Resources Center. She described the challenge of managing Yellowstone National Park—a coupled system—with respect to climate change as follows: "Knowing this [warming of climate systems] is true does not necessarily help us understand how these changes are affecting Yellowstone National Park and the surrounding area. … It is even harder to go out on a limb and say how complex natural systems, with all their fuzzy feedback mechanisms, might react to a changing climate" (Rodman, 2015). Rodman's statement is apropos to all coupled systems, not just Yellowstone National Park. In general, managers of coupled systems and scientists whose research informs management decisions for coupled systems face the daunting task of understanding how coupled systems are likely to respond to management actions and uncertain future changes in climate and human use. Examples of management actions for alleviating potential adverse impacts of climate change on coupled systems include assisted migration, preserving genetic diversity for foundation species, and conserving nature's stage (Carswell, 2015).

Fuzzy decision rules are based on fuzzy logic, which is a mathematical way of representing the imprecise or approximate nature of decision-making under uncertainty (Zadeh, 1965; Bellman and Zadeh, 1970; Bass and Kwakernaak, 1977; Barrett and Pattanaik, 1989; Klir and Yuan, 1995; Carlsson and Fuller, 1996; Phillis and Andriantiatsaholiniaina, 2001; Andriantiatsaholiniaina et al., 2004; Prato, 2007). Fuzzy logic is well suited to formulating decision rules for managing coupled systems when there is uncertainty about future changes in system driver(s) (e.g., Chen, 2003; Adriaenssens, 2004; Svoray, 2004; Prato, 2009, 2012). Most fuzzy logic-based decision rules involve the use of complex mathematical operations (e.g., Prato, 2005, 2009) that are difficult for managers of coupled systems to understand and apply. In contrast, the fuzzy logic-based decision rules proposed here can be understood by coupled system managers.

This paper describes a fuzzy logic-based, adaptive management (FLAM) model for adapting coupled systems to future climate change when there is uncertainty about the extent of that change and the efficacy of management actions in achieving desired outcomes. The FLAM model provides three kinds of results: (1) the preferred management action for each climate change scenario and time period determined by applying the fuzzy Technique for Order Preference by Similarity of Ideal Solution (fuzzy TOPSIS) to a manager's ratings of the estimated attributes of system responses to climate change and management actions (or attributes for short) and the relative importance of those attributes; (2) the preferred management action for each time period determined by applying the minimax regret criterion to the first result; and (3) the best strategy for adapting management actions to future climate change across time periods determined based on the second result.

10.2 METHODS

This section describes a FLAM model for managing the backcountry of Cascadia National Park (CNP), a hypothetical coupled system. A hypothetical coupled system is used because the FLAM model has not been applied to a real coupled system. In addition, using a hypothetical coupled system makes the description of the model less abstract.

10.2.1 FLAM Model for CNP

CNP's backcountry contains environmentally sensitive alpine and subalpine areas that are vulnerable to natural resource degradation and loss of biodiversity from human use and climate change. In recent years, the demand for backcountry camping permits has increased. For that reason, park managers want to determine preferred management actions over time for increasing the number of backcountry campsites while minimizing degradation of backcountry areas from increased human use and climate change. Increasing the number of backcountry campsites would allow CNP managers to issue more backcountry camping permits thereby increasing the number of backcountry campers.

An overview of the FLAM model for CNP is illustrated in Fig. 10.1. The model assumes that, in each of four, five-year time periods, a different climate change scenario can occur, a different management action can be implemented, and different system responses can result. The hypothetical FLAM model for CNP contains four climate change scenarios (C_1, C_2, C_3, and C_4), four management actions (A_1, A_2, A_3, and A_4), and four system responses to climate change and management actions (R_1, R_2, R_3, and R_4). Climate change scenarios are equivalent to the Intergovernmental Panel on Climate Change's (IPCC's) four representative concentration pathways (RCPs), namely RCP2.6, RCP4.5, RCP6, and RCP8.5 (IPCC, 2014). The number associated with each RCP (e.g., 2.6) refers to radiative forcing in the tropopause measured in watts per square meter of the Earth's surface. The higher the radiative forcing, the greater the climate change.

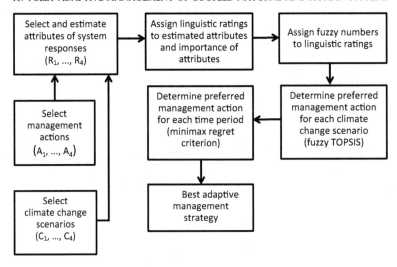

FIGURE 10.1   Diagram of FLAM model for CNP.

The four management actions correspond to four percentage increases in the number of backcountry campsites chosen by CNP managers: 0% increase for A_1; 5% increase for A_2; 10% increase for A_3; and 15% increase for A_4. System responses are discussed in the next section.

Preferred management actions for the CNP are determined taking into account two kinds of uncertainty. First, there is uncertainty about future climate change, which implies the probabilities of the climate change scenarios are unknown in all time periods. Climate change uncertainty is realistic because the IPCC's Fifth Assessment Report does not assign probabilities to the RCPs.

Second, there is uncertainty about system responses to climate change scenarios and management actions, which implies the probabilities of R1 through R4 are unknown. Consequently, risk-based decision frameworks that require knowledge of the probabilities of different responses to system drivers cannot be used to determine the preferred management actions and best adaptive management strategy for CNP. Examples of such frameworks include maximizing the expected value of responses and Bayesian belief networks, both of which require knowledge of the $p(R_j)$s, and Monte Carlo simulation that requires knowledge of the probability distributions for climate variables (i.e., precipitation and temperature) for each climate change scenario. The FLAM model does not require knowledge of $p(C_i)$s and $p(R_j)$s, or the probability distributions for climate variables.

10.2.2 Attributes

Coupled systems are typically managed for multiple attributes. Consequently, the FLAM model evaluates system responses to climate change and management actions in terms of multiple attributes. Other studies have used multiple attributes to characterize system responses. For example, Mackinson (2000) developed an adaptive fuzzy expert system for predicting structure, dynamics, and distribution (responses) of herring shoals. Responses were evaluated in terms of 26 attributes. Prato (2005) developed a FLAM model for determining whether or not management of an ecosystem is strongly sustainable based on three attributes: regional income, biodiversity, and water quality.

The FLAM model for CNP evaluates system responses in terms of four attributes: two attributes for backcountry campers' satisfaction and two attributes for resource protection. Camper satisfaction attributes are the percent of backcountry camping permittees that believe: (1) the number of parties encountered on backcountry trails is excessive (PT); and the number of backcountry campsites is sufficient (PC). Conservation attributes are the percent of backcountry areas with favorable habitat for threatened and endangered species (PF) and the percent of backcountry trails with unacceptable soil erosion rates (SE).

10.2.3 Adaptive Management

Adaptive management (AM) has been proposed and/or used to adapt management actions over time in response to changes in system drivers (Holling, 1978; Walters, 1996; Parma, 1998; Prato, 2007). AM is a form of integrated learning that acknowledges and accounts for the surprising and unpredictable nature of system responses due to uncertainty about the temporal changes in drivers. Kohm and Franklin state that "adaptive management is the only logical approach under the circumstances of uncertainty" (Kohm and Franklin, 1997). Baron et al. (2009) assert that AM is the best way to manage natural protected areas for future climate change and variability.

Adaptive management can be active or passive. There are different opinions about the distinction between active and passive AM (Bormann et al., 1996; Schreiber et al., 2004). Based on Walters (1996), Williams (2011) defines active AM as an approach that evaluates management alternatives for reducing uncertainty about ecological processes and how those processes are influenced by management actions, and passive AM as an approach that focuses on resource management objectives with less emphasis on learning about the efficacy of management actions on ecological processes. Nyberg (1998) and Prato (2012) define active AM as a management approach that designs and conducts experiments to test hypotheses about the efficacy of management alternatives and adapts management alternatives over time when warranted based on test results, and passive AM as a management approach that does not involve experiments and hypothesis testing. For the results of active AM to be statistically reliable, the experiments must incorporate replicated, randomized, and independent treatments and controls; the latter refer to management actions.

For the hypothetical coupled system, active AM experiments are conducted on the four management actions to generate experimental data on the attributes of system responses to those actions and climate change. Experiments are based on the following experimental design. The backcountry region of CNP is divided into several biophysical zones having dissimilar biophysical characteristics. Each biophysical zone is divided into several areas that serve as experimental units. At the beginning of the first time period, the four management actions are randomly assigned to the areas within each zone, such that the number of areas treated with the same management action (i.e., number of replications per zone) is approximately the same.

A_1	A_4	A_3	A_2	A_4
A_1	A_2	A_4	A_3	A_1
A_4	A_3	A_1	A_2	A_4
A_2	A_1	A_3	A_2	A_3

FIGURE 10.2   Example of random allocation of management actions to areas of a biophysical zone.

Fig. 10.2 illustrates a random allocation of management actions to areas within a single zone. For simplicity, the figure illustrates a square zone containing square areas. In reality, zones and areas are not square. At the end of each time period, backcountry satisfaction attributes are measured for each zone using data obtained from backcountry camping surveys and conservation attributes are measured using field surveys. If the backcountry area of CNP is divided into m biophysical zones and each zone has five replications of management actions, then there are 5m data points per attribute per time period for each management action.

10.2.4 Estimating Attributes

The FLAM model estimates the attributes using one of two methods. First, the attributes of responses to experimental combinations of climate change scenarios and management actions (i.e., combinations for which experiments have been performed) are estimated using the experimental data. For example, if climate change scenario C_2 occurs during the first time period, then the attributes under C_2 are estimated using the experimental data for the four management actions under C_2. Valid experimental data are available for estimating attributes when the experiments have replicated, randomized, and independent treatments and controls (i.e., management actions). Second, the attributes of responses to nonexperimental combinations of climate change scenarios and management actions (i.e., combinations for which experimental data are not available) are estimated using expert judgment (Linstone and Turoff, 2002) and/or simulation models. For example, it would be necessary to use the second method to estimate the attributes of responses under C_1, C_3, and C_4 if climate change scenario C_2 occurs during the first time period. The combined effect of climate change and management actions on PF can be estimated using habitat suitability models (e.g., Store and Jokimäki, 2003) and on SE can be estimated using the Variable Cross-Sectional Area method (Olive and Marion, 2009).

10.2.5 Determining Preferred Management Actions

A preferred management action is determined at the beginning of each time period using a two-step procedure. In the first step, fuzzy TOPSIS is used to rank the four management actions for each climate change scenario. The preferred management action for each climate change scenario is the top-ranked action for that scenario. In the second step, the preferred management action for each time period is determined by applying the minimax regret criterion to maximum loss indices for the preferred management actions identified in the first step. This section explains both steps.

10.2.5.1 Determining Preferred Management Action for Climate Change Scenarios

The preferred management action for each climate change scenario within time periods is determined by applying fuzzy TOPSIS to the estimated attributes of responses for each combination of management action and climate change scenario. For example, at the beginning of the first time period, fuzzy TOPSIS is applied to the estimated attributes of responses to C_1 and A_1, C_1 and A_2, C_1 and A_3, and C_1 and A_4 to determine which of the four management actions is preferred under C_1. This procedure is repeated to determine the preferred management actions under C_2, C_3, and C_4.

As an example, consider using fuzzy TOPSIS to calculate the distances of the estimated attributes of responses to C_1 and A_1, C_1 and A_2, C_1 and A_3, and C_1 and A_4 from the fuzzy positive-ideal solution (d_i^+) and the fuzzy negative-ideal solution (d_i^-) for the attributes and the closeness coefficient (E_i) for each attribute. These metrics are calculated as follows:

$$d_i^+ = \sum_j d\left(w_j r_{ij}, v_j^+\right)$$ (10.1)

$$d_i^- = \sum_j d\left(w_j r_{ij}, v_j^-\right)$$ (10.2)

$$E_i = d_i^- / \left(d_i^+ + d_i^-\right) \quad (0 \le E_i \le 1)$$ (10.3)

where:

i = C_1 and A_1, C_1 and A_2, C_1 and A_3, and C_1 and A_4;
j = PT, PC, PF, and SE;
w_j = normalized triangular fuzzy number corresponding to the linguistic variable chosen by managers to rate the relative importance of attribute j;
r_{ij} = normalized triangular fuzzy number corresponding to the linguistic variable chosen by managers to rate the estimated effect of C_1 and A_i on attribute j;
$d(w_j r_{ij}, v_j^+)$ = vertex distance between the weighted normalized fuzzy effect of management action i and climate change scenario C_1 on attribute j and the positive-ideal solution for attribute j; and
$d(w_j r_{ij}, v_j^-)$ = vertex distance between the weighted normalized fuzzy effect of management action i and climate change scenario C_1 on attribute j and the negative-ideal solution for attribute j.

Various linguistic variables can be used to rate the estimated attributes and their relative importance. An example of linguistic variables and corresponding triangular fuzzy numbers is given in Table 10.1. Each triangular fuzzy number (a, b, c) defines a triangular probability distribution like the one illustrated in Fig. 10.3. The triangular probability distribution for a random variable x is $T(a, b, c) = [2(x - a)/(c - a)(b - a)]$ for $a \le x \le b$ and $T(a, b, c) = [2(c - x)/(c - a)(c - b)]$ for $b < x \le c$, where a is the minimum value, b is the modal

TABLE 10.1   Linguistic Rating Scale for Estimated Attributes and Relative Importance of Attributes, and Corresponding Triangular Fuzzy Numbers

Linguistic Rating	Triangular Fuzzy Number[a]
Very low	$(0.05, 0.05, 1)$[b]
Low	$(0.05, 1, 3)$
Moderate	$(3, 5, 7)$
High	$(7, 9, 10)$
Very high	$(9, 10, 10)$

[a]*Adapted from Chen (2000) and Prato (2012).*
[b]*The first number is the minimum value, the second number is the mode, and the third number is the maximum value for a triangular probability distribution.*

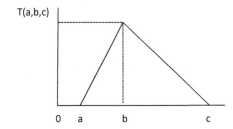

FIGURE 10.3   Triangular probability distribution.

value, and c is the maximum value of x. Fuzzy numbers can be defined based on other probability distributions besides the triangular.

The vertex distance between two triangular fuzzy numbers $z_1 = (e_1, e_2, e_3)$ and $z_2 = (k_1, k_2, k_3)$ is $d(z_1, z_2) = \{0.33[(e_1 - k_1)^2 + (e_2 - k_2)^2 + (e_3 - k_3)^2]\}^{0.5}$, where, for Eqs. (10.1) and (10.2), $z_1 = w_j r_{ij}$ and $z_2 = v_j^+$ or v_j^-.

A positive attribute is one for which more of the attribute is desirable and a negative attribute is one for which less of the attribute is desirable from the viewpoint of the decision maker. For CNP managers, PC and PF are positive attributes and PT and SE are negative attributes. The fuzzy positive- and negative-ideal solutions for the four attributes are:

$$v_j^+ = (1, 1, 1) \text{ for } j = PC \text{ and } PF;$$

$$v_j^- = (0, 0, 0) \text{ for } j = PC \text{ and } PF;$$

$$v_j^+ = (0, 0, 0) \text{ for } j = PT \text{ and } SE; \text{ and}$$

$$v_j^- = (1, 1, 1) \text{ for } j = PT \text{ and } SE.$$

E_i approaches 0 (or 1) as the triangular fuzzy numbers for the attributes of the response to C_1 and A_i move farther away from (or closer to) the fuzzy positive-ideal solution and closer to (or farther away from) the attributes for the fuzzy negative-ideal solution for attributes. Because the desirability of a response decreases (or increases) as the closeness coefficient for the attributes of that response approaches zero (or one), the four responses to C_1 and A_1, C_1 and A_2, C_1 and A_3, and C_1 and A_4 are ranked based on the values of the closeness coefficients. The rank order for these four responses implies a rank order for the four management actions. For example, if the rank order of the responses indicate C_1 and A_3 is preferred to C_1 and A_2 is preferred to C_1 and A_1 is preferred to C_1 and A_4, then the rank order of management actions under C_1 is A_3 is preferred to A_2 is preferred to A_1 is preferred to A_4. Therefore, A_3 is the preferred management action under C_1. This ranking procedure is repeated for responses involving C_2, C_3, and C_4 and each of the four management actions to determine the preferred management actions under each of those climate change scenarios. Suppose the preferred management actions under C_2, C_3, and C_4 for the first time period are A_4 under C_2 and C_3, and A_2 under C_4.

10.2.5.2 *Example of Preferred Management Action for a Climate Change Scenario*

This section uses a numerical example to illustrate how fuzzy TOPSIS is used to determine the preferred management action for one climate change scenario in one time period, namely climate change scenario C_1 in the first time period. Applying fuzzy TOPSIS involves several steps. In the first step, the manager must linguistically rate the estimated attributes of responses to combinations of C_1 and each of the four management actions and the relative of importance of the attributes. For the hypothetical example, the CNP managers determine the linguistic ratings shown in Table 10.2. If there is a management team for CNP and the team collectively rates the estimated attributes of responses and the relative of importance of attributes, then the triangular fuzzy numbers corresponding to the collective linguistic ratings are used. If managers on the team independently rate the estimated attributes of responses and the relative of importance of attributes, then the triangular fuzzy numbers corresponding to the linguistic ratings made by individual members of the team are averaged to obtain collective fuzzy numbers. If there are multiple estimated values of an attribute for the same management action, climate change scenario, and time period, then the triangular

TABLE 10.2   Linguistic Ratings of the Estimated Values and Relative Importance of the Four Attributes of Responses Under Climate Change Scenario C_1 in the First Time Period

Management Action	PT	PC	SE	PF
A_1	Very low	Very low	Very low	Very high
A_2	Low	Moderate	Moderate	High
A_3	Moderate	High	High	Moderate
A_4	High	Very high	High	Low
Importance	Moderate	Moderate	High	Very high

fuzzy numbers corresponding to the linguistic ratings for the multiple estimated values are averaged.

In the second step, the fuzzy effects matrix (see Table 10.3) is created by assigning the triangular fuzzy numbers in Table 10.1 to the corresponding linguistic ratings in Table 10.2. In the third step, the normalized fuzzy effects matrix is formed (see Table 10.4). Each element of the normalized fuzzy effects matrix is formed by applying the following formula to the corresponding element of the fuzzy effects matrix:

$$\text{Positive attributes}: r_{ij} = \left(a_{ij}/c_j^+, b_{ij}/c_j^+, c_{ij}/c_j^+\right) \text{ where } c_j^+ = \max_i c_{ij} \ (j = PC, PF); \text{ and}$$

$$\text{Nagative attributes}: r_{ij} = \left(a_j^-/c_{ij}, a_j^-/b_{ij}, a_j^-/a_{ij}\right) \text{ where } a_j^- = \min_i a_{ij} \ (j = PT, SF).$$

In the fourth step, the weighted normalized fuzzy effects matrix is determined (see Table 10.5) by multiplying the normalized weight for an attribute and the corresponding normalized fuzzy effect for that attribute. For example, the weighted normalized fuzzy effect of A_1 on PT is $w_{PT} \times r_{A1PT}$, where w_{PT} is the triangular fuzzy number for the relative importance of PT and r_{A1PT} is the normalized fuzzy number for the effect of A_1 on PT. The numerical weighted normalized fuzzy effect of A_1 on PT is $w_{PT} \times r_{A1PT} = (0.3, 0.5, 0.7)(0.05, 1, 1) = (0.015, 0.5, 0.7)$. Other weighted normalized fuzzy effects are calculated in a similar manner.

TABLE 10.3   Fuzzy Effects Matrix for Climate Change Scenario C_1 and Four Management Actions in the First Time Period and Fuzzy Numbers for Relative Importance of Attributes

Management Action	PT	PC	SE	PF
A_1	(0.05, 0.05, 1)	(0.05, 0.05, 1)	(0.05, 0.05, 1)	(9, 10, 10)
A_2	(0.05, 1, 3)	(3, 5, 7)	(3, 5, 7)	(7, 9, 10)
A_3	(3, 5, 7)	(7, 9, 10)	(7, 9, 10)	(3, 5, 7)
A_4	(7, 9, 10)	(9, 10, 10)	(7, 9, 10)	(0.05, 1, 3)
Importance	(3, 5, 7)	(3, 5, 7)	(7, 9, 10)	(9, 10, 10)

TABLE 10.4   Normalized Fuzzy Effects Matrix for Climate Change Scenario C_1 and Four Management Actions in the First Time Period

Management Action	PT	PC	SE	PF
A_1	(0.05, 1, 1)	(0.9, 1, 1)	(0.05, 1, 1)	(0.005, 0.005, 0.11)
A_2	(0.017, 0.05, 1)	(0.7, 0.8, 0.9)	(0.007, 0.01, 0.017)	(0.33, 0.55, 0.77)
A_3	(0.007, 0.01, 0.017)	(0.3, 0.5, 0.7)	(0.005, 0.006, 1)	(0.77, 0.88, 1)
A_4	(0.005, 0.006, 0.007)	(0.005, 0.1, 0.3)	(0.005, 0.005, 0.005)	(0.77, 0.88, 1)

TABLE 10.5 Weighted Normalized Fuzzy Effects Matrix for Climate Change Scenario C_1 and Four Management Actions in the First Time Period

Management Action	PT	PC	SE	PF
A_1	(0.015, 0.5, 0.7)	(0.81, 1, 1)	(0.015, 0.5, 0.7)	(0.004, 0.004, 0.1)
A_2	(0.005, 0.025, 0.7)	(0.63, 0.8, 0.9)	(0.002, 0.005, 0.012)	(0.23, 0.44, 0.7)
A_3	(0.002, 0.005, 0.012)	(0.27, 0.5, 0.7)	(0.002, 0.003, 0.7)	(0.54, 0.71, 0.9)
A_4	(0.002, 0.003, 0.005)	(0.004, 0.1, 0.3)	(0.001, 0.002, 0.003)	(0.54, 0.71, 0.9)

10.2.5.3 Determining Preferred Management Actions for Time Periods

The preferred management action for a time period is determined by applying the minimax regret criterion to the preferred management actions for the climate change scenarios for that time period. With the minimax regret criterion, the preferred management action for a time period is the one that minimizes the maximum loss index (MLI) over the four preferred management actions under C_1, C_2, C_3, and C_4 for that time period. The MLI for the preferred management action for a climate change scenario is a weighted average of the expected maximum losses in the attributes that would occur if that action was implemented. Expected maximum loss for a single attribute with the preferred management action for a climate change scenario is the estimated value of that attribute with that action and no future climate change minus the estimated value of that attribute with that action and the climate change scenario for which that action is preferred. Losses in individual attributes without and with climate change are estimated using biophysical simulation models, visitor surveys, and/or mental models. Construction of the MLI requires managers to assign weights to the four attributes, such that the weights sum to one.

The MLIs for the preferred management actions for each climate change scenario in the first time period are given in Table 10.6. A_3 is the preferred management action for the first time period because it has the lowest MLI across the four climate change scenarios. Therefore, A_3 is implemented at the beginning of the first time period.

The preferred management actions for the second, third, and fourth time periods are determined using a similar procedure to the one used for the first time period with two major differences. First, the attributes for the second, third, and fourth time periods are measured using a combination of experimental data (provided active AM is implemented at the

TABLE 10.6 MLIs for Preferred Management Actions Under Four Climate Change Scenarios in the First Time Period

Climate change scenario	C_1	C_2	C_3	C_4
Preferred management action	A_3	A_4	A_4	A_2
MLI	45	58	65	72

beginning of the first time period) and nonexperimental data. Second, if active AM is implemented, then there are multiple observations on the attributes of responses to C_{it} and A_1, C_{it} and A_2, C_{it} and A_3, and C_{it} and A_4, where C_{it} is the climate change scenario that occurred during time period t for t ≥ 2. In this case, distances from the fuzzy positive-ideal solution $(d_i{}^+)$ and fuzzy negative-ideal solution $(d_i{}^-)$ for those attributes and the closeness coefficient (E_i) for each management action are estimated for each observation and the average value of E_i over the multiple observations is used to rank management actions.

10.2.5.4 *Determining Best Adaptive Management Strategy*

The previous section describes how the minimax regret criterion is used to determine that A_3 is the preferred management action for the first time period. Suppose the minimax regret criterion indicates that the preferred management actions for the second through fourth time periods are A_3 in the second time period, A_2 in the third time period, and A_1 in the fourth time period. Based on these results, the best AM strategy for CNP is to implement A_3 at the beginning of the first time period and continue using it through the end of the second time period, implement A_2 at the beginning of the third time period, and implement A_1 at the beginning of the fourth time period. This strategy assumes it is feasible to change management actions across time periods. If that is not the case, then the best strategy would have to be altered to accommodate any limitations on altering management actions across time periods for CNP.

10.3 DISCUSSION

This section discusses the steps a coupled system manager would have to take in order to implement the FLAM model. Those steps include: (1) selecting the length and number of time periods for evaluating management actions; (2) picking future climate change scenarios; (3) choosing management actions; (4) selecting the multiple attributes of system responses; (5) if active AM is used, designing and conducting AM experiments in each time period to determine how experimental combinations of climate change scenarios and management actions influence the attributes; (6) using expert judgment, surveys, and/or simulation models to estimate how nonexperimental combinations of climate change scenarios and management actions influence the attributes; (7) using fuzzy TOPSIS to determine the preferred management action for each climate change scenario within time periods; (8) applying the minimax regret criterion to determine the preferred management action for each time period; and (9) determining the best AM strategy.

The time periods should span a sufficiently enough long period of time to allow climate change and management actions to fully influence the attributes. The slower the responses to climate change and management actions, the longer the time periods should be.

The most commonly used climate change scenarios are the ones developed by the IPCC (see Section 10.2.1). The IPCC scenarios project changes in climate through 2100. The FLAM model requires climate projections for a portion of or the entire area of a coupled system. Such projections can be developed using the high-resolution (800-m), bias-corrected, ensemble climate projections for monthly average maximum temperature, minimum temperature, and precipitation developed by NASA Earth Exchange (NEX)

(Thrasher et al., 2013). Such projections should be sufficient to evaluate the responses of a coupled system to climate change. The choice of management actions and attributes depends to a large extent on the mission of the agency managing the coupled system and the management challenges facing that system. For example, the hypothetical CNP is a national park. National parks in the United States are managed by the National Park Service whose goal is "… to conserve the scenery and natural and historic objects and the wildlife therein and to provide for the enjoyment of the same by such manner and by such means as will leave them unimpaired for the enjoyment of future generation" (National Park Service Organic Act of 1914). For the hypothetical CNP, the four management actions and the two camper satisfaction attributes address enhancing visitor enjoyment by increasing the number of backcountry campsites, and the two conservation attributes address conserving scenery, natural objects, and wildlife.

Designing and conducting AM experiments requires applying the principles of experimental design, a well-established branch of statistics. In some cases, it may not be feasible for managers to conduct AM experiments due to technical and/or financial reasons. For example, from a technical viewpoint, if the management actions pertain to alleviating adverse impacts of climate change on grizzly bears, it is unlikely that the treatments (i.e., management actions) applied to biophysical zones will be independent because the home range for grizzly bears is likely to exceed the size of a biophysical zone. From a financial viewpoint, managers may not be able to afford long-term AM experiments due to limited operating budgets.

Using expert judgment, such as the Delphi method, surveys, and/or simulation models, to estimate how nonexperimental combinations of climate change scenarios and management actions influence the attributes is likely to be the most difficult step in implementing the FLAM model. While these estimation methods are commonly used by researchers, they are not typically used by coupled system managers. Managers can overcome this potential limitation by enlisting the assistance of technical experts that are knowledgeable about the application of these methods.

Applying fuzzy TOPSIS requires the coupled system manager to assign linguistic ratings to the estimated values and relative importance of the attributes, such as the ones shown in Table 10.2 and perform the fuzzy mathematical operations needed to generate Tables 10.2−10.5. The hypothetical CNP generates numerous estimated attributes, which means that managers would have to do a large number of linguistic ratings of estimated attributes. The number of linguistic ratings can be substantially reduced by developing a look-up table that indicates which linguistic variables the manager assigns to different percentage ranges of the estimated attributes. An example of a look-up table for an attribute is 0−15% is very low, 16−45% is low, 46−65% is moderate, 66−84% is high, and greater than 85% is very high. A look-up table is not be needed to rate the relative importance of the attributes because it only requires one rating per attribute. If the linguistic ratings vary across time periods, then it would be necessary to develop separate look-up tables for each time period.

The fuzzy mathematical operations required to generate Tables 10.3−10.5 can be performed using a spreadsheet developed by the author. That spreadsheet requires the user to create a table, like Table 10.1, that assigns numerical codes (e.g., 1 through 5) to the five linguistic ratings/triangular fuzzy numbers, and develop a matrix for each climate change scenario that contains the corresponding codes for all combinations of management actions and attributes and the relative importance of the attributes. The spreadsheet automatically generates Tables 10.3—10.5 based on those codes.

Using the minimax regret criterion is relatively straightforward once the MLIs have been calculated (i.e., Table 10.6). Defining the MLI requires the manager to estimate the attributes without future climate change for each time period, which can be done using biophysical simulation models, visitor surveys, and/or mental models, and assign weights to the attributes.

Determining the best AM strategy is straightforward once the preferred management actions have been determined for all time periods.

10.4 CONCLUSIONS

This chapter describes a FLAM model for adaptively managing a hypothetical coupled system when the manager is uncertain about the extent of future changes in system drivers and system responses to both climate change and management actions. The FLAM model demonstrates two advantages of fuzzy decision rules. First, the fuzzy decision rule used in the FLAM model allows for sampling and measurement errors in attribute data and stochastic variability in attributes. Second, the decision rules used in the model do not require managers to specify probability distributions for system attributes. This feature is particularly advantageous when climate change is one of the drivers of system attributes because probability distributions for climate change scenarios have not been specified.

In general, the FLAM model can be applied to any coupled system. The mathematical operations required by most fuzzy logic-based decision rules are complex. In contrast, the mathematical operations required to implement the FLAM model have been programmed into a spreadsheet that is relatively easy to use. However, other elements of the model, notably using expert judgment, surveys, and/or simulation models to estimate how nonexperimental combinations of climate change scenarios and management actions influence the attributes, are more challenging to apply. For that reason, managers who want to use the FLAM model would most likely have to enlist the assistance of individuals who are familiar with applying those elements.

References

Adriaenssens, V., De Baets, B., Goethals, P.L.M., De Pauw, N., 2004. Fuzzy rule-based models for decision support in ecosystem management. Science of the Total Environment 319, 1—12.

Andriantiatsaholiniaina, L.A., Kouikoglou, V.S., Phillis, Y.A., 2004. Evaluating strategies for sustainable development: fuzzy logic reasoning and sensitivity analysis. Ecological Economics 48, 149—172.

Baron, J.S., Gunderson, L., Allen, C.D., Fleishman, E., McKenzie, D., Meyerson, L.A., Oropeza, J., Stevenson, N., 2009. Options for national parks and reserves for adapting to climate change. Environmental Management 44, 1033—1042.

Barrett, C.R., Pattanaik, P.K., 1989. Fuzzy sets, preference and choice: some conceptual issues. Bulletin of Economic Research 41, 229—253.

Bass, S.M., Kwakernaak, H., 1977. Rating and ranking of multiple-aspect alternatives using fuzzy sets. Automatica 13, 47—58.

Bellman, R.E., Zadeh, L.A., 1970. Decision-making in a fuzzy environment. Management Science 17, B-141—B-164.

Bormann, B.T., Cunningham, P.G., Gordon, J.C., 1996. Best management practices, adaptive management, or both?. In: Proceedings, National Society of American Foresters Convention, Portland, ME.

Carlsson, C., Fuller, R., 1996. Fuzzy multiple criteria decision making: recent developments. Fuzzy Sets and Systems 78, 139–153.

Carswell, C., 2015. Tree of life. High Country News 47, 13–19.

Chen, C.-T., 2000. Extensions to the TOPSIS for group decision–making under fuzzy environment. Fuzzy Sets and Systems 114, 1–9.

Chen, Q., Mynett, A.E., 2003. Integration of data mining techniques and heuristic knowledge in fuzzy logic modelling of eutrophication in Taihu Lake. Ecological Modelling 162, 55–67.

Thrasher, B., Xiong, J., Wang, W., Melton, F., Michaelis, A., Nemani, R., 2013. Downscaled climate projections suitable for resource management–summary. Eos 94, 321–323.

Holling, C.S., 1978. Adaptive Environmental Assessment and Management. Wiley, Chichester, England.

Intergovernmental Panel on Climate Change (IPCC), 2014. Fifth Assessment Report (AR5). http://ipcc.ch/report/ar5/.

Klir, G.J., Yuan, B., 1995. Fuzzy Sets and Fuzzy Logic: Theory and Applications. Prentice-Hall, Inc., Upper Saddle River, NJ.

Kohm, K.A., Franklin, J.F., 1997. Introduction. In: Kohm, K.A., Franklin, J.F. (Eds.), Creating Forestry for the 21st Century: The Science of Ecosystem Management. Island Press, Washington, DC, pp. 1–5.

Linstone, H.A., Turoff, M. (Eds.), 2002. The Delphi Method: Techniques and Applications. http://is.njit.edu/pubs/delphibook/.

Liu, J., Dietz, T., Carpenter, S.R., Alberti, M., Folke, C., Moran, E., Pell, A.N., Deadman, P., Kratz, T., Lubchenco, J., Ostrom, E., Ouyang, Z., Provencher, W., Redman, C.L., Schneider, S.H., Taylor, W.W., 2007. Complexity of coupled human and natural systems. Science 317, 1513–1516.

Mackinson, S., 2000. An adaptive fuzzy expert system for predicting structure, dynamics and distribution of herring shoals. Ecological Modelling 126, 155–178.

Nyberg, J.B., 1998. Statistics and the practice of adaptive management. In: Sit, V., Taylor, B. (Eds.), Statistical Methods for Adaptive Management Studies, Land Management Handbook No. 42. Ministry of Forests Research Program, Victoria, BC, pp. 1–8.

Olive, N.D., Marion, J.L., 2009. The influence of use-related, environmental, and managerial factors on soil loss from recreational trails. Journal of Environmental Management 90, 1483–1493.

Parma, A.M., 1998. NCEAS Working Group on Population Management. What can adaptive management do for our fish, forest, food, and biodiversity? Integrative Biology 1, 16–26.

Phillis, Y.A., Andriantiatsaholiniaina, L.A., 2001. Sustainability: an ill-defined concept and its assessment using fuzzy logic. Ecological Economics 37, 435–456.

Prato, T., 2005. A fuzzy logic approach to ecosystem sustainability. Ecological Modelling 187, 361–368.

Prato, T., 2007. Assessing ecosystem sustainability and management using fuzzy logic. Ecological Economics 61, 171–177.

Prato, T., 2009. Fuzzy adaptive management of social and ecological carrying capacities for protected areas. Journal of Environmental Management 90, 2551–2557. http://dx.doi.org/10.1016/j.jenvman.2009.01.015.

Prato, T., 2012. Increasing resilience of natural protected areas to future climate change: a fuzzy adaptive management approach. Ecological Modelling 242, 46–53. http://dx.10.1016/j.ecolmodel.2012.05.014.

Rodman, A., 2015. Fear is not the answer. Ecological implications of climate change on the greater yellowstone ecosystem (special issue). Yellowstone Science 23, 2.

Schreiber, E.S.G., Bearlin, A.R., Nicol, S.J., Todd, C.R., 2004. Adaptive management: a synthesis of current understanding and effective application. Ecological Management and Restoration 5, 177–182.

Store, R., Jokimäki, J., 2003. A GIS-based multi-scale approach to habitat suitability modeling. Ecological Modelling 169, 1–15.

Svoray, T., Gancharski, S.-Y., Henkin, Z., Gutman, M., 2004. Assessment of herbaceous plant habitats in water-constrained environments: predicting indirect effects with fuzzy logic. Ecological Modelling 180, 537–556.

Walters, C., 1996. Adaptive Management of Renewable Resources. Blackburn Press, Caldwell, New Jersey.

Williams, B.K., 2011. Passive and active adaptive management: approaches and an example. Journal of Environmental Management 92, 1371–1378.

Zadeh, L.A., 1965. Fuzzy sets. Information and Control 8, 338–353.

CHAPTER

11

Coastal Ecosystem Modeling in the Context of Climate Change: An Overview With Case Studies

D. Justic[*,1], S.M. Duke-Sylvester[§], J.M. Visser[§], Z. Xue[*], J. Liang[*]

[*]Louisiana State University, Baton Rouge, LA, United States and [§]University of Louisiana, Lafayette, LA, United States

[1]Corresponding author: E-mail: djusti1@lsu.edu

OUTLINE

Ecological Model Types, Volume 28
http://dx.doi.org/10.1016/B978-0-444-63623-2.00011-6

11.1 MODELS AS TOOLS FOR ASSESSING THE EFFECTS OF CLIMATE CHANGE ON COASTAL ECOSYSTEMS

Future climatic conditions are expected to include a warmer atmosphere, warmer and more acidic oceans, higher sea levels, and altered precipitation and freshwater discharge patterns, all of which will profoundly influence coastal ecosystems and people living near the coasts. In order to adapt to continuing climate change, it is important to understand how different climatic drivers affect ecosystem processes and species distribution in coastal habitats. Climatic drivers do not act alone but rather in combination with other ecosystem stressors, such as habitat degradation, pollution, eutrophication, and overharvesting of commercially important species. To project ecosystem responses over long temporal and large spatial scales and to separate the influence of climatic drivers from other ecosystem stressors, numerical simulation models have emerged as a major research tool to study the effects of climate change on coastal ecosystems. The complexity and size of these models have been steadily increasing over the past 20 years due to continuing developments in computer technologies and computational techniques but also in response to increased concerns about the sustainability of coastal ecosystems under future climatic conditions.

Numerical models used in the assessment of effect of climate change on coastal ecosystems can be broadly divided into two groups. General circulation models or global climate models (GCMs), such as those featured in the Intergovernmental Panel on Climate Change reports (e.g., IPCC, 2014), are primarily aimed at modeling Earth's climate system at relatively coarse scales (~ 100 km). In contrast, regional models rely on downscaled GCM products and simulate climate and associated ecosystem processes at much finer scales (~ 10 km). Numerous regional models have been developed to predict the extent and the potential future effects of climate change on coastal ecosystems, including the rising sea levels (e.g., Craft et al., 2009; McLeod et al., 2010; Storlazzi et al., 2011; Rogers et al., 2012), storm surges (e.g., Wang et al., 2008; Tebaldi et al., 2012), altered precipitation and freshwater discharge patterns (e.g., Najjar et al., 2010; Sperna Weiland et al., 2012; Gochis et al., 2014; Ren et al., 2015), acidification and deoxygenation (e.g., Lerman et al., 2011; Bendtsen and Hansen, 2013), harmful algal blooms (e.g., Glibert et al., 2014), and fisheries impacts (e.g., Hare et al., 2010; Diamond et al., 2013; Rose et al., 2015; Ruzicka et al., 2016). Many of the models used in climate change impact studies have originally been developed for other applications, e.g., to investigate the influences of specific ecosystem stressors or explain the environmental effects of various human actions and subsequently adapted for climate research applications as regional climate change scenarios became available.

The four case studies presented below discuss modeling studies from ecologically diverse coastal ecosystems of the northern Gulf of Mexico, California Coastal Current system, and the Western Pacific (Fig. 11.1). The approach was not intended to systematically evaluate the use of numerical models in coastal ecosystems but rather to provide examples of how coastal ecosystem modeling is being carried out in the context of climate change.

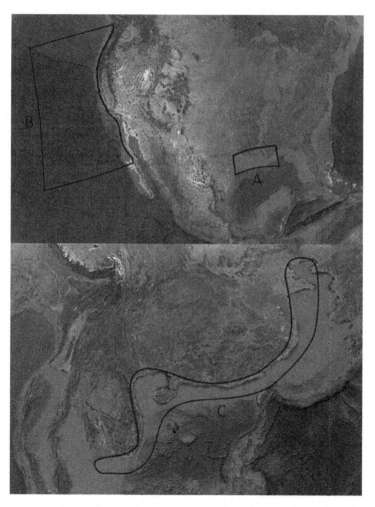

FIGURE 11.1   A map showing the locations of sites in (A) the northern Gulf of Mexico, (B) California Coastal Current system, and (C) the Western Pacific where the selected case studies have been conducted.

11.2 CASE STUDIES

11.2.1 Modeling the Effects of Climate Change on Louisiana's Coastal Wetland Plant Communities

Plant communities play a central role in shaping wetland ecosystems. Both the species composition and plant growth forms define habitat conditions for a diverse collection of arthropods, birds, reptiles, fish, mammals, and a host of other organisms. The plant species that comprise a wetland also shape hydrology and edaphic conditions through processes such as evapotranspiration and frictional resistance to water flow, thus affecting sedimentation and erosion rates. For these reasons, vegetation models are an integral part of ecosystem modeling projects (Craft et al., 2009; Davis and Ogden, 1994; DeAngelis et al., 1998; Peyronnin et al., 2013). The LAVegMod vegetation model is an integral part of a suite of models developed for the Louisiana Coastal Master Plan (LCMP). An earlier version of the model is described in Visser et al. (2013). This section describes an updated and expanded version of the LAVegMod model that is currently used in the development of the 2017 LCMP (Visser et al., 2015). For a more comprehensive view of the overall LCMP modeling effort, see http://coastal.la.gov/a-common-vision/2017-master-plan-update/technical-analysis/.

The objectives of LAVegMod model are twofold. First, it is designed to provide a landscape-scale assessment of potential changes in the response of Louisiana coastal plant communities to natural and anthropogenic perturbations that may occur in coming decades. Second, it is a tool for assessing the effectiveness of management and restoration projects aimed at preserving and enhancing wetlands for their ecological values as well as for sustainable commercial use. This section focuses on the first objective and explores the potential shifts in coastal plant communities in response to various future climate change scenarios.

11.2.1.1 Development and Implementation of the LAVegMod Model

The LAVegMod vegetation model is a component of an integrated compartment model (ICM) that also includes a hydrodynamic box model, a wetland geomorphology model, a barrier island model, and several habitat suitability models. These models are linked together and allow for feedback between the processes captured by each model. The hydrology model simulates the movement of water over the landscape and does so based in part on the distribution of soils and the distribution of plants. Information from the hydrology model is passed to the soil morphology model and the vegetation model. The wetland geomorphology model uses the distribution of water provided, as well as the distribution of plant species, to determine rates of soil accumulation and erosion. The vegetation model receives information from the morphology model about projected additions and/or losses, as well as the distribution of water depth and salinity in space and time provided by the hydrology model, and uses this information to update the distribution of plant species on the landscape.

11.2.1.1.1 MODEL STRUCTURE

The model is spatially explicit and simulates changes in the area covered by each of the 36 plant species (Table 11.1) over time. The species included in our model represent many of the most common dominant plants found in coastal Louisiana. They are associated with the marsh and forest habitats that characterize most of Louisiana's coastland. Each species is

TABLE 11.1 List of Species Included in the LAVegMod Model

Model Type	Habitat	Species	Species Code
Hardwood forest	Bottomland hardwood forest	*Quercus lyrata* Walter	QULE
		Quercus texana Buckley	QUTE
		Quercus laurifolia Michx.	QULA3
		Ulmus americana L.	ULAM
		Quercus nigra L.	QUNI
		Quercus virginiana Mill.	QUVI
Swamp forest	Swamp forest	*Salix nigra* Marshall	SANI
		Taxodium distichum (L.) Rich.	TADI2
		Nyssa aquatica L.	NYAQ2
Floating marsh	Fresh floating marsh	*Panicum hemitomon* Schult.	PAHE2
		Eleocharis baldwinii (Torr.) Chapm.	ELBA2
		Hydrocotyle umbellata L.	HYUM
Emergent marsh	Fresh attached marsh	*Morella cerifera* (L.) small	MOCE2
		Panicum hemitomon Schult.	PAHE2
		Eleocharis baldwinii (Torr.) Chapm.	ELBA2
		Hydrocotyle umbellata L.	HYUM
		Sagittaria latifolia Willd.	SALA2
		Zizaniopsis miliacea (Michx.) Döll & Asch.	ZIMI
		Cladium mariscus (L.) Pohl	CLMA10
		Typha domingensis Pers.	TYDO
		Schoenoplectus californicus (C.A. Mey.) Palla	SCCA11
	Intermediate marsh	*Sagittaria lancifolia* L.	SALA
		Phragmites australis (Cav.) Trin. ex Steud.	PHAU7
		Iva frutescens L.	IVFR
		Baccharis halimifolia L.	BAHA
	Brackish marsh	*Spartina patens* (Aiton) Muhl.	SPPA
		Paspalum vaginatum Sw.	PAVA
	Saline marsh	*Juncus roemeriaunus* Scheele	JURO
		Distichlis spicata (L.) Greene	DISP
		Spartina alterniflora Loisel.	SPAL
		Avicennia germinans (L.) L.	AVGE

(Continued)

TABLE 11.1   List of Species Included in the LAVegMod Model—cont'd

Model Type	Habitat	Species	Species Code
Barrier Island	Dune	*Uniola paniculata* L.	*UNPA*
		Panicum amarum Elliott	*PAAM2*
		Sporobolus virginicus (L.) Kunth.	*SPVI3*
	Swale	*Spartina patens* (Aiton) Muhl.	*SPPABI*
		Distichlis spicata (L.) Greene	*DISPBI*
		Solidago sempervirens L.	*SOSE*
		Strophostyles helvola (L.) Elliott	*STHE9*
		Baccharis halimifolia L.	*BAHABI*
Submerged aquatic vegetation	Water		*SAV*

characterized by the environmental conditions that allow its establishment and persistence. The model also includes the information on the conditions that may lead to plant senescence.

The spatial domain of the model is the entire Louisiana coast, extending from the Texas border to the Mississippi border. The southernmost extent of the model falls within the Gulf of Mexico and is located at approximately the 30 m isobath which corresponds to a distance between 40 and 100 km from the coast (Fig. 11.2). This domain was selected so that the hydrology along the border was no longer influenced by freshwater and estuarine processes.

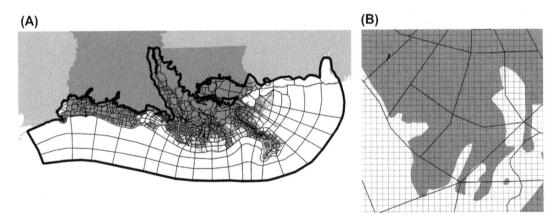

FIGURE 11.2   (A) The spatial domain of the LAVegMod model. Gray areas denote land, with Louisiana shown in dark gray and the neighboring Gulf coast states shown in lighter gray. The heavy black line denotes the overall model domain. The partitions used by the hydrology model are shown as thinner black lines. The red box shows the location and extent of the area show in panel B. (B) A subset of the model domain. Areas that are wetland marshes are shown in gray and open water areas are shown in white. The irregular polygons are the boxes used by the hydrology model. The regular lattice is the 500 × 500 m grid used by the vegetation model.

Instead the hydrology dynamics at the southern boundary are primarily driven by tidal and wave processes. The northern boundary of the model was selected to correspond to the 10 m elevation contour with the Atchafalaya basin drainage added up to the stream flow gage located at Simmesport, Louisiana (Fig. 11.2).

The 48,722 km^2 spatial domain is divided into a structured grid composed of 500 × 500 m cells. Within each cell, the model only tracks the fraction of the area that is covered by each species as well as the area that is covered by water. The model does not track the spatial structure of species within a cell.

The temporal domain of the model is a 50-year planning horizon. The model updates the plant species distributions at the end of the growing season (October) on a yearly time step. In each time step, the fraction of each plot that is covered by each species is evaluated and updates in the vegetation cover are performed on a plot-by-plot basis. Changes in the vegetation cover within a plot are based on the local hydrology and morphology, as well as on the range of hydrologic conditions each species can tolerate and dispersal of propagules from neighboring locations.

The 36 plant species are grouped into six functional groups that are represented by different vegetation submodels: emergent wetlands, hardwood forest, swamp forest, floating marsh, submerged aquatics (SAV), and barrier island vegetation (Table 11.1). For each species, the probability of establishment and senesce are specified. At each time step, the model uses this information in a two-step process. In the first step, the model computes the probability of senescence for each species based on the hydrology and salinity conditions. The cover of each species within a cell is decreased in proportion to its probability of senescence. The reduction in cover of species i within a plot is calculated as:

$$C_i'(t+1) = \left[1 - P_{senescence,i}\{H(t), S(t)\}\right] C_i(t) \tag{11.1}$$

where t is time, $C_i(t)$, is the cover of species i at time t, $H(t)$ is local hydrology conditions, $S(t)$ is salinity, $P_{senescence,i}\{H(t),S(t)\}$ is the probability of senescence under the local hydrology and salinity conditions and $C_i'(t+1)$ is the cover of species i at time $t+1$ after the effects of senescence have been assessed, but before the effects of establishment have been assessed. The $H(t)$ and $S(t)$ values are different for different functional groups and are defined in the description of each submodel. The location index is not included in the above notation to minimize the notational clutter. However, Eq. (11.1) is applied to each 500 × 500 m cell of the model, where $H(t)$, $S(t)$, $C_i(t)$ and $C_i'(t+1)$ represent local quantities for each cell.

The second model step is used to determine what species become established in any open area within a plot. The open area within a plot consists of the area made vacant by the senescing of species from the first step plus any area that was not previously occupied. The available area is divided among the species in proportion to their probability of establishment under the local hydrology and salinity conditions. Establishment is also governed by the ability of species to disperse from the area surrounding each 500 × 500 m cell. The increase in area covered by species i is evaluated as:

$$C_i(t+1) = \left[\left(A - \sum_{j=1}^{K} C_j(t)\right) + \sum_{j=1}^{K}\left\{C_j(t) - C_j'(t+1)\right\}\right] \frac{P_{establish,i}(H(t), S(t))P_{disp,i}}{\sum_{j=1}^{K} P_{establish,j}(H(t), S(t))P_{disp,j}}$$

$$\tag{11.2}$$

where t is time, and i and j are species indices. A is the total area in the plot, $C_j(t)$ is the cover of species j, the sum of $C_j(t)$ is the total area covered by species at time step t, and the difference of this sum and A is the area that was unoccupied at time t. $C'_j(t+1)$ is the cover of species j after the effects of senescence have been assessed, the difference between $C_j(t)$ and $C'_j(t+1)$ is the area lost by species j, and the sum of these differences is the total area vacated as a result of senescence. The sum of the first two terms on the right-hand side is the total area that is unoccupied and is available for species to become established. This quantity is multiplied by the relative probability of establishment by species i, where $P_{establish,i}(H(t),S(t))$ is the probability of species i becoming established under conditions, $H(t)$ and $S(t)$. $P_{disp,i}$ is the probability of species i dispersing into the local patch from the surrounding area. The product of $P_{establish,i}$ and $P_{disp,i}$ is normalized by the total probability of establishment summed over all species. As in Eq. (11.1), the spatial indexing is omitted to make the equation more readable. The probability of dispersing into a plot by species i ($P_{dist,i}$) varies according to the description of each species group as different groups of species are adapted to different hydrology and salinity conditions.

11.2.1.1.2 EMERGENT WETLANDS

The model includes 20 species that represent emergent wetlands (Table 11.1). Included in this list are species that are among the most common inhabitants of freshwater, intermediate, brackish, and saline marshes. For these species, the environmental conditions that govern establishment and senescence are the annual standard deviation in water depth, $H(t) = H_{stdev}(t)$ and the annual mean salinity, $S_{mean}(t)$. The probability of senescence, $P_{senescence,i}(H_{stdev}(t),S_{mean}(t))$ is a bivariate function obtained by applying bilinear interpolation to a table of probabilities. The probability of establishment, $P_{establish,i}(H_{stdev}(t),S_{mean}(t))$, is also obtained by applying bilinear interpolation to a table of probabilities (Table 11.2).

The use of annual mean salinity and the annual standard deviation in stage height as the factors defining a species niche was developed through an analysis of the data collected by the Coast-wide Reference Monitoring System (CRMS) (Folse et al., 2012; Visser et al., 2015). There are over 350 CRMS monitoring stations distributed throughout Louisiana's coastal wetlands, with most stations having collected data from 2009 to the present. Data from these stations include continuous monitoring of environmental parameters including stage height, salinity, water temperature, and air temperature. CRMS data also include annual surveys of vegetation cover. We used this data to estimate the niche width of each of our emergent wetland species with respect to a number of potential factors including hydroperiod, annual mean and median water depth, and salinity. Salinity and the standard deviation in water depth were the only two factors where different species showed different niche ranges (Snedden and Steyer, 2013). Salinity is a well-known factor governing the distribution of coastal wetland species (Mitsch and Gosselink, 2000; Baldwin and Batzer, 2012) and its inclusion here reflects the stress that plants experience in the presence of saltwater and the adaptations some mesohaline and polyhaline species have evolved.

The standard deviation in water depth is a less obvious factor distinguishing species. This factor was used as a measure of the rate of nutrient exchange where areas with high variation in stage experience rapid exchange and frequent replenishment of nutrients while areas with low variation in stage are nutrient limited. This factor also reflects the differential responses

TABLE 11.2   Probability of establishment of *Spartina patens* as a function of annual mean salinity and standard deviation in stage height

Salinity	0	0.04	0.08	0.12	0.16	0.2	0.24	0.28	0.32	0.36	0.4	0.44	0.48	0.52	0.56	0.6	0.64	0.68	0.72	0.76	10
								Water level variability													
0	0	0	0	0	0	0	0	0	0	0	0	0	0	0	0	0	0	0	0	0	0
0.2	0	0	0	0	0	0	0	0	0	0	0	0	0	0	0	0	0	0	0	0	0
0.4	0	0	0	0	0	0	0	0	0	0	0	0	0	0	0	0	0	0	0	0	0
0.6	0	0	0	0	0	0	0	0	0	0	0	0	0	0	0	0	0	0	0	0	0
0.8	0.1	0.2	0.3	0.05	0.1	0.1	0.1	0.05	0	0	0	0	0	0	0	0	0	0	0	0	0
1	0.2	0.3	0.4	0.2	0.25	0.3	0.25	0.15	0.1	0.05	0	0	0	0	0	0	0	0	0	0	0
1.2	0.3	0.4	0.5	0.35	0.4	0.4	0.4	0.35	0.3	0.25	0.2	0.1	0.05	0	0	0	0	0	0	0	0
1.4	0.4	0.5	0.6	0.5	0.55	0.6	0.55	0.5	0.45	0.4	0.4	0.3	0.25	0.15	0.1	0.05	0	0	0	0	0
1.6	0.5	0.6	0.7	0.65	0.7	0.7	0.7	0.65	0.6	0.55	0.5	0.45	0.4	0.35	0.3	0.25	0.15	0.1	0.05	0	0
1.8	0.6	0.7	0.8	0.8	0.85	0.9	0.85	0.8	0.75	0.7	0.7	0.55	0.5	0.45	0.4	0.35	0.3	0.25	0.2	0.15	0
2	0.7	0.8	0.9	0.95	1	1	1	0.95	0.9	0.85	0.8	0.75	0.7	0.65	0.55	0.5	0.45	0.4	0.35	0.3	0
3	0.7	0.8	0.9	0.95	1	1	1	0.95	0.9	0.85	0.8	0.75	0.7	0.65	0.55	0.5	0.45	0.4	0.35	0.3	0
4	0.7	0.8	0.9	0.95	1	1	1	0.95	0.9	0.85	0.8	0.75	0.7	0.65	0.55	0.5	0.45	0.4	0.35	0.3	0
5	0.7	0.8	0.9	0.95	1	1	1	0.95	0.9	0.85	0.8	0.75	0.7	0.65	0.55	0.5	0.45	0.4	0.35	0.3	0
6	0.5	0.6	0.7	0.75	0.8	0.8	0.8	0.75	0.7	0.65	0.6	0.55	0.5	0.45	0.4	0.35	0.3	0.25	0.15	0.1	0
7	0.3	0.4	0.5	0.55	0.6	0.6	0.6	0.55	0.5	0.45	0.4	0.35	0.3	0.25	0.15	0.1	0.05	0	0	0	0
8	0.1	0.2	0.3	0.35	0.4	0.4	0.4	0.35	0.3	0.25	0.2	0.15	0.1	0.05	0	0	0	0	0	0	0
9	0	0	0.1	0.15	0.2	0.2	0.2	0.15	0.1	0.05	0	0	0	0	0	0	0	0	0	0	0
10	0	0	0	0	0	0	0	0	0	0	0	0	0	0	0	0	0	0	0	0	0
12	0	0	0	0	0	0	0	0	0	0	0	0	0	0	0	0	0	0	0	0	0
14	0	0	0	0	0	0	0	0	0	0	0	0	0	0	0	0	0	0	0	0	0
100	0	0	0	0	0	0	0	0	0	0	0	0	0	0	0	0	0	0	0	0	0

For intermediate values of salinity and standard deviation in stage height, the probability is calculated based on linear interpolation of the nearest neighboring values.

of species to nutrient availability, with some species being adapted to nutrient stress and others being adapted to take advantage of available nutrients.

Propagules of emergent wetland species disperse into a cell in proportion to the fraction of area covered by a species in the neighboring cells. The probability of species i dispersing its propagules into a given cell is evaluated based on the average cover in the neighboring cells:

$$P_{dist,i} = \frac{1}{N} \sum_{k=1}^{N} C_i(t;k) \tag{11.3}$$

where k is the location index for the neighbors. The number of neighbors varies from 1 to 8, depending whether a cell is on the border of the model domain or is situated in the interior. Note that the number of neighboring cells can be less than three because the model domain is not rectilinear. If species i completely covers the surrounding neighboring cells, i.e., $C_i(t;k) = 1$ for all k, then $P_{dist,i}$ is also equal to 1. In all other cases $P_{dist,i}$ is smaller than 1.

11.2.1.1.3 HARDWOOD FOREST

Six of the species included in our model represent hardwood forest species. These are primarily oak species that commonly grow near the coast but the forest group also includes American elm (*Ulmus americana*) (Table 11.1). The probability that a hardwood species will become established depends on the duration of flooding within the year, the elevation of the land above annual mean stage height, and salinity. For these species, salinity ($S(t)$) is expressed as the annual average salinity. Hydrology is summarized in two ways for hardwood species, as an index of flooding conditions, $H_{flood}(t)$, and as the land elevation above the annual mean stage height, $H_{height}(t)$.

In order for seedlings to become established, there must be a continuous two-week period with no surface water present, followed immediately by a two-week period where the water never exceeds 0.1 m in depth. These criteria are common to all six of the hardwood species we have included. The equation for the flooding index is:

$$H_{flood}(t) = \begin{cases} 1 & H_{daily}(d) < 0 \quad \text{for} \quad d = d_0...d_0 + 15 \\ & \text{and} \\ & H_{daily(d)} < 10 \text{ cm} \quad \text{for} \quad d = d_0 + 15 \ldots d_0 + 29 \\ & \text{for any } d_0 \text{ in year } t \\ 0 & \text{otherwise} \end{cases} \quad (11.4)$$

where d and d_0 denote time in days within a year t and $H_{daily}(d)$ is the stage height relative to land surface elevation (e.g., $H_{daily}(d) < 0$ indicates subsurface water).

The probability of establishment is calculated as:

$$P_{establish,i}(H_{height}(t), H_{flood}(t), S(t)) = \begin{cases} P_{elv,i}(H_{height}(t)) & \text{if } S(t) < 1 \text{ ppt} \quad \text{and} \quad H_{flood}(t) = 1 \\ 0 & \text{otherwise} \end{cases}$$

$$(11.5)$$

where t is time in years. In Eq. (11.5), $P_{elv,i}(H_{height}(t))$ is the probability of establishment for species i expressed as a function of the annual average height of the ground above the water surface, $H_{height}(t)$. The establishment probability is a piecewise linear function whose shape is parameterized by a table of values relating elevation to establishment probability (Table 11.3). *Quercus lyrata*, *Quercus texana*, and *Quercus laurifolia* occupy the lowest range of elevations, followed by *U. americana*, *Quercus nigra*, and finally *Quercus virginiana* at the highest elevations.

Senescence conditions are based on annual average salinity ($S(t)$) and land elevation above annual mean stage height, $H_{height}(t)$:

$$P_{senescence,i}(H_{height}(t), S(t)) = \begin{cases} P'_{senescence,i}(H_{height}(t)) & \text{if } S(t) < 1 \text{ ppt} \\ 1 & \text{otherwise} \end{cases} \quad (11.6)$$

TABLE 11.3 Probability of Establishment for Six Selected Upland Forest Species

Elevation (m)	Quercus lyrata	Quercus texana	Quercus laurifolia	Ulmus americana	Quercus nigra	Quercus virginiana
−0.1525	0	0	0	0	0	0
0	0.1	0.1	0.1	0	0	0
0.1525	0.2	0.2	0.2	0	0	0
0.305	0.3	0.3	0.3	0.1	0.1	0
0.475	0.4	0.4	0.4	0.2	0.2	0
0.61	0.7	0.7	0.7	0.3	0.3	0
0.7625	0.8	0.8	0.8	0.4	0.4	0
0.915	1	1	1	0.6	0.6	0.2
1.0675	1	1	1	0.7	0.7	0.4
1.22	0.8	0.8	0.8	1	1	0.6
1.3725	0.7	0.7	0.7	1	1	0.8
1.525	0.4	0.4	0.4	0.6	0.6	1
1.6775	0.2	0.2	0.2	0.4	0.4	1
1.83	0	0	0	0.2	0.2	1
1.9825	0	0	0	0.1	0.1	1
2.135	0	0	0	0	0	1

where $P'_{senescence,i}(H_{height}(t))$ is the probability of senescence for species i. $P'_{senescence,i}(H_{height}(t))$ is a piecewise linear function obtained by applying linear interpolation to parameters in Table 11.3. The conditions defined by the establishment table are the opposite of those for senescent. For example, Q. lyrata, Q. texana, and Q. laurifolia will begin to senesce if the elevation above mean water height either exceeds 1.22 m or falls below 0.76 m.

The establishment tables for these species were derived based on the literature data on the distribution of these species along the elevation gradient (Wall and Darwin, 1999; Denslow and Battaglia, 2002; Natural Resource Professionals, 2001; Theriot, 1993).

11.2.1.1.4 SWAMP FOREST

The model includes three swamp forest species, *Salix nigra*, *Taxodium distichum*, and *Nyssa aquatica* (Table 11.1). Senescence for these species is described in the same way as for the emergent wetland species. The environmental conditions that govern senescence and establishment are the annual standard deviation of stage height ($H(t)$) and the annual mean salinity ($S(t)$). The probability of senescence, $P_{senescence,i}(H(t),S(t))$, is a function obtained by applying bilinear interpolation to a table of probabilities. The establishment tables for these species are also based on the analysis of the CRMS data (Visser et al., 2015).

The establishment model for swamp forest species is based on annual mean salinity, $S(t)$, and on two summaries of the hydrology data: the annual standard deviation in water depth, $H_{stdev}(t)$, and the flooding index used for the upland hardwood species, $H_{flood}(t)$. The probability of establishment is calculated as:

$$P_{establish,i}(H_{stdev}(t), H_{flood}(t), S(t)) = \begin{cases} P'_{establish,i}(H_{stdev}(t), S(t)) & \text{if} \quad H_{flood}(t) = 1 \\ 0 & \text{otherwise} \end{cases} \quad (11.7)$$

where $P'_{establish,i}(H_{stdev}(t), S(t))$ is a function obtained by applying bilinear interpolation to a species-specific probability table similar to Table 11.2. Dispersal for the swamp forest species is evaluated in the same way as described for the emergent wetland species (Eq. (11.3)).

11.2.1.1.5 FLOATING MARSH

Our model includes three floating marsh species (Table 11.1). Floating marshes form when soils within the rooting zone of the plants become detached from lower soil strata and the resulting vegetation mat becomes buoyant (Sasser et al., 1996). In Louisiana, these marshes are typically associated with three emergent wetland species, *Panicum hemitomon*, *Eleocharis baldwinii*, and *Hydrocotyle umbellata*. These species are already included in the model and are modeled using the emergent wetlands model described above. They are grouped as a separate model category because floating marshes in which these species dominate are the result of different processes and have a different fate when they senesce. The formation of these marshes is thought to take place on a much slower time scale than the 50-year time horizon used in this model. For this reason, the model does not include an establishment process for these species and so $P_{establish,i}(H(t),S(t)) = 0$ for all three emergent wetland species. The initial distribution of floating marsh is determined by the model's initial conditions and the marsh can only experience senescence during a simulation. Senescence of floating marshes is governed by the same rules as used for the emergent wetland species and so the senescence table for each of the species associated with floating marshes is the same one used for the emergent wetland species.

When floating marshes senesce, the area they previously occupied becomes open water. This is different from the emergent marshes, where area lost by one emergent marsh species can be reoccupied by another species. The floating mat of soil and plants is often separated from the soil surface by a half a meter or more of water (Sasser et al., 1996). If a floating marsh is destroyed, the resulting water is too deep for emergent species to become established and the area becomes open water. The floating marshes are included in the model to capture this process and the location and extent of open water created by the death of floating marsh is communicated to the hydrology and soil morphology models that are linked to the LAVeg-Mod model.

11.2.1.1.6 SUBMERGED AQUATICS

Submerged aquatic vegetation (SAV) is a common feature in Louisiana's wetlands. SAV can influence hydrology by increasing surface friction and soil morphology as they stabilize soils with their roots. In the LAVegMod model, SAV is the only plant category that is not modeled at the species level. The model uses a different set of summaries of hydrology and salinity to estimate cover. This approach is dictated by the limited availability of data for the

SAV species that occupy Louisiana's wetlands. For example, the CRMS data set that was the basis for our emergent wetlands parameters does not include SAV species.

The SAV model is based on a linear regression that relates the cover of SAV to summer mean water depth, water temperature, and salinity (Visser et al., 2013):

$$C_{SAV}(t) = C_{Water}(t)\left[1.83 - 3.7 \times 10^2\, T_{summer}(t) - 7.7 \times 10^2\, S_{summer}(t) - 2.6\right.$$
$$\left. \times 10^2\, H_{summer}(t)\right] \tag{11.8}$$

where $C_{water}(t)$ is the fraction of a cell covered by water in year t, $T_{summer}(t)$, $S_{summer}(t)$, and $H_{summer}(t)$ are the summer mean water temperature, summer mean salinity, and the summer mean water depth in year t, respectively.

11.2.1.1.7 BARRIER ISLAND VEGETATION

The model includes eight species that are commonly found on Louisiana's coastal barrier islands. Three of these species are also included in our emergent wetland species (*Baccharis halimifolia*, *Spartina patens*, and *Distichlis spicata*). We include these species separately for the barrier islands because different processes govern their establishment and senescence on barrier islands. The remaining five species are typically found only on Louisiana's barrier islands.

The establishment of barrier island species is determined by height above mean sea level. Height above mean sea level is a proxy for a number of related factors, including exposure to salt from sea spray, exposure to scouring from wind, and exposure to tidal energy and inundation. Low elevation areas are exposed to high wave energy, frequent inundation, high salinity, and relatively low scouring from wind. As elevation increases wave energy and inundation decrease because these locations are farther from the shore, salinity decreases because rain water tends to flush salts downhill, and scouring from wind increases as plants become more exposed. The cumulative effect of these processes is such that each species is found only within a characteristic band at a distinctive elevation above the shoreline. The probability of establishment is a piecewise linear function of elevation that is obtained by applying linear interpolation to a species-specific probability table similar to Table 11.2.

11.2.1.2 Model Inputs

The vegetation model takes input from a hydrology model and a soil morphology model. The hydrology model provides daily estimates of water stage height, salinity, and water temperature for the entire model domain at a 500 × 500 m resolution. This information is then summarized into the annual and summer season statistics required by the various submodels (Table 11.4). The model also records a map of this distribution of land and water from the soil morphology model. This map is updated every year, noting the fraction of each 500 × 500 m cell that is classified as land and the fraction classified as open water for the entire model domain. This map is used to determine where plants can become established. Only the portion of each cell that is classified as "land" can be occupied by terrestrial species, while only areas classified as "open water" can be occupied by SAV. These quantities are updated by the soil morphology model based on changes in hydrology, the transport of sediments, and the distribution of plants. At the start of each yearly time step, the vegetation model updates the cover of plants in each 500 × 500 m cell to reflect the input from the soil morphology

TABLE 11.4   Hydrology and Soil Morphology Model Inputs to the LAVegMod Submodels

Input	Description	LAVegMod Submodels
$H_{flood}(t)$	Index of flooding used for tree establishment	Hardwood forests
		Swamp forest
$H_{height}(t)$	Height of land surface above mean water surface	Hardwood forests
		Barrier islands
$H_{stdev}(t)$	Annual standard deviation in stage height	Emergent wetlands
		Swamp forest
$S(t)$	Annual mean salinity	Hardwood forest
		Swamp forest
		Emergent wetlands
$T_{summer}(t)$	Summer mean water temperature	Submerged aquatic vegetation
$S_{summer}(t)$	Summer mean salinity	Submerged aquatic vegetation
$D_{summer}(t)$	Summer mean water depth	Submerged aquatic vegetation

model. If a cell experiences no change in the proportion of land and water, there is no adjustment needed to the plant species cover. If the soil morphology model predicts an increase in land fraction (and a corresponding decrease in water fraction), the model computes the difference between the area currently covered by plants and the area of "new" land provided by the soil morphology model. The difference is classified as unoccupied and will be included in the next update of the vegetation dynamics.

If the soil morphology model predicts a reduction in the fraction of a cell that is covered by land, the reduction in the cover of terrestrial species is computed as:

$$C_i'(t) = C_i(t)\frac{L(t)}{\sum C_i(t)} \tag{11.9}$$

where $L(t)$ is the new fraction of a cell predicted to be land by the soil morphology model and the sum of the loss is taken over all terrestrial plant species (i.e., all species but SAV). $C_i'(t)$ is the cover of species i after the adjustment, which is subsequently used to update cover based on hydrology and salinity (Eqs. (11.1) and (11.2)).

The initial conditions for the model are based on a habitat classification map formulated from satellite telemetry data for the year 2010. The map provides individual species cover for each 30×30 m cell. This information is then converted into a coarser 500×500 m map that is used in the LAVegMod model.

11.2.1.3 Climate Change Scenarios

We considered three hypothetical climate change scenarios (Table 11.5). All three scenarios assume increased flooding, both in extent and duration, and associated increases in salinity.

TABLE 11.5 Parameters Used to Configure the Hydrology Dynamics for the Three Hypothetical Climate Change Scenarios

Scenario	Precipitation	Evapotranspiration	ESLR (m)	Subsidence
A	>Historical	<Historical	0.43	20%
B	>Historical	Historical	0.63	20%
C	Historical	Historical	0.83	50%

ESLR is eustatic sea level rise, given as the total increase in sea level over the 50-year modeling period. Subsidence is given as a percentage into the range of subsidence rates that occur in different parts of coastal Louisiana (Meselhe et al., 2015).

The extent of future eustatic sea level rise (ESLR) varies among scenarios, where each successive scenario assumes progressively higher ESLR. The scenarios are parameterized in terms of ESLR, soil subsidence, and precipitation. The values for precipitation and evaporation for different scenarios were chosen based on downscaled global climate models (Meselhe et al., 2015). Subsidence rates for Louisiana's coast were based on available data and expert panel input (Meselhe et al., 2015). Rates of ESLR are based on the IPCC predictions adjusted based on Boesch et al. (2013) local adaptive approach (Meselhe et al., 2015). Increases in ESLR and subsidence result in increased coastal inundation. Water depth, duration of flooding events, and spatial extent of flooding are all positively correlated with increases in ESLR and subsidence. The extent to which salt water penetrates into the coastal marshes and swamps is controlled by the sea level variations in the Gulf of Mexico. Increases in rainfall also increase flooding but have the opposite effect on salinity.

Scenario A (Table 11.5) assumes the lowest rate of ESLR and the lowest rate of soil subsidence. Accordingly, this scenario results in the least extensive flooding and the least extensive intrusion of saltwater relative to the other scenarios considered. Scenario B represents a ~50% increase in the rate of ESLR. Scenario C assumes a doubling of ESLR and a 30% increase in subsidence compared to scenario A.

11.2.1.4 Model Results

All three climate change scenarios show similar patterns over the first 20 years of the simulation (Fig. 11.3). During the first 20 years, land area remains remarkably stable, as does the distribution of the estuarine gradient. However, in the first 20 years, the model also predicts some changes within the different salinity zones. In the brackish zone, *D. spicata* expands at the expense of *S. patens*. In the fresh zone *Sagittaria lancifolia* and *P. hemitomon* decline, while *E. baldwinii* and *H. umbellata* increase.

In the last 30 years of the simulation, the effects of climate change become more apparent, with drastic losses of marsh in scenario C, and smaller but significant losses in scenario A. Under all scenarios, land loss occurs (Fig. 11.4) and fresh water marshes are invaded by brackish and saline marshes as the estuarine gradient shifts up the estuaries. Under scenarios A and B, only *D. spicata* and *Paspalum vaginatum* increase their cover, while all other species decline in the last 30 years of the simulation. Under scenario C, the increases of *D. spicata* and *P. vaginatum* start around year 15 and continue for about 10 years, while after year 25 all species decline in cover and only 5% of the model domain remains vegetated.

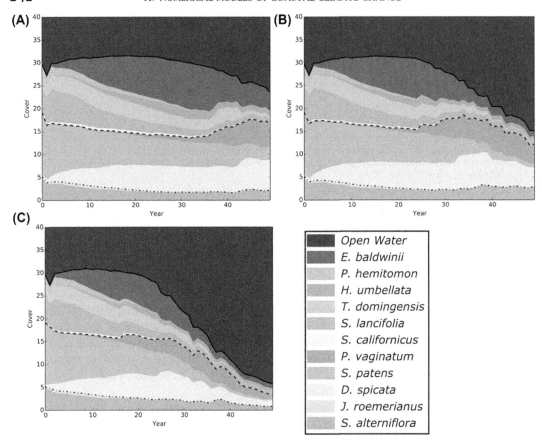

FIGURE 11.3 A summary of simulated regional changes in species composition under scenarios A–C (panels A–C) for 13 of the 19 emergent wetland species with the cover >1% of the model domain at any point during the 50-year simulation period. Species that are not shown had very low cover values. The six most abundant species are freshwater species *Eleocharis baldwinii, Hydrocotyle umbellata, Morella cerifera, Panicum hemitomon, Typha domingensis,* and *Sagittaria lancifolia*. The next in abundance are the intermediate salinity species *S. lancifolia* and *Schoenoplectus californicus*, followed by brackish species *Paspalum vaginatum* and *Spartina patens*. Polyhaline marsh species *Distichlis spicata, Spartina alterniflora, Juncus roemeriaunus,* and *Avicennia germinans* have the lowest abundance.

As apparent sea-level rise increases, the model predicts a shift and shortening of the estuarine gradient upslope (Fig. 11.4). In many areas that are occupied by swamp forest today, the model predicts a change to fresh marsh (Scenario A), brackish marsh (Scenario B), or open water (Scenario C) at the end of the 50-year simulation period. The only region that is relatively stable is the central part of the coast, where the Atchafalaya River maintains coastal freshwater marshes in its delta. In the Atchafalaya River flood plain, bottomland hardwood forest is to a great extend replaced by the more flood tolerant swamp forest species. Along the western part of the coast the areas currently occupied by fresh and brackish marsh species convert to more salt-tolerant species under scenario A, saline marsh species, and open water under scenario B and almost complete conversion to open water under

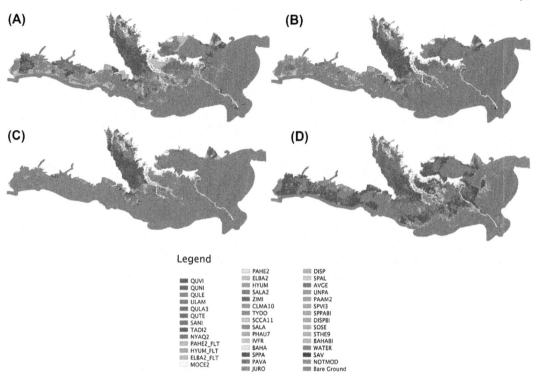

FIGURE 11.4 The forecasted change in species composition in year 50 of the simulation under scenarios A–C (panels A–C) and at the start of the simulation (panel D). Species codes in the legend are provided in Table 11.1.

scenario C. Importantly, these model results suggest that future land loss in the study region could be much faster compared to predictions from previous studies (e.g., Visser et al., 2013). Under the higher ESLR and slightly lower subsidence rates used in this study, the model predicts a near-total loss of the Louisiana coastal wetlands within the next 50 years (Fig. 11.4). This is significantly faster than proposed by Blum and Roberts (2009), who predicted a similar collapse by the year 2100.

11.2.2 Modeling the Impacts of Climate Change on Gulf of Mexico Hypoxia

Hypoxia refers to conditions in the water column where dissolved oxygen falls below levels that can support most marine life (i.e., 2 mg O_2 L^{-1}). The number of coastal hypoxic zones, often referred to as "dead zones," has been increasing at an exponential rate since the 1960s and there are currently over 500 documented coastal hypoxic zones (Diaz and Rosenberg, 2008). Hypoxia causes large-scale spatial population displacement and often mass mortalities of marine organisms that live on or near the bottom (Craig and Crowder, 2005; Breitburg et al., 2009). Hypoxia can also negatively affect the growth and reproduction of commercially important species (Rose et al., 2009).

The northern Gulf of Mexico shelf (NGOM) is the site of one of the world's largest coastal hypoxic zones (up to 22,000 km^2, Rabalais et al., 2007). Hypoxia typically occurs from March through October in waters below the pycnocline (~ 10 m) and extends from the depths of ~ 5 m near the shore to as deep as 60 m at the offshore boundary of the hypoxic zone (Rabalais et al., 2007). Historical reconstruction based on dated sediment cores (Turner and Rabalais, 1994; Osterman et al., 2008; Osterman et al., 2009) and model hindcasts (Justic et al., 2002) indicate that bottom-water hypoxia started to develop in the early part of the twentieth century and has become more frequent and widespread since the 1960s. The temporal increase in the extent and severity of hypoxia has been largely attributed to the increased nitrogen inputs by the Mississippi River and more balanced nutrient ratios in fresh waters (Turner et al., 2012). However, it was also recognized that development of hypoxia in the NGOM is highly sensitive to interannual variability in the Mississippi River discharge. During the drought of 1988 (a 52-year low discharge record of the Mississippi River), for example, the areal extent of midsummer hypoxia was minimal. In contrast, during the flood of 1993 (a 62-year maximum discharge for August and September) the hypoxic zone doubled in size, relative to the 1985–1990 average (Rabalais et al., 2007).

Hypoxia in the NGOM develops as a synergistic product of high surface primary production and high stability of the water column. High surface primary production fuels the water column and benthic respiration by increasing the vertical flux of organic carbon while high stability of the water column inhibits mixing and oxygen resupply to bottom waters. Consequently, climate change could affect the extent and the severity of hypoxia in two major ways. First, changes in temperature and riverine freshwater inputs (Sperna Weiland et al., 2012; IPCC, 2014) would likely modify the spatial and temporal patterns of productivity and respiration. Also, the physical characteristics of the environment (i.e., stratification) would likely change, resulting in more favorable conditions for hypoxia development (Justic et al., 2005).

11.2.2.1 Hypoxia Models

A number of different hypoxia models have been developed for NGOM over the past 20 years, ranging from simple statistical models to complex three-dimensional (3D) coupled hydrodynamic–biogeochemical models (Justic et al., 2007). Recent advances in three-dimensional (3D) coupled hydrodynamic-water quality models for the NGOM hypoxic zone (e.g., Hetland and DiMarco, 2008; Wang and Justic, 2009; Fennel et al., 2013; Justic and Wang, 2014) allow for a more detailed assessment of changes in the spatial patterns of hypoxia in response to climate change.

The effects of climate change on hypoxia in the NGOM have been modeled previously using relatively simple models. For example, Donner and Scavia (2007) examined climate change impacts on hypoxia using a modified Streeter–Phelps model. In a related study, Justic et al. (1996, 2003) used a two-box model to evaluate the impacts of double CO_2 scenario on dissolved oxygen dynamics at a site within the core of the NGOM hypoxic zone. The model assumes uniform properties for the layers above and below the average depth of the pycnocline. The oxygen concentration in the upper water column changes as a result of biological oxygen production and consumption, oxygen transport in the horizontal and vertical direction, and atmospheric exchanges. The oxygen concentration in the lower water column reflects the balance between oxygen uptake due to the benthic and water column respiration and oxygen resupply from the upper water column via turbulent diffusion (Justic et al., 2002). Model scenarios included a historical baseline conditions for the period 1955–2000 and six hypothetical future scenarios that were based on observed and projected changes in the Mississippi River discharge, Mississippi River nitrate concentrations, and ambient water temperatures (Justic et al., 2003).

Here, the predictions of the Justic et al. (2003) model are supplemented by the new results from the high resolution, coupled hydrodynamic-biogeochemical FVCOM-WASP model (Justic and Wang, 2014). FVCOM is a 3D, primitive equation coastal ocean circulation model that features an unstructured triangular grid to resolve complex coastlines and a terrain following coordinate system to convert irregular bottom topography into a regular computational domain (Chen et al., 2003). The current implementation for the NGOM uses 14,740 triangular nodes and 28,320 triangular elements, and has a variable 1–10 km horizontal resolution. The vertical grid domain consisted of 31 uniformly distributed σ layers. The WASP water quality model simulates nitrogen and phosphorous kinetics, carbonaceous biological oxygen demand in the water column and sediments, phytoplankton biomass (expressed as carbon and chlorophyll), and dissolved oxygen dynamics (Justic and Wang, 2014).

11.2.2.2 Model Results

Box-model run for historical baseline conditions (1955–2000) predicted the occurrence of hypoxia in 19 out of 45 simulated years, 16 of which were years with severe hypoxia when the dissolved oxygen concentrations decreased below 1 mg/L (Fig. 11.5). The model results also suggested that hypoxia first appeared in the mid-1970s and has become more severe over time. The predicted onset of hypoxia is in good agreement with the timing of first reports documenting hypoxia in this region (Rabalais and Turner, 2001) and the incidence of hypoxia proxies in dated sediment cores from the hypoxia region (e.g., Turner and Rabalais, 1994). For a future hypothetical climate scenarios characterized by 4°C temperature increase and a 20% increase in the average annual Mississippi River discharge (Justic et al., 2003), the model predicted 31 years with hypoxia, 26 of which were years with severe hypoxia (Fig. 11.5).

The FVCOM-WASP model simulations for 2002 showed a good agreement between simulated and observed areal extent of hypoxia (Fig. 11.6). The simulated area of hypoxia during July 2002 was 18,550 km^2, which was about 15% lower than the area of hypoxia measured during the concurrent hypoxia monitoring cruise (22,000 km^2, www.gulfhypoxia.net). The measured area of hypoxia during July 2002 was the highest since the monitoring program started in 1985 (Rabalais et al., 2007) and provides a good reference for the baseline model scenario. Assuming a 4°C increase in temperature and a 20% increase in the average annual Mississippi River discharge, the FVCOM-WASP model predicts an approximate doubling in the area of hypoxia compared to the baseline model scenario, along with severely reduced bottom oxygen levels over much of the study area and expansion of hypoxia into deeper waters on the continental shelf (Fig. 11.6).

11.2.3 Modeling the Impacts of Climate Change on Coastal Ecosystems in the Western Pacific

The Western Pacific is a vast marginal ocean stretching from 6°N to 42°N and consisting of several marginal seas including the Bohai Sea, Yellow Sea, East China Sea, and South China Sea. The circulation in these marginal seas is controlled by a strong M2 tide and two monsoon systems, the East Asian Monsoon from the east and the Indian Monsoon from the south. Meanwhile, the Western Pacific receives discharges from several large river systems,

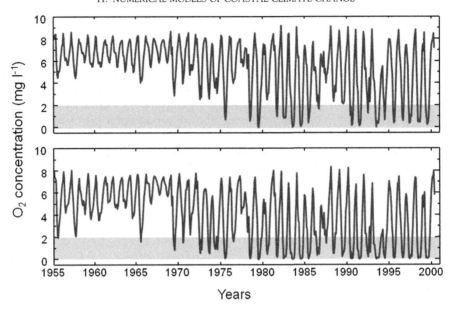

FIGURE 11.5   Simulated changes in the average bottom oxygen concentrations at a station within the core of the northern Gulf of Mexico hypoxic zone for the 1955–2000 baseline scenario (top panel), and for a hypothetical future scenario that assumes a 4°C increase in temperature and a 20% increase in the Mississippi River discharge (bottom panel). Shaded areas denote hypoxic conditions (<2 mg O_2 L^{-1}). *Adapted from Justic, D., Rabalais, N.N., Turner, R.E., 2003. Simulated responses of the Gulf of Mexico hypoxia to variations in climate and anthropogenic nutrient loading. Journal of Marine Systems 42, 115–126.*

including the Yellow River (Huanghe), Yangtze River (Changjiang), Pearl River (Zhujiang), Red River (Songhong), and Mekong River (Lancang), all of which originate from the Qinghai-Tibetan Plateau. These large rivers travel thousands of kilometers before entering the ocean and deliver large amounts of sediments along with nutrients and other particulate and dissolved constituents. Annual and decadal changes in the constituent fluxes of these large rivers reflect the climatic and anthropogenic influences in their massive watersheds and provide an excellent research framework to examine the potential consequences of future climate change.

The marginal seas in the Western Pacific are also bordered by some of the most densely populated coastal regions in the world. Human activities, mainly through dam construction and deforestation, have dramatically reduced sediment delivery to the coastal ocean. The total annual sediment flux from these rivers has decreased from 20 to 6 million tons per year over the past century (Wang et al., 2011), which when combined with the rising sea level, seriously threatens the sustainability of the coastal communities. On these heavily populated deltaic coasts, the blooming economic activities are further contributing to land subsidence via ground water extraction, various coastal engineering projects, and changes in land use practices. The Mekong Delta, for example, was listed as the most endangered delta in the world by the Intergovernmental Panel on Climate Change (IPCC, 2007).

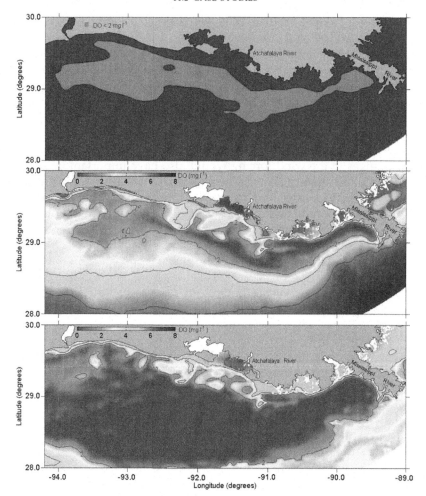

FIGURE 11.6   Measured area of bottom water hypoxia (<2 mg O_2 L^{-1}) in the northern Gulf of Mexico during July 21–26, 2002 (top panel, www.gulfhypoxia.net). Also shown are the simulated bottom oxygen concentrations during July 21–26, 2002, for the baseline scenario (middle panel, adapted from Justic and Wang, 2014), and for a hypothetical future scenario that assumes a 4°C increase in temperature and a 20% increase in the Mississippi River discharge (lower panel). The area of hypoxia is denoted by the 2 mg O_2 L^{-1} isoline.

11.2.3.1 Monsoon Dynamics

Climate in the Western Pacific is dominated by the East Asian Monsoon and South Asian Monsoon (Indian Monsoon). Monsoons impact to the Western Pacific region in various ways. In the ocean, monsoons are affecting the circulation and the associated biogeochemical processes via wind, pressure, heat, and freshwater flux. On land, monsoons are controlling the snow formation and melting on the Tibetan Plateau, and so they affect riverine discharge and constituent transport to the coastal ocean. Potential impacts of climate change on monsoon dynamics were explored using various global coupled ocean–atmosphere general circulation models (e.g., Kripalani et al., 2006; Kripalani et al., 2007; Turner and Annamalai, 2012). Results from a multimodel ensemble projected a significant increase in summertime precipitation under future climate scenarios for both the East Asian (7.8%) and the Indian Monsoons (8.0%). The dynamics of the East Asian Monsoon is more complex compared to the Indian Monsoon due to its subtropical origin and close coupling to ENSO (Wang et al., 2000). For the Indian monsoon, the model results point to a longer monsoon season (June—September) and more variable precipitation pattern, in response to intensified heat flow over northwest India (Kripalani et al., 2007).

While most of existing modeling studies were focused on changes in the precipitation pattern, potential changes in the wind forcing would also have important implications for the coastal ecosystems in the Western Pacific. The pattern of coastal surface winds during winter monsoon controls the resuspension and alongshore transport of river-derived materials (e.g., Xue et al., 2012). Coastal circulation in the vicinity of large river deltas along the Western Pacific margin is largely dominated by the geostrophic balanced currents, which will be enhanced by the downwelling favorable northeasterly winds. The results from a multimodel ensemble predict an increase in wind speed in the global monsoon regions and a reduction in wind speed over most of the tropical oceans (Hsu et al., 2013). For instance, the northwesterly winds along the coastal region of north Western Pacific would increase significantly in response to global warming (Xu et al., 2015).

11.2.3.2 Sediment Transport

Over the past 9000 years, the five large river deltas (Yellow, Yangtze, Pearl, Red, and Mekong) have been prograding toward the Western Pacific because of the annual supply of millions of tons of sediments together with a relative steady sea level. However, this process will likely be reversed under projected future climate scenarios due to rising sea level. For instance, by the end of this century, sea-level rise in the South China Sea is projected to increase by 0.64 m (Huang and Qiao, 2015).

With advances of high performance computing and algorithm development, sediment transport models have become an important tool in assessing the potential impacts of climate change on continental shelf processes. These sediment models are generally used in conjunction with ocean circulation models, such as the Regional Ocean Modeling System (ROMS, e.g., Shchepetkin and McWilliams, 2005; Haidvogel et al., 2008), and allow for characterization of bottom boundary layers and description of cohesive sediment behavior. One notable development was the Coupled Ocean Atmosphere Wave and Sediment Transport modeling system (COAWST, Warner et al., 2010). The COAWST couples sediment transport with wave and atmospheric forcings and includes the ROMS, the Weather Research Forecast model (WRF, Skamarock et al., 2005), and the Simulate WAve Nearshore model (SWAN, Booij et al., 1999). This coupled modeling system allows for efficient exchange of information (air pressure, heat flux, sea surface temperature, sea level, etc.) among different models. Also, the close coupling between waves and ocean circulation allows the estimation of wave-induced enhancement of surface roughness, water column mixing, and bottom stress, all of which are critical for sediment transport modeling.

Recently the COAWST was applied to the Western Pacific to simulate the sediment transport and deposition on the continental shelf. In one of the COAWST applications Xue et al. (2012) simulated coupled ocean, wave, and sediment transport dynamics. The model

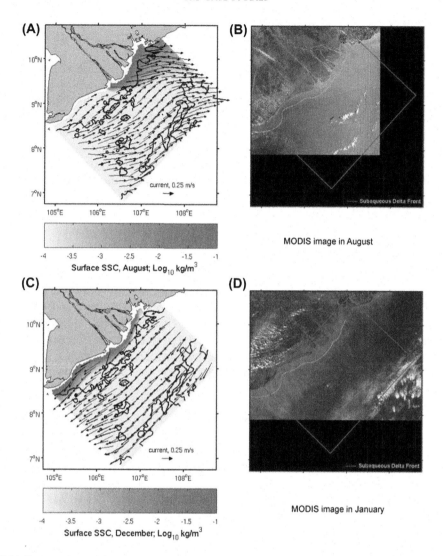

FIGURE 11.7 Monthly mean surface suspended sediment concentration (SSC, color shading) and surface current fields in (A) summer (August) and (C) winter (December), 2005. Also shown are Moderate Resolution Imaging Spectroradiometer (MODIS) images in August 2002 (B) and January 2005 (D). Simulated surface SSC values (kg/m^3) are shown on the log scale (color). *From Xue, Z., He, R., Liu, J.P., Warner, J.C., 2012. Modeling transport and deposition of the Mekong River sediment. Continental Shelf Research 37, 66–78; reprinted with permission from Elsevier.*

accurately described the seasonal shift of monsoonal winds and associated sediment transport on the Mekong Shelf and confirmed the previously proposed "summer deposit–winter transport" mechanism for large river-derived sediments in the Western Pacific (Fig. 11.7). Using the COAWST, Bian et al. (2013) examined the transport pathways of shelf sediments in the Bohai-Yellow-East China Seas under climatological mean circulation and wind forcing.

Further, the modeling results suggested that most of the sediments supplied by the Yellow River remain within the Bohai Sea, whereas the Yangtze-derived sediments are transported to the Yellow and East China Seas (Zeng et al., 2015).

While the current state-of-the-art sediment transport models show reasonable skill in quantifying sediment fluxes and reproducing the patterns of sediment distribution, much remains to be learned regarding the response of sediment transport to combined effects of sea-level rise and changes in monsoonal regime. Further, ongoing human activities in river basins and along the coast make the forecasting of future conditions even more challenging (Syvitski et al., 2009). For instance, projected increases of summer monsoon precipitation will likely intensify inland erosion and thus increase riverine-suspended sediment load. However, because a large fraction of suspended sediment load is deposited behind dams, it remains unclear whether the altered monsoon dynamics would significantly affect the overall sediment delivery to the Western Pacific.

11.2.3.3 Coastal Ecosystem Processes

Ecosystems of the Western Pacific marginal seas are highly sensitive to climate-driven variability in monsoon dynamics and changes in sediment and nutrient fluxes at land-ocean interface (e.g., Liu et al., 2014). For example, model estimates by Liu et al. (2010) suggest that primary productivity in the East China Sea increased by 17% between 1970 and 2002. This increase in primary productivity was caused by a 2.4-fold increase in the dissolved inorganic nitrogen load of the Yangtze River over the same period of time (Liu et al., 2015). While in the coastal regions primary productivity is predominantly controlled by the riverine nutrient inputs, in the open ocean it is sustained largely by the nutrients supplied by the monsoon-induced upwelling. A coupled physical—biological model developed by Liu et al. (2002) successfully simulated the existence of three distinct high-chlorophyll regions in the South China Sea, which correspond to the monsoon-influenced upwelling regions that develop northwest of Luzon and north of the Sunda Shelf in winter and off the east coast of Vietnam in summer. Chai et al. (2009) used the ROMS CoSINE ecosystem model to examine the carbon budget of the Western Pacific. They concluded the South China Sea is overall a weak source CO_2 for the atmosphere.

In a related study, Glibert et al. (2014) used the Global Coastal Ocean Modelling System (GCOMS, Holt et al., 2009) to examine the changes in harmful algal bloom distribution in response to climate change. The model was forced by atmospheric and oceanic boundary conditions from the IPCC climate models and river inputs were derived from the NEWS model (Global Nutrient Export from Watersheds, Seitzinger et al., 2005). Projections from this coupled model suggested that HAB potential in the Yellow and East China seas will not change substantially by the end of this century. However, it should be pointed out that our understanding of the linkages between climate variability and HABs in this region is still very limited.

11.2.4 Modeling the Impacts of Climate Change on the California Current System

Four major coastal upwelling systems, including the California upwelling system, the Humboldt upwelling system, the Canary upwelling system, and the Benguela upwelling system, account for less than 2% of the ocean, yet support more than 20% of the global fish catch (Pauly and Christensen, 1995). Located at the eastern boundaries of the four subtropical gyres in both the Pacific and the Atlantic oceans, coastal upwelling systems are driven by seasonal equatorward winds that push surface water offshore and subsequently bring subsurface water closer to the coast. The subsurface water is distinctly different from surface water in its physical and chemical properties. It is low in temperature, enriched in nutrients, deficient in dissolved oxygen and enriched in dissolved inorganic carbon (DIC). The upwelled nutrients fuel the surface ocean productivity, while the shallow hypoxia depth and aragonite saturation depth limit the vertical distribution of marine organisms. The abundance of DIC in upwelled water also makes the region a source of atmospheric CO_2.

11.2.4.1 *Modeling of the California Current System*

Observational programs in the California Current system (CCS), most notably the California Cooperative Oceanic Fisheries Investigations (CalCOFI) time series that started at 1949 (Bograd et al., 2003), have significantly advanced our understanding of the physics, chemistry, ecology, and fisheries in the CCS. Early ecosystem modeling studies relied on physical models of reduced complexity, such as one-dimensional (e.g., Moisan and Hofmann, 1996) and two-dimensional models (e.g., Spitz et al., 2003). Ecosystem in CCS is highly modulated by three-dimensional circulation; however, those simplified physical models provided a useful basis for the development of ecological models of the CCS. Subsequently, Gruber et al. (2006) embedded a nitrogen-based Nutrient-Phytoplankton-Zooplankton-Detritus (NPZD) model within the 3D Regional Oceanic Modeling System (ROMS) and ran the coupled model for 10 years so that a climatological equilibrium was reached. The NPZD model simulates the dynamics of two macronutrients (nitrate and ammonia), one phytoplankton group (i.e., diatoms), chlorophyll-to-carbon ratio of phytoplankton, one zooplankton group, and two detritus pools with different sinking speeds. By a comprehensive comparison with different satellite products and in situ measurements, the authors showed that this relatively simple ecological model was able to skillfully reproduce the spatial and temporal distributions of temperature, nutrients, and chlorophyll in the CCS.

Superimposed on the large-scale equatorward eastern boundary currents and the upwelling flows are mesoscale eddies arising from the unstable sheared currents and tilting thermocline. Mesoscale activities are important source of nutrients in the open ocean, particularly in oligotrophic ocean gyres (McGillicuddy et al., 1998; Levy et al., 2001). Cyclonic eddies uplift thermocline/nutricline during their passage and pump nutrients from subsurface into the eutrophic zone. In the eastern boundary upwelling systems, however, eddy activities are negatively correlated with upper ocean productivity (Gruber et al., 2011). By contrasting models that resolve eddies with models that do not, Gruber et al. (2011) showed that mesoscale eddies bury upwelled nutrients during their offshore propagation. Since thermocline shoals toward the coast and material transport by eddies is largely along isopycnals, there is a downward component in material transport by westward offshore eddy propagation. The reduction in nearshore production also leads to lower nutrient content in source water for upwelling by reducing the amount of sinking organic matters. Subsequent studies (e.g., Combes et al., 2013; Nagai et al., 2015) further demonstrated that anticyclonic eddies have a larger contribution for offshore downward material transport than cyclonic eddies. Using a Lagrangian particle tracking module in the 3D circulation model, Lachkar and Gruber (2011) showed that strong eddy activity contributes to relatively low nutrient residence time and nutrient utilization rate in the CCS compared to the Canary current system.

11.2.4.2 *Future Evolution of the California Current System*

Due to its prime importance in climate, environment, and fishery, the future evolution of coastal upwelling systems has received considerable attention. Bakun (1990) proposed a mechanism demonstrating that the equatorward upwelling-favorable winds strengthen under rising atmospheric CO_2 concentrations. The equatorward upwelling-favorable wind is a geostrophic wind driven by large-scale land-sea pressure difference. Each spring and summer when the hemisphere warms up, the land warms up faster than the ocean because of its smaller heat capacity, and a low-pressure cell develops above the land. With the build-up of CO_2 and other greenhouse gases in the atmosphere, more heat is trapped and further enhances the land—ocean temperature and pressure gradients, leading to stronger geostrophic upwelling favorable winds. The intensification of upwelling favorable winds has been observed in both historical wind measurements (e.g., Sydeman et al., 2014) and in simulated future climate scenarios (Wang et al., 2015).

The ecological consequences of the wind strengthening embedded in a warming, souring, and deoxygenating ocean, is less certain. Biological production increases as a result of upwelling intensification that enhances nutrient supply to the euphotic zone. The increase in biological production, however, is not proportional to the increase in wind stress. Lachkar and Gruber (2013) showed that a doubling of wind stress results in less than 50% increase in biological production in the California current system while the same increase in wind stress doubles production in the Canary current system. The relatively low increase in the productivity of the California current system is a combined result of low phytoplankton growth under nutrient replete conditions and strong offshore transport. The authors also looked at influence of doubling wind stress on air—sea CO_2 flux (with a constant atmospheric CO_2 concentration). They showed that the doubling of wind stress increases air—sea CO_2 efflux by 4—6 times in central California and southern Canary current and yet has little effect on other upwelling regions. The doubling of wind stress not only increases the upwelled supply of DIC and biological consumption but also enhances the wind-dependent gas transfer rate. Regional differences in the relative importance of the DIC supply, biological consumption, and gas transfer lead to the differences in the variability of CO_2 flux.

The California current system is a hotspot of ocean acidification which has a negative impact on calcifying marine organisms and aquaculture (e.g., Barton et al., 2012). Coastal water in the CCS (and in other eastern boundary upwelling systems) is naturally more acidic than other oceanic regions because upwelling brings DIC-enriched and carbonate-depleted deeper water to the surface. The large subsurface remineralization of sinking organic matter further makes the source water for upwelling even more acidic. Feely et al. (2008) observed seawater with pH values as low as 7.75, and aragonite undersaturation starting at depths varying from the 120 m to the surface in the northern US pacific coast. Hauri et al. (2009) reproduced the observed pattern of pH and aragonite undersaturation with an NPZD model coupled with an equilibrium carbon chemistry module embedded in ROMS. Using climatological forcing conditions except a rising atmospheric CO_2 concentration, Gruber et al. (2012)

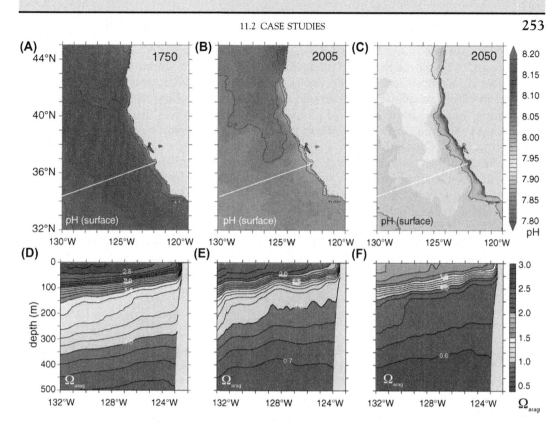

FIGURE 11.8 Temporal evolution of ocean acidification in the California Current system from 1750 until 2050. Maps A to C show the modeled pH values for 1750, 2005, and 2050. Maps D to F illustrate decrease in the annual mean saturation state of seawater with respect to aragonite (Ω_{arag}) and the shoaling of the aragonite saturation depth ($\Omega_{arag} = 1$). The white lines in panels A to C indicate the position of the offshore transect. *From Gruber, N., Hauri, C., Lachkar, Z., Loher, D., Frolicher, T.L., Plattner, G.K., 2012. Rapid progression of ocean acidification in the California Current system. Science 337, 220–223; reprinted with permission from AAAS.*

and Hauri et al. (2013a,b) projected that more than half of the water in eutrophic zone would be undersaturated with respect to aragonite by the year 2050 and bottom water would become year-round undersaturated within the next 20–30 years (Fig. 11.8). Lachkar (2014) included the upwelling intensification in the acidification projection simulations and showed that upwelling intensification strengthens acidification in the California current and weakens acidification in the Canary current. In the California current, stronger upwelling of acidic water dominates while in the Canary current, reduced remineralization due to strengthening of offshore organic matter transport and increased biological production appears more important.

Hypoxia occurs in the CCS during summer upwelling seasons. Unlike other coastal systems where nutrient loading from human activities play a significant role (e.g., Rabalais et al., 2009, 2010), hypoxia in the CCS and other upwelling systems is primarily controlled by upwelling intensity and oxygen contents in upwelled source water (Grantham et al., 2004). Observations have shown the decline in oxygen content in southern California waters from 1984 to 2006 (Bograd et al., 2008). Using an NPZD model coupled with ROMS, Lachkar and Gruber (2012) showed that wind intensification increases the amount of hypoxic water in the California current system while decreases the amount of hypoxic water in the Canary current system. Similar to the above-illustrated mechanisms controlling the variability in ocean acidification, increase in advection of oxygen-poor water controls the volume of hypoxic water in the California current. In contrast, the increase in biological oxygen production lessens hypoxia in the Canary current.

In addition to changes in the upwelling regime, variability in large-scale circulation associated with global warming also play significant role in the CCS by modulating the properties of the upwelled source water. By analyzing solutions from NOAA's coupled biogeochemical–physical general circulation model, Rykaczewski and Dunne (2010) suggested that source waters for the upwelling could be more enriched in nitrate and more depleted in oxygen under future climate conditions. This is because surface water warms up faster than deep water, resulting in stronger stratification. Consequently, the ventilation of deep water capped by a stronger thermocline slows down, resulting in stronger remineralization, nutrient accrual, and oxygen depletion.

11.3 CHALLENGES IN PREDICTING THE EFFECTS OF CLIMATE CHANGE ON COASTAL ECOSYSTEMS

The four case studies presented above illustrate the power of simulation modeling in projecting future conditions in the coastal zone and the wealth of information and details they can provide. However, while significant advances in numerical modeling have been made, there are also continuing challenges in modeling the effects of climate change on coastal ecosystems (e.g., Rose and Allen, 2013). For example, most global climate models (e.g., IPCC/CMIP5) have a relative coarse resolution ($\sim 1°$) and do not account for many coastal processes such as inundation, estuarine-shelf exchanges, and riverine freshwater, sediment, and nutrient inputs. Further, wind drop-off toward the coast, coastal upwelling, and formation of eddies are all poorly resolved in global models, which decreases the accuracy of ocean–atmosphere momentum and heat flux estimates. Also, using products from global models to drive regional models is not trivial and the computational expense of dynamically downscaling a 100-year global atmospheric forcing to a regional scale remains formidable obstacle for many researchers working in coastal ecosystems. Statistical downscaling, while computationally much less intensive, largely depends on the existing predictor–predictand relationship and cannot incorporate future climatic changes. The existing regional ecosystem models also need improvement. For example, there are still large uncertainties in the numerical formulations of biogeochemical and food web processes. Model formulations are often semiempirical with parameters tuned to best match historical observations. It is uncertain how reliable those models can be for future conditions that will likely include different environmental conditions and altered coastal food webs. Finally, as model spatial and temporal domains increase due to developments in computer technologies and computational techniques, much work remains to be done to enhance data collection in support of model calibration and validation.

11.4 CONCLUSIONS

Numerical simulation models play important roles in assessing the potential effects of climate change on coastal ecosystems and developing management strategies aimed at minimizing risks to sensitive habitats, species, and people living along the coasts. Numerical models also provide a quantitative framework to disentangle the synergistic influences of multiple factors and isolate the effects of individual stressors. This is important because climatic drivers do not act alone, but rather in conjunction with other ecosystem stressors, such as habitat degradation, pollution, eutrophication, and overharvesting of commercially important species. Undoubtedly, coastal ecosystem models will continue to evolve and will play even larger roles as computational power increases and time-series of observed climatic and environmental variables become longer and more informative. There is every reason to be optimistic that continued model improvement will lead to better understanding of coastal ecosystems, more instructive climate impact assessments, and better informed management decisions.

Acknowledgments

Hypoxia modeling effort carried out by D. Justic was funded in part by the NOAA/CSCOR Northern Gulf of Mexico Ecosystems and Hypoxia Assessment Program under award NA09NOS4780230 to Louisiana State University. Drs. Duke-Sylvester and Visser, who contributed the section on modeling the effects of climate change on Louisiana's coastal wetland plant communities, were supported by the Louisiana Coastal Protection and Restoration Authority through The Water Institute of the Gulf under project award number CPRA-2013-T03-EM, as part of a larger effort to support of the development of Louisiana's 2017 Coastal Master Plan. The views expressed in this publication are those of the authors and do not necessarily represent the views of the Coastal Protection and Restoration Authority or The Water Institute of the Gulf.

References

Bakun, A., 1990. Global climate change and intensification of coastal ocean upwelling. Science 247, 198–201.
Baldwin, A.H., Batzer, D.P., 2012. Wetland Habitats of North America: Ecology and Conservation Concerns. University of California Press, Oakland, California.
Barton, A., Hales, B., Waldbusser, G.G., Langdon, C., Feely, R.A., 2012. The Pacific oyster, Crassotrea gigas, shows negative correlation to naturally elevated carbon dioxide levels: implications for near-term ocean acidification effects. Limnology and Oceanography 57, 698–710.
Bendtsen, J., Hansen, J.L.S., 2013. Effects of global warming on hypoxia in the Baltic Sea–North Sea transition zone. Ecological Modelling 264, 17–26.
Bian, C., Jiang, W., Greatbatch, R.J., 2013. An exploratory model study of sediment transport sources and deposits in the Bohai Sea, Yellow Sea, and East China Sea. Journal of Geophysical Research: Oceans 118, 5908–5923.
Blum, M.D., Roberts, H.H., 2009. Drowning of the Mississippi Delta due to insufficient sediment supply and global sea-level rise. Nature Geoscience 2 (7), 488–491.
Boesch, D.F., Atkinson, L.P., Boicourt, W.C., Boon, J.D., Cahoon, D.R., Dalrymple, R.A., Ezer, T., Sommerfield, C.K., 2013. Updating Maryland's SLR Projections. Special Report of the Scientific and Technical Working Group to the Maryland Climate Change Commission. University of Maryland Center for Environmental Science, Cambridge, Maryland.
Bograd, S.J., Checkley, D.A., Wooster, W.S., 2003. CalCOFI: a half century of physical, chemical, and biological research in the California current system. Deep-Sea Research 50 (14–16), 2355–2370.
Bograd, S.J., Castro, C.G., Di Lorenzo, E., Palacios, D.M., Bailey, H., Gilly, W., Chavez, F.P., 2008. Oxygen declines and the shoaling of the hypoxic boundary in the California Current. Geophysical Research Letters 35, L12607.

Booij, N., Ris, R., Holthuijsen, L.H., 1999. A third-generation wave model for coastal regions: 1. Model description and validation. Journal of Geophysical Research: Oceans 104, 7649–7666.

Breitburg, D.L., Hondorp, D.W., Davias, L.A., Diaz, R.J., 2009. Hypoxia, nitrogen, and fisheries: integrating effects across local and global landscapes. Annual Review of Marine Science 1, 329–349.

Chai, F., Liu, G., Xue, H., Shi, L., Chao, Y., Tseng, C.-M., et al., 2009. Seasonal and interannual variability of carbon cycle in South China Sea: a three-dimensional physical-biogeochemical modeling study. Journal of Oceanography 65, 703–720.

Chen, C., Liu, H., Beardsley, R.C., 2003. An unstructured grid, finite-volume, three-dimensional, primitive equations ocean model: application to coastal ocean and estuaries. Journal of Atmospheric and Oceanic Technology 20, 159–186.

Combes, V., Chenillat, F., Di Lorenzo, E., Riviere, P., Ohman, M.D., Bograd, S.J., 2013. Cross-shore transport variability in the California Current: Ekman upwelling vs. eddy dynamics. Progress in Oceanography 109, 78–89.

Craft, C., Clough, J., Ehman, J., Joye, S., Park, R., Pennings, S., Guo, H., Machmuller, M., 2009. Forecasting the effects of accelerated sea-level rise on tidal marsh ecosystem services. Frontiers in the Ecology and Environment 7, 73–78.

Craig, J.K., Crowder, L.B., 2005. Hypoxia-induced habitat shifts and energetic consequences in Atlantic croaker and brown shrimp on the Gulf of Mexico shelf. Marine Ecology Progress Series 294, 79–94.

Davis, S., Ogden, J.C., 1994. Everglades: The Ecosystem and Its Restoration. CRC Press, Boca Raton, Florida.

DeAngelis, D.L., Gross, L.J., Huston, M.A., Wolff, W.F., Fleming, D.M., Comiskey, E.J., Sylvester, S.M., 1998. Landscape modeling for Everglades ecosystem restoration. Ecosystems 1 (1), 64–75.

Denslow, J.S., Battaglia, L.L., 2002. Stand composition and structure across a changing hydrologic gradient: Jean Lafitte National Park, Louisiana, USA. Wetlands 22 (4), 738–752.

Diamond, S., Murphy, C.A., Rose, K.A., 2013. Simulating the effects of global climate change on Atlantic croaker population dynamics in the mid-Atlantic Region. Ecological Modelling 264, 98–114.

Diaz, R.J., Rosenberg, R., 2008. Spreading dead zones and consequences for marine ecosystems. Science 321, 926–928.

Donner, S.D., Scavia, D., 2007. How climate controls the flux of nitrogen by the Mississippi River and the development of hypoxia in the Gulf of Mexico. Limnology and Oceanography 52, 856–861.

Feely, R.A., Sabine, C.L., Hernandez-Ayon, J.M., Ianson, D., Hales, B., 2008. Evidence for upwelling of corrosive "acidified" water onto the continental shelf. Science 320, 1490–1492.

Fennel, K., Hu, J., Laurent, A., Marta-Almeida, M., Hetland, R., 2013. Sensitivity of hypoxia predictions for the Northern Gulf of Mexico to sediment oxygen consumption and model nesting. Journal of Geophysical Research: Oceans 118, 990–1002.

Folse, T.M., West, J.L., Hymel, M.K., Troutman, J.P., Sharp, L.A., Weifenbach, D., McGinnis, T., Rodrigue, L.B., Boshart, W.M., Richardi, D.C., Miller, C.M., Wood, W.B., 2012. A Standard Operating Procedures Manual for the Coast-wide Reference Monitoring System-wetlands: Methods for Site Establishment, Data Collection, and Quality Assurance/Quality Control. Louisiana Coastal Protection and Restoration Authority, Office of Coastal Protection and Restoration, Baton Rouge, Louisiana.

Glibert, P.M., Allen, J.I., Artioli, Y., Beusen, A., Bouwman, L., Harle, J., et al., 2014. Vulnerability of coastal ecosystems to changes in harmful algal bloom distribution in response to climate change: projections based on model analysis. Global Change Biology 20, 3845–3858.

Gochis, D.J., Yu, W., Yates, D.N., 2014. The NCAR WRF-Hydro Technical Description and User's Guide v2.0. http://www.ral.ucar.edu/projects/wrf_hydro/.

Grantham, B.A., Chan, F., Nielsen, K.J., Fox, D.S., Barth, J.A., Huyer, A., Lubchenco, J., Menge, B.A., 2004. Upwelling-driven nearshore hypoxia signals ecosystem and oceanographic changes in the northeast Pacific. Nature 429, 749–754.

Gruber, N., Frenzel, H., Doney, S.C., Marchesiello, P., McWilliams, J.C., Moisan, J.R., Oram, J.J., Plattner, G.K., Stolzenbach, K.D., 2006. Eddy-resolving simulation of plankton ecosystem dynamics in the California current system. Deep-Sea Research 53, 1483–1516.

Gruber, N., Lachkar, Z., Frenzel, H., Marchesiello, P., Munnich, M., McWilliams, J.C., Nagai, T., Plattner, G.K., 2011. Eddy-induced reduction of biological production in eastern boundary upwelling systems. Nature Geosciences 4, 787–792.

Gruber, N., Hauri, C., Lachkar, Z., Loher, D., Frolicher, T.L., Plattner, G.K., 2012. Rapid progression of ocean acidification in the California current system. Science 337, 220–223.

Haidvogel, D.B., Arango, H., Budgell, W.P., Cornuelle, B.D., Curchitser, E., Di Lorenzo, E., et al., 2008. Ocean fore-casting in terrain-following coordinates: formulation and skill assessment of the regional ocean modeling system. Journal of Computational Physics 227, 3595–3624.

Hare, J.A., Alexander, M.A., Fogarty, M.J., Williams, E.H., Scott, J.D., 2010. Forecasting the dynamics of a coastal fishery species using a coupled climate-population model. Ecological Applications 20, 452–464.

Hauri, C., Gruber, N., Plattner, G.K., Alin, S., Feely, R.A., Hales, B., Wheeler, P.A., 2009. Ocean acidification in the California current system. Oceanography 22, 60–71.

Hauri, C., Gruber, N., McDonnell, A.M.P., Vogt, M., 2013a. The intensity, duration, and severity of low aragonite saturation state events on the California continental shelf. Geophysical Research Letters 40, 3,424–3,428.

Hauri, C., Gruber, N., Vogt, M., Doney, S.C., Feely, R.A., Lachkar, Z., Leinweber, A., McDonnell, A.M.P., Munnich, M., Plattner, G.K., 2013b. Spatiotemporal variability and long-term trends of ocean acidification in the California current system. Biogeosciences 10 (1), 193–216.

Hetland, R.D., DiMarco, S.F., 2008. How does the character of oxygen demand control the structure of hypoxia on the Texas-Louisiana continental shelf? Journal of Marine Systems 70 (1–2), 49–62.

Holt, J., Harle, J., Proctor, R., Michel, S., Ashworth, M., Batstone, C., et al., 2009. Modelling the global coastal ocean. Philosophical Transactions of the Royal Society of London A: Mathematical, Physical and Engineering Sciences 367, 939–951.

Hsu, P.-C., Li, T., Murakami, H., Kitoh, A., 2013. Future change of the global monsoon revealed from 19 CMIP5 models. Journal of Geophysical Research: Atmospheres 118, 1247–1260.

Huang, C., Qiao, F., 2015. Sea level rise projection in the South China Sea from CMIP5 models. Acta Oceanologica Sinica 34, 31–41.

IPCC, 2007. In: Parry, M.L., Canziani, O.F., Palutikof, J.P., Van Der Linden, P.J., Hanson, C.E. (Eds.), Contribution of Working Group II to the Fourth Assessment Report of the Intergovernmental Panel on Climate Change. Cambridge University Press, Cambridge & New York.

IPCC, 2014. In: Barros, V.R., Field, C.B., Dokken, D.J., Mastrandrea, M.D., Mach, K.J., Bilir, T.E., Chatterjee, M., Ebi, K.L., Estrada, Y.O., Genova, R.C., Girma, B., Kissel, E.S., Levy, A.N., MacCracken, S., Mastrandrea, P.R., White, L.L. (Eds.), Climate Change 2014: Impacts, Adaptation, and Vulnerability. Part B: Regional Aspects. Contribution of Working Group II to the Fifth Assessment Report of the Intergovernmental Panel on Climate Change. Cambridge University Press, Cambridge & New York.

Justic, D., Rabalais, N.N., Turner, R.E., 1996. Effects of climate change on hypoxia in coastal waters: a doubled CO_2 scenario for the northern Gulf of Mexico. Limnology and Oceanography 41, 992–1003.

Justic, D., Rabalais, N.N., Turner, R.E., 2002. Modeling the impacts of decadal changes in riverine nutrient fluxes on coastal eutrophication near the Mississippi River Delta. Ecological Modelling 153, 33–46.

Justic, D., Rabalais, N.N., Turner, R.E., 2003. Simulated responses of the Gulf of Mexico hypoxia to variations in climate and anthropogenic nutrient loading. Journal of Marine Systems 42, 115–126.

Justic, D., Rabalais, N.N., Turner, R.E., 2005. Coupling between climate variability and marine coastal eutrophication: Historical evidence and future outlook. Journal of Sea Research 54, 25–35.

Justic, D., Bierman, J.V., Scavia, D., Hetland, R., 2007. Forecasting Gulf's hypoxia: the next 50 years? Estuaries and Coasts 30, 791–801.

Justic, D., Wang, L., 2014. Assessing temporal and spatial variability of hypoxia over the inner Louisiana-upper Texas shelf: application of an unstructured-grid three-dimensional coupled hydrodynamic-water quality model. Continental Shelf Research 72, 163–179.

Kripalani, H.R., Oh, H.J., Chaudhari, S.H., 2006. Response of the East Asian summer monsoon to doubled atmospheric CO_2: coupled climate model simulations and projections under IPCC AR4. Theoretical and Applied Climatology 87, 1–28.

Kripalani, R.H., Oh, J.H., Kulkarni, A., Sabade, S.S., Chaudhari, H.S., 2007. South Asian summer monsoon precipitation variability: coupled climate model simulations and projections under IPCC AR4. Theoretical and Applied Climatology 90, 133–159.

Lachkar, Z., Gruber, N., 2011. What controls biological production in coastal upwelling systems? Insights from a comparative modeling study. Biogeosciences 8, 2,961–2,976.

Lachkar, Z., Gruber, N., 2012. Exploring the future evolution of multiple stressors in eastern boundary upwelling systems. Ocean Carbon Biogeochemistry News 5 (2), 5–9.

Lachkar, Z., Gruber, N., 2013. Response of biological production and air-sea CO_2 fluxes to upwelling intensification in the California and Canary Current systems. Journal of Marine System 109–110, 149–160.

Lachkar, Z., 2014. Effects of upwelling increase on ocean acidification in the California and Canary Current systems. Geophysical Research Letters 41, 90–95.

Lerman, A., Guidry, M., Andersson, A.J., Mackenzie, F.T., 2011. Coastal ocean last glacial maximum to 2100 CO_2-carbonic acid-carbonate system: a modeling approach. Aquatic Geochemistry 17, 749–773.

Levy, M., Klein, P., Treguier, A.M., 2001. Impact of sub-mesoscale physics on production and subduction of phytoplankton in an oligotrophic regime. Journal of Marine Research 59, 535–565.

Liu, K.K., Chao, S.Y., Shaw, P.T., Gong, G.C., Chen, C.C., Tang, T.Y., 2002. Monsoon-forced chlorophyll distribution and primary production in the South China Sea: observations and a numerical study. Deep Sea Research Part I: Oceanographic Research Papers 49, 1387–1412.

Liu, K.-K., Chao, S.-Y., Lee, H.-J., Gong, G.-C., Teng, Y.-C., 2010. Seasonal variation of primary productivity in the East China Sea: A numerical study based on coupled physical-biogeochemical model. Deep Sea Research Part II: Topical Studies in Oceanography 57, 1762–1782.

Liu, Z., Zhang, L., Cai, W.-J., Wang, L., Xue, M., Zhang, X., 2014. Removal of dissolved inorganic carbon in the Yellow River Estuary. Limnology and Oceanography 59, 413–426.

Liu, K.-K., Yan, W., Lee, H.-J., Chao, S.-Y., Gong, G.-C., Yeh, T.-Y., 2015. Impacts of increasing dissolved inorganic nitrogen discharged from Changjiang on primary production and seafloor oxygen demand in the East China Sea from 1970 to 2002. Journal of Marine Systems 141, 200–217.

McGillicuddy, D.J., Robinson, A.R., Siegel, D.A., Jannasch, H.W., Johnson, R., Dickey, T.D., McNeil, J., Michaels, A.F., Knap, A.H., 1998. Influence of mesoscale eddies on new production in the Sargasso Sea. Nature 394, 263–266.

McLeod, E., Poulter, B., Hinkel, J., Reyes, E., Salm, R., 2010. Sea-level rise impact models and environmental conservation: a review of models and their applications. Ocean and Coastal Management 53 (9), 507–517.

Meselhe, E., Reed, D.J., Grace, A.O., 2015. 2017 Coastal Master Plan: Appendix C: Modeling, Version I. Coastal Protection and Restoration Authority, Baton Rouge, Louisiana. http://coastal.la.gov/wp-content/uploads/2016/02/Appendix-C-Ch123_021716.pdf.

Mitsch, W.J., Gosselink, J.G., 2000. Wetlands. John Wiley & Sons Inc., New York.

Moisan, J.R., Hofmann, E.E., 1996. Modeling nutrient and plankton processes in the California coastal transition zone 1. A time- and depth-dependent model. Journal of Geophysical Research 101, 22647–22676.

Nagai, T., Gruber, N., Frenzel, H., Lachkar, Z., McWilliams, J.C., Plattner, G.K., 2015. Dominant role of eddies and filaments in the offshore transport of carbon and nutrients in the California Current System. Journal of Geophysical Research 120. http://dx.doi.org/10.1002/2015JC010889.

Najjar, R., Pyke, C., Adams, M.B., Breitburg, D., Hershner, C., Kemp, M., Howarth, R., Mulholland, M., Paolisso, M., Secor, D., Sellner, K., Wardrop, D., Wood, R., 2010. Potential climate-change impacts on the Chesapeake Bay. Estuarine, Coastal and Shelf Science 86, 1–20.

Natural Resource Professionals, 2011. Spanish Lake Wetland Data Report. USACE, Vicksburg, Mississippi.

Osterman, L.E., Poore, R.Z., Swarzenski, P.W., 2008. The last 1000 years of natural and anthropogenic low oxygen bottom water on the Louisiana Shelf Gulf of Mexico. Marine Micropaleontology 66, 291–303.

Osterman, L.E., Poore, R.Z., Swarzenski, P.W., Senn, D.B., DiMarco, S.F., 2009. The 20th-century development and expansion of Louisiana shelf hypoxia. Geo-Marine Letters 29, 405–414.

Pauly, D., Christensen, V., 1995. Primary production required to sustain global fisheries. Nature 374, 225–257.

Peyronnin, N., Green, M., Richards, C.P., Owens, A., Reed, D., Chamberlain, J., Groves, D.G., Rhinehart, W.K., Belhadjali, K., 2013. Louisiana's 2012 coastal master plan: overview of a science-based and publicly informed decision-making process. Journal of Coastal Research 67 (sp1), 1–15.

Rabalais, N.N., Turner, R.E., 2001. Hypoxia in the northern Gulf of Mexico: description, causes and change. In: Rabalais, N.N., Turner, R.E. (Eds.), Coastal Hypoxia: Consequences for Living Resources and Ecosystems, Coastal and Estuarine Studies, vol. 58. American Geophysical Union, Washington, D.C, pp. 1–36.

Rabalais, N.N., Turner, R.E., Sen Gupta, B.K., Boesch, D.F., Chapman, P., Murrell, M.C., 2007. Hypoxia in the northern Gulf of Mexico: does the science support the plan to reduce, mitigate, and control hypoxia? Estuaries and Coasts 30, 753–772.

Rabalais, N.N., Turner, R.E., Diaz, R.J., Justic, D., 2009. Global change and eutrophication of coastal waters. ICES Journal of Marine Science 66, 1528–1537.

Rabalais, N.N., Diaz, R.J., Levin, L.A., Turner, R.E., Gilbert, D., Zhang, J., 2010. Dynamics and distribution of natural and human-caused hypoxia. Biogeosciences 7, 585–619.

Ren, W., Tian, H., Tao, B., Yang, J., Pan, S., Cai, W.J., Lohrenz, S.E., He, R., Hopkinson, C.S., 2015. Large increase in dissolved inorganic carbon flux from the Mississippi River to Gulf of Mexico due to climatic and anthropogenic changes over the 21st century. Journal of Geophysical Research: Biogeosciences 120, 724–736.

Rogers, K., Saintilan, N., Copeland, C., 2012. Modelling wetland surface elevation dynamics and its application to forecasting the effects of sea-level rise on estuarine wetlands. Ecological Modelling 244, 148–157.

Rose, K.A., Adamack, A.T., Murphy, C.A., Sable, S.E., Kolesar, S.E., Craig, J.K., Breitburg, D.L., Thomas, P., Brouwer, M.H., Cerco, C.F., Diamond, S., 2009. Does hypoxia have population-level effects on coastal fish? Musings from the virtual world. Journal of Experimental Marine Biology and Ecology 381, S188–S203.

Rose, K.A., Allen, J.I., 2013. Modeling marine ecosystem responses to global climate change: where are we now and where should we be going? Ecological Modelling 264, 1–6.

Rose, K.A., Fiechter, J., Curchitser, E.N., Hedstrom, K., Bernal, M., Creekmore, S., Haynie, A., Ito, S., Lluch-Cota, S., Megrey, B.A., Edwards, C., Checkley, D., Koslow, T., McClatchie, S., Werner, F., 2015. Demonstration of a fully-coupled end-to-end model for small pelagic fish using sardine and anchovy in the California Current. Progress in Oceanography 138, 348–380.

Ruzicka, J.J., Brink, K.H., Gifford, D.J., Bahr, F., 2016. A physically coupled end-to-end model platform for coastal ecosystems: Simulating the effects of climate change and changing upwelling characteristics on the Northern California Current ecosystem. Ecological Modelling. http://dx.doi.org/10.1016/j.ecolmodel.2016.01.018.

Rykaczewski, R.R., Dunne, J.P., 2010. Enhanced nutrient supply to the California current ecosystem with global warming and increased stratification in an earth system model. Geophysical Research Letters 37, L21606.

Sasser, C.E., Gosselink, J.G., Swenson, E.M., Swarzenski, C.M., Leibowitz, N.C., 1996. Vegetation, substrate and hydrology in floating marshes in the Mississippi river delta plain wetlands, USA. Vegetatio 122 (2), 129–142.

Seitzinger, S.P., Harrison, J.A., Dumont, E., Beusen, A.H.W., Bouwman, A.F., 2005. Sources and delivery of carbon, nitrogen, and phosphorus to the coastal zone: an overview of Global Nutrient Export from Watersheds (NEWS) models and their application. Global Biogeochemical Cycles 19, GB4S01. http://dx.doi.org/10.1029/2005GB002606.

Shchepetkin, A.F., Mcwilliams, J.C., 2005. The Regional Ocean Modeling System (ROMS): a split-explicit, free-surface, topography-following coordinates ocean model. Ocean Modelling 9, 347–404.

Skamarock, W.C., Klemp, J.B., Dudhia, J., Gill, D.O., Barker, D.M., Wang, W., et al., 2005. A Description of the Advanced Research WRF Version 2. Technical Note, NCAR/TN-468+STR. NCAR.

Snedden, G.A., Steyer, G.D., 2013. Predictive occurrence models for coastal wetland plant communities: Delineating hydrologic response surfaces with multinomial logistic regression. Estuarine. Coastal and Shelf Science 118, 11–23.

Sperna Weiland, F.C.,L.P.H., van Beek, L.P.H., Kwadijk, J.C.J., Bierkens, M.F.P., 2012. Global patterns of change in discharge regimes for 2100. Hydrology and Earth System Sciences 16, 1047–1062.

Spitz, Y.H., Newberger, P.A., Allen, J.S., 2003. Ecosystem response to upwelling off the Oregon coast: Behavior of three nitrogen-based models. Journal of Geophysical Research 108 (C3), 3062.

Storlazzi, C.D., Elias, E., Field, M.E., Presto, M.K., 2011. Numerical modeling of the impact of sea-level rise on fringing coral reef hydrodynamics and sediment transport. Coral Reefs 30, 83–96.

Sydeman, W.J., Garcia-Reyes, M., Schoeman, D.S., Rykaczewski, R.R., Thompson, S.A., Black, B.A., Bograd, S.J., 2014. Climate change and wind intensification in coastal upwelling ecosystems. Science 345, 77–80.

Syvitski, J.P.M., Kettner, A.J., Overeem, I., Hutton, E.W.H., Hannon, M.T., Brakenridge, G.R., et al., 2009. Sinking deltas due to human activities. Nature Geosciences 2, 681–686.

Tebaldi, C., Strauss, B.H., Zervas, C.E., 2012. Modeling sea level rise impacts on storm surges along US coasts. Environmental Research Letters 7, 014032. http://dx.doi.org/10.1088/1748-9326/7/1/014032.

Theriot, R.F., 1993. Flood Tolerance of Plant Species in Bottomland Hardwood Forests of the Southeastern United States. Wetland Research Program Technical Report WRP-DE-6.

Turner, R.E., Rabalais, N.N., 1994. Evidence for coastal eutrophication near the Mississippi river delta. Nature 368, 619–621.

Turner, A.G., Annamalai, H., 2012. Climate change and the South Asian summer monsoon. Nature Climate Change 2, 587–595.

Turner, R.E., Rabalais, N.N., Justic, D., 2012. Predicting summer hypoxia in the northern Gulf of Mexico: Redux. Marine Pollution Bulletin 64, 319–324.

Visser, J.M., Duke-Sylvester, S.M., Carter, J., Broussard III, W.P., 2013. A computer model to forecast wetland vegetation changes resulting from restoration and protection in coastal Louisiana. Journal of Coastal Research 67 (sp1), 51–59.

Visser, J.M., Duke-Sylvester, S.M., Shaffer, G.P., Hester, M.W., Couvillion, B., Broussard III, W.P., Willis, J., Beck, H., 2015. 2017 Coastal Master Plan: Model Improvement Plan, Additional Vegetation Communities (Subtask 4.4), Version I. Coastal Protection and Restoration Authority, Baton Rouge, Louisiana.

Wall, D.P., Darwin, S.P., 1999. Vegetation and elevation gradients within a bottomland hardwood forest of southeastern Louisiana. American Naturalist 142 (1), 17–30.

Wang, B., Wu, R., Fu, X., 2000. Pacific-East Asia teleconnection: how does ENSO affect East Asian Climate. Journal of Climate 13, 1517–1536.

Wang, S., McGrath, R., Hanafin, J., Lynch, P., Semmler, T., Nolan, P., 2008. The impact of climate change on storm surges over Irish waters. Ocean Modelling 25 (1–2), 83–94.

Wang, L., Justic, D., 2009. A modeling study of the physical processes affecting the development of seasonal hypoxia over the inner Louisiana-Texas shelf: Circulation and stratification. Continental Shelf Research 9, 1464–1476.

Wang, H.J., Saito, Y., Zhang, Y., Bi, N.S., Sun, X.X., Yang, Z.S., 2011. Recent changes of sediment flux to the western Pacific Ocean from major rivers in East and Southeast Asia. Earth-Science Reviews 108, 80–100.

Wang, D., Gouhier, T.C., Menge, B.A., Ganguly, A.R., 2015. Intensification and spatial homogenization of coastal upwelling under climate change. Nature 518, 390–394.

Warner, J.C., Armstrong, B., He, R., Zambon, J.B., 2010. Development of a Coupled Ocean-Atmosphere-Wave-Sediment Transport (COAWST) modeling system. Ocean Modeling 35, 230–244.

Xu, M., Xu, H., Ma, J., 2015. Responses of the East Asian winter monsoon to global warming in CMIP5 models. International Journal of Climatology. http://dx.doi.org/10.1002/joc.4480.

Xue, Z., He, R., Liu, J.P., Warner, J.C., 2012. Modeling transport and deposition of the Mekong River sediment. Continental Shelf Research 37, 66–78.

Zeng, X., He, R., Xue, Z., Wang, H., Wang, Y., Yao, Z., et al., 2015. River-derived sediment suspension and transport in the Bohai, Yellow, and East China Seas: A preliminary modeling study. Continental Shelf Research 111 (Part B), 112–125.

Index

Note: Page numbers followed by "f" indicate figures, "t" indicate tables and "b" indicate boxes.

Ecological law of thermodynamics (ELT), 112
Ecological magnification factor (EMF), 162
Ecological modelling, 13, 33–37, 34f, 103–104
  application, 1–4
    classification and types, 4–7
    ecosystem as object for modeling, 9–10
    environmental problems, 2f
    model types, 7–9, 8t
  development, 20
  MLP, 132
  MLP usage
    contribution of variables, 136–138
    data, 134, 135t
    data preparation, 134
    model training, 135, 135f
    problem, 134
    results, 135–136, 136f
  recommendations, 134
Ecological thermodynamic theory, 66
Ecopath model, 66–67
  development
    construction and parameter, 73–74
    determination of TL, 76
    evaluation of ecosystem functioning, 74–76
  for large Chinese lake, 68–85
    natural system, 68–69
    study site, 70
    TLs, 69
  results and discussion
    basic model performance, 76–79
    changes in ecosystem functioning, 79–81
    collapse in food web, 81–83
    immature state, 83–84
    lake fishery and restoration, 85
    potential driving factors and underlying
      mechanisms, 84–85
Ecosystem(s), 103–104
  functioning
    changes, 79–81
    evaluation, 74–76
  model, 142–143
  as object for modeling, 9–10
  properties, 104–108
Ecotox, 159
Ecotoxicological models, 4, 6–7, 154, 203–205. See also
    Fugacity models
  application, 141–144
  application of models in, 154
  characteristics, 144–153
    dynamic model of toxic substance, 147–148
    with effect components, 149–153
    food chain, 146–147
    in population dynamics, 148–149
    static models of toxic substance mass flows, 147

    contamination of agricultural products, 166–172
    ecotoxicological properties, 159t
    environmental properties of organic compounds, 158t
    estimation of ecotoxicological parameters, 154–166
    examples of toxic substance models, 155t–157t
    as experimental tools, 172–176
    heavy metal model, 172f
    STELLA diagram of model, 175f
    submodels of total ecotoxicological model, 151f
EEP. See Estimation of ecotoxicological parameters (EEP)
Effects models, 141–144
ELT. See Ecological law of thermodynamics (ELT)
Emergent wetlands, 234–236
EMF. See Ecological magnification factor (EMF)
Environmental
  biogeochemical models, 33–37, 34f, 36t
  factors, 136
  management, 3, 3f
  problems, 35, 37t
Environmental risk assessment (ERA), 142, 160
Equivalent indexes (EQIs), 92–93
ERA. See Environmental risk assessment (ERA)
ESLR. See Eustatic sea level rise (ESLR)
Estimation methods, 153
Estimation of ecotoxicological parameters (EEP),
    165–166
  network of estimation methods in, 166f
Euclidean motion, 92
Eustatic sea level rise (ESLR), 240–241
Experimental mathematics models, 171
Experimental tools, 4
  ecotoxicological models as, 172–176
External variables, 16–18

F
Fate models, 141–142
Fate-transport-effect models (FTE models), 143, 203–205
FCI. See Finn's cycling index (FCI)
Fick's laws, 14–15
Finn's cycling index (FCI), 74–76
Finn's mean path length (FML), 74–76
First law of population dynamics, 40–43
Fla. See Fluoranthene (Fla)
FLAM. See Fuzzy logic-based, adaptive management
    (FLAM)
Flo. See Fluorene (Flo)
Floating marsh, 238
Fluoranthene (Fla), 187–188
Fluorene (Flo), 187–188
FML. See Finn's mean path length (FML)
Food chain(s), 56–58, 56f, 146–147
  bioaccumulation, 146f
  model, 117, 118f
  with two trophic levels and cycling of nutrients, 60–61